PRINCIPLES AND PRACTICE

Springer
Berlin
Heidelberg
New York
Barcelona
Budapest
Hong Kong
London
Milan
Paris
Santa Clara
Singapore
Tokyo

G. Isenberg (Ed.)

Modern Optics, Electronics and High Precision Techniques in Cell Biology

With 120 Figures, 13 in Color

Springer

PROFESSOR
DR. DR. GERHARD ISENBERG
Technical University of Munich
Dept. of Biophysics E-22
James-Franck-Strasse
85747 Garching/Munich
Germany

ISBN-13:978-3-642-80372-7

CIP Data applied for
Die Deutsche Bibliothek – CIP-Einheitsaufnahme
Modern optics, electronics and high precision techniques in cell biology / Gerhard Isenberg (ed.).
– Berlin; Heidelberg; New York; Santa Clara; Singapore; Tokyo: Springer, 1997
ISBN-13:978-3-642-80372-7 e-ISBN-13:978-3-642-80370-3
DOI: 10.1007/978-3-642-80370-3 DBN 95.083379.7

Softcover reprint of the hardcover 1st edition 1998

Typesetting: K+V Fotosatz GmbH, Beerfelden
Cover design: D&P, Heidelberg

SPIN 10533885 271/3137 5 4 3 2 1 0 – Printed on acid-free paper

Preface

In spite of tremendous scientific progress over the past years, cell biologists do not yet understand the fundamental processes that determine the life cycle of a cell. Such are: cell movement and cell spreading, cell division, cell communication, cell signaling, cell regeneration and cell death. Biochemistry has enabled us to recognize and to isolate an overwhelming number of new proteins. In vitro assays and the reinjection of proteins into cells and tissues have provided insights into molecular functions and cellular mechanisms. The renaissance of the genetic approach by applying restriction enzymes and vectors, PCR and antisense technology has enabled us to overexpress certain cellular products, to make altered constructs of cell components or to create "knock-out" mutants that entirely lack the factor of interest. Amazingly enough, all these molecular toys have led to a stream of information but not, in a comparable degree, to a better understanding. Has the puzzle become too complex to get solved; or are the windows too small that we are looking through?

As an attempt to answer both questions, the aim of the present monograph *Modern Optics, Electronics and High Precision Techniques in Cell Biology* is first to provide cell and molecular biologists with a whole new scope of easily applicable techniques including brand-new optical, biophysical, physicochemical and biosensoric devices. Secondly, these newly developed techniques allow us to look at cells and biological systems as a whole. This means that cells and molecular systems can be analyzed in real life, i.e. at a real place, in real time and in real dimensions.

The optical section begins with a description of the principles and application of atomic force microscopy (AFM) by Drs. C.-A. Schoenenberger, D.J. Müller and A. Engel. Imaging of native biological structures, culminating in monitoring and changing molecular conformations of individual molecules, is the main topic in this "state-of-the-art" article. Three-dimensional imaging by optical sectioning through a cell is possible by the newly developed Confocal Laser Scanning Microscopy at the light microscope level (Drs. Borlinghaus and Gröbler).

Cellular neurobiology contributes with two revolutionary novel techniques to this volume: Drs. Dodt and Zieglgänsberger have developed an infrared illumination which, in combination with a gradient contrast system and video enhancement, allows monitoring of development and plasticity of neuronal networks in their natural environment.

Exciting time-resolved imaging of membrane potentials and ion fluxes at the cellular level may be obtained by a new fiber array photodiode camera, developed and described by Drs. Hosoi, Tsuchiya, Takahashi, Kashiwasake-Jibu, Sakatani and Hayakawa.

Both approaches open up quite new dimensions for analyzing molecular nerve cell functions.

Drs. Schliwa and Käs with their collaborators look at individual biopolymers with different techniques: Optical Laser Tweezers allow the manipulation of single microtubules in order to measure their rigidity and stiffness. Polymer dynamics of semi-flexible actin rods and a corresponding theory are derived from high resolution single protein imaging by fluorescence.

Neutron reflection and its application for investigating protein-lipid interactions in cell membranes are covered in the chapter by Drs. Bayerl and Maierhofer. This non-invasive technique is dependent on large nuclear reactor facilities and therefore is of increasing interest worldwide. Utilization of physical techniques for cell science results from the attempt to find a common scientific language among biophysicists and biologists. An example may also be the following chapter by Drs. Goldmann, Guttenberg, Ezzel and Isenberg, introducing the stopped-flow technique for measuring fast kinetics of biological reactions.

Primarily derived from a physical technique, namely, surface plasmon resonance, is the BIA-CORE system, a biosensor-based analytical technology for monitoring interactions in real time without using any labeling (Drs. Markey and Schindler). Both techniques have a wide application range and fill a gap within the spectrum of cell biological methods.

An intriguing new way of producing micromachined substrates is illustrated by Drs. Galbraith and Sheetz. Micromachining combined with two-dimensional micro-electronic sensors gives new insights into the signal processing of neuronal networks. Here, the device is used to record traction forces of individual cell contacts.

A final example that the new age of cell science significantly relates to bioengineering is given by Dr. Sackmann, who describes the operation of high precision viscometers designed to measure the viscoelasticity of biological polymers.

Each of the new technologies presented here is directly correlated with an actual field of cell biological research. One can envisage for the next decades that cell biological research will be determined not only by genetics but will necessarily have to include modern optics, electronics and high precision techniques to open up the new era of Molecular Cell Physiology. I dare to predict that knowledge and understanding are guaranteed when one is open-minded towards new approaches.

Munich, January 1997 G. Isenberg

Contents

CHAPTER 1

Atomic Force Microscopy Provides Molecular Details of Cell Surfaces

Cora-Ann Schoenenberger[1] · Daniel J. Müller[1,2] · Andreas Engel[1]*

Contents

[1] M.E. Müller Institute for Microscopy, Biozentrum, University of Basel, CH-4056 Basel, Switzerland
[2] IBI-2:Structural Biology, Forschungszentrum Jülich, D-52425 Jülich, Germany
* Corresponding address: M.E. Müller Institute, Biozentrum, University of Basel, Klingelbergstrasse 70, CH-4056 Basel, Switzerland, Tel.: +41-61-267-2261, Fax: +41-61-267-2109, e-mail: aengel@ubaclu.unibas.ch

1.1 Introduction

Ten years ago, Binnig, Quate, and Gerber presented the microscopy world with the atomic force microscope (AFM; Binnig et al. 1986). As a member of the scanning probe microscopes, the AFM provides detailed maps of surface topographies by raster scanning a fine tip over the sample surface and recording the changes in tip-sample interactions. Initially, the AFM found its applications in material sciences, for example, in the imaging of magnetic fields above thin-film recording heads (Martin and Wickramasinghe 1987) or of polymers (Albrecht et al. 1988). It took as little as one year to achieve atomic resolution on hard flat surfaces (Binnig et al. 1987). Unlike its ancestor, the scanning tunneling microscope (STM), the AFM does not require the specimen to be either electron or ion conductive. In addition, it is operational in liquid as well as in air or in vacuo. The possibility of imaging nonconductive samples in aqueous solutions has launched this technology into biological research. With the AFM, the imaging of native biological structures in a physiological environment at molecular resolution had moved into the realm of reality. That biological compounds could indeed be imaged in aqueous solutions was first reported by Drake and colleagues in 1989 who showed images of rehydrated polyalanine adsorbed to glass and of fibrin polymers adsorbed to mica in phosphate-buffered saline (Drake et al. 1989). Compared to these early images, the resolution obtained by AFM has come a long way in a very short time. A number of technical developements were vital for improving resolution. The optical lever significantly increased the sensitivity of the detection of the deflecting cantilever (Meyer and Amer 1988). Together with the microfabrication of cantilevers with integrated tips that have a defined geometry and high-aspect ratios, a more reproducible operation at forces as low as a few hundred picoNewtons was achieved. Further on in this chapter, we will present examples from our own work where submolecular details of native membrane proteins are resolved. In its infancy, the resolution obtained with biomolecules was limited, but the potential of AFM to monitor dynamic biological processes was already recognized early on. The recent observation of reversible conformational changes of bacteriorhodopsin (Müller et al. 1995b) and of single pores of the hexagonally packed intermediate (HPI) layer (Müller et al. 1996b) have spurred the hope that structure–function relationships can be directly monitored with the AFM.

In this review we will briefly introduce the principle of AFM and then discuss the most important aspect of sample preparation which is its immobilization. With respect to the AFM applications in biology, space limitations oblige us to focus on the imaging of biomolecular assemblies and the structural information contained in these images. We will do so by illustrating the different degrees of resolution obtained with examples from our own research. Other exciting applications, such as the measurement of intermolecular forces (Florin et al. 1994; Lee et al. 1994; Moy et al. 1994; Dammer et al. 1995, 1996) and the measurement of visco-elastic cell surface properties (Weisenhorn et al. 1993; Hoh and Schoenenberger 1994; Radmacher et al. 1994) merit being discussed independently.

1.2 Principles of Atomic Force Microscopy

The physical principle of the AFM (Binnig et al. 1986) is simple (Fig. 1.1). An ultrasharp tip mounted on a cantilever is scanned over the surface of an object. The force applied to the sample is detected by the bending of the cantilever. A feedback loop connected between the detection system and the piezoceramic elements controls the vertical movement of the piezoelement to keep the applied force constant (constant-force mode). Thus, the vertical movement of the piezoelements is locked to the lateral movement by the feedback loop during scanning, and the sample topography is contoured. As the feedback system can only read after the cantilever bending changes, a small error is inevitable. The corresponding error signal produces an image akin to differential interference contrast in light microscopy (error mode; Putman et al. 1992). The vertical information of the image is due to structural height differences, but can exhibit additional information based on the specific interactions between tip and sample. These can be of electrostatic, magnetic or chemical nature. If a step of a homogeneous material is monitored in constant-force mode, the height difference is only of structural origin. This is not the case if the upper surface of the step is of different material from the lower surface, because the tip–surface interactions are then different. Therefore, the height detected can also show a contribution due to forces caused by these specific interactions.

The microscope can be operated in various environments such as vacuum, gaseous phases, and liquids. Of particular interest for biological applications is the operation in aqueous solution (Drake et al. 1989). Under all of these environments the surfaces of solid-state materials can be imaged with atomic resolution (Binnig et al. 1987; Ohnesorge and Binnig 1993; Sugawara et al. 1995). Forces between stylus and sample depend on the environment in

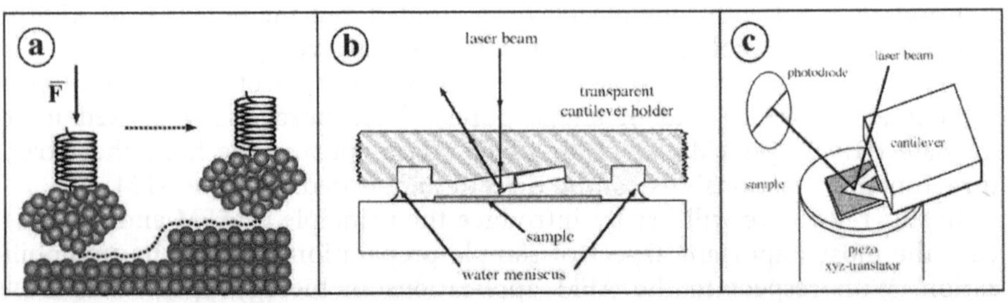

Fig. 1.1 a–c. Principles of atomic force microscopy. **a** A stylus with a very small tip radius is mounted on a cantilever spring and scanned over the sample surface. At sufficiently small forces, the corrugations of the scanning lines represent the surface topography of the sample. **b** The imaging process can be performed in a liquid cell filled with buffer solution. This warrants the preservation of biomolecules in their native state. **c** The vertical deflections of the cantilever are detected by the displacement of the laser beam reflected onto a two-segment photodiode. The difference signal of the photodiode is used to drive a feedback loop which controls the movement of the piezo xyz-translator. In aqueous solution, the applied force between the stylus and the sample can be kept constant in the range of ~10–25 pN

which the measurements are performed. In humid air, hydrophilic surfaces are covered by an ultrathin water layer. When the tip approaches the sample these water layers coalesce to build a meniscus, and tip and sample are then pulled together by capillary forces (Weisenhorn et al. 1992). This makes it impossible to operate the AFM in air at applied forces smaller than 100 nN (Weisenhorn et al. 1993). This limitation is overcome by imaging in an aqueous environment (Weisenhorn et al. 1992). Most biological systems are soft, comparable to a water-filled sponge. Depending on the force applied during imaging, sample deformation can occur (Weisenhorn et al. 1993). This has been observed on living cells (Hoh and Schoenenberger 1994) as well as on single membrane proteins (Müller et al. 1995b). Forces above 1 nN not only influence the native biological structure, but also increase the contact area between tip and sample, thereby limiting the achievable lateral resolution (Weihs et al. 1991). To minimize the sample perturbation and to achieve high resolution, small forces are required when acquiring surface topographs of biological samples with the AFM. The sensitivity of the microscope depends critically on the feedback system, the detection system and the cantilever spring constant. Therefore, only a very sensitive AFM that can reproducibly be operated at forces of 0.1–0.5 nN is suitable for biological aplications.

The liquid environment changes the physical properties of the cantilever. Its movement is damped, while its effective mass is increased (Butt et al. 1993). Therefore, the scan speed has to be reduced when imaging in liquid. The scan speed limit to image biomolecules with a resolution of 1 nm is about 2.2 µm/s (191 µm long cantilever, $k=0.057$ N/m). Taking the scan speed limit and the required small forces of about 100 pN for nonperturbative imaging of biological systems into account, it was possible to image macromolecular systems with submolecular resolution (Schabert et al. 1995; Müller et al. 1995a, 1996a; Mou et al. 1995, 1996).

The most commonly used imaging modes in biological atomic force microscopy are the constant-force mode described above, and the recently invented tapping mode (Hansma et al. 1994; Putman et al. 1994a). While in the constant-force mode the cantilever is in continuous contact with the surface, the tip–sample separation is modulated while the sample is scanned in the tapping mode. Because the sample is only tapped at each minimum of the tip oscillation, the contact time is reduced. To reduce the contact time between the tip and the sample is one way to minimize the friction which can easily displace samples that were not sufficiently attached to the support. Other possibilities are to minimize the applied force or to modify the stylus surface to prevent specific interaction with the sample.

1.3 Sample Immobilization for Imaging in Physiological Environment

An important prerequisite for the imaging of native biological systems is the immobilization of the sample on a solid, flat support so that it cannot be displaced by the scanning probe. For most biological applications, this is the only preparation step needed or wanted. The major consideration in establishing the optimal immobilization conditions is the integrity of sample structure and function. Since it is beyond the scope of this article to discuss the details of immobilization, we will concentrate on the general principles and refer the reader to more specific reviews (Amrein and Müller 1997).

1.3.1
Supports

The support itself is preferably flat with little surface corrugation in order to minimize contributions of the support topography to the final object image. In addition, the support should be chemically rather inert to provide consistent adsorption conditions that do not interfere with sample integrity. Glass coverslips and freshly cleaved mica are frequently used supports. Compared to freshly cleaved mica or silicon wafers, which are atomically flat, glass surfaces have a texture with a peak to peak corrugation of approximately 1 nm (Karrasch et al. 1993). However, the optical transparency makes glass a suitable support for combined AFM and light microscopy (Putman et al. 1993a; Schabert et al. 1993). Mica minerals are characterized by their layered crystal structure. Basal cleavage results in an atomically flat, negatively charged surface. A number of native membranes and other supramolecular assemblies adhere strongly to freshly cleaved mica in the presence of the appropriate counterions. Molecular resolution of gap junctions has been obtained on purified membrane fragments adsorbed to mica under physiological conditions (Hoh et al. 1993). Another approach is to reconstitute purified membrane proteins into a supported planar lipid bilayer prepared by a Langmuir trough on a mica surface (for review see Peachy and Eckhardt 1994). When the glycosphingolipid receptor for cholera toxin B (CTX-B) was included in the bilayer, the pentameric CTX-B protein was bound to the bilayer with high affinity and its five-fold symmetrical structure was resolved by AFM at 1 to 2 nm lateral resolution (Mou et al. 1995). Not only proteins that are laterally supported by lipids but also a number of soluble proteins adsorb well to mica and can be imaged under near physiological conditions albeit at lower resolution (Yang et al. 1994; Mou et al. 1996). The adhesion of intact living cells, representing the opposite end of supramolecular complexity, on mica or glass surfaces largely depends on the intrinsic properties of the specific cell type.

1.3.2
Sample Immobilization

The major immobilization principles that overlap to some extent are (a) adsorption, (b) chemical cross-linking, and (c) physical immobilization.

(a) Adsorption

The most widely applied immobilization technique is the adsorption of biological samples from solution to a flat support. Long range (3–30 Å) weak Van der Waals attractions, the repulsive electrostatic double-layer force (i.e., the interaction between charged surfaces), as well as hydrophobic and hydrophilic interactions are effective in adsorption. A net attractive force between the surface of the object in solution and the planar support is required for initial attachment. To strengthen the attachment, the interactions can be modified by varying the pH or the electrolyte concentration of the solution. It has been shown for a number of biological macromolecular structures (e.g., actin, purple membrane, hexagonally packed intermediate layer, laminin, DNA) that by adjusting the pH and the concentration of the counterions, the repulsive electrical double layer interactions can be reduced. Consequently, the attractive van der Waals forces predominate, thus improving adsorption (Müller et al. 1997). Hydrophobic and hydrophilic interactions are intricately related to the structure of water and its interaction with solutes. As a general rule, hydrophobic molecules do not attach to hydrophilic surfaces and vice versa. The hydrophilic and hydrophobic interactions can result in an oriented adsorption of molecular assemblies. For example, the HPI layer of *Deinoccocus radiiodurans* has a more hydrophilic outer surface and a more hydrophobic inner surface (Baumeister et al. 1986). On a hydrophilic surface, the HPI layer preferentially adsorbs with its outer surface (Müller et al. 1996b) while on a hydrophobic substrate, the inner surface attaches more readily (Karrasch et al. 1994).

A number of additional parameters may affect adsorption. For one, the binding energy may be too low to prevent single molecules or filaments diffusing over the support. Adequate modification of the support surface may help to overcome this difficulty. A variety of molecular coatings have been shown to improve the adhesive properties of the support. The basic side chains of polylysine render coated surfaces positively charged and help to firmly adsorb negatively charged objects (Mazia et al. 1975). A mussel adhesive protein extracted from *Mytilus edulis* has been shown to mediate a tight nonspecific binding of a wide range of samples, including proteins and nucleic acids (D'Costa and Hoh 1995). The coating of glass coverslips with extracellular matrix components promotes the adhesion of living cells (Henderson et al. 1992).

(b) Chemical Cross-Linking

For biomolecules that can not be sufficiently immobilized by mere adsorption, a covalent chemical cross-linking of the sample to the solid support may prove helpful. A rich literature for covalent immobilization of biomolecules on solid substrates emerged from solid-phase peptide synthesis and immuno-binding assays (Aebersold et al. 1986). Unmodified double-stranded DNA could be repeatedly imaged in water when it was bound to chemically functionalized mica (Lyubchenko et al. 1993). A variety of supramolecular assemblies could be covalently bound to a silanized glass surface via a photoreactive cross-linker and subsequently imaged by AFM (Karrasch et al. 1993). In another approach, activation of a silanized silicon substrate with glutaraldehyde was used to covalently link microtubules via their external amino groups (Vinckier et al. 1995). Similar to a chemical funtionalization of the substrate, molecular recognition forces can be exploited to attach receptor molecules to a matrix of ligands. Biotin-derivatized surfaces were shown to immobilize streptavidin in a geometrically defined manner through high-affinity binding (Mazzola and Fodor 1995). Specific chemical modification of the sample may result in a better defined spatial orientation. DNA molecules were anchored on gold surfaces at one end only by introducing thiol groups at the 5′ phosphates (Hegner et al. 1996). However, chemical cross-linking of the sample can alter its molecular structure and interfere with function.

(c) Physical Immobilization

An alternative strategy for immobilizing a sample is to physically hold it in place. In analogy to the micropipettes in patch clamp experiments, glass pipettes with an opening of about 2–4 mm were used to fix individual, otherwise nonadherent cells by suction (Hörber et al. 1992). The free cell surface protruding from the open end of the micropipette was then scanned with the probe mounted in a fixed position. In this experimental set-up, the inner pressure of the cell can be increased with the effect that the plasma membrane stiffens and consequently the resolution improves. In addition, other types of measurements (e.g., all classical electrophysiological techniques) can be carried out on the cell in parallel to AFM. However, refined experimental set-ups require customized instrumentation. In contrast, the trapping of yeast cells in the pores of filter membranes is one example of yet another technique to immobilize large particles (Holstein et al. 1994; Kasas and Ikai 1995) and it is applicable to conventional microscopes. As with the micropipette, only the protruding sample surface is accessible to the probe. This restriction is not necessarily a disadvantage, since the AFM is not designed to contour deep corrugations by the scanning tip.

The immobilization procedures discussed here represent only a selection and are far from complete. Despite the variety of approaches it still remains a challenge to fix biological samples to a substrate under physiological conditions and reproducibly image proteins at adequate resolution. In particular, high lateral forces exerted by the scanning tip in contact mode AFM can impede imaging of single molecules and filamentous assemblies in solution with few exceptions (Karrasch et al. 1993; Mou et al. 1996). The recently developed tapping mode AFM (Hansma et al. 1994; Putman et al. 1994a) abates the disruptive influence of the lateral forces applied in contact mode. With tapping mode AFM it became possible to image globular and filamentous biomolecules that are not strongly attached to the substrate (Hansma et al. 1994, 1996; Fritz et al. 1995). The helical structure of DNA (Hansma et al. 1996) and of filamentous actin (Fritz et al. 1995) could be resolved. However, the subnanometer resolution achieved on two-dimensional crystalline arrays of membrane proteins (see below) in contact mode remains unparallelled. With the better understanding of the fundamental aspects of the imaging process, the tapping technique is likely to be optimized for achieving high resolution topographies of both solid state and biological materials.

1.4
Imaging of Biomolecular Assemblies

1.4.1
Living Cells

The ability to operate the AFM in aqueous environments kindled the interest of cell biologists in this technology. The generation of image contrast by a mechanism that is different from light microscopy and the unprecedented resolution of native biomolecules indicated that new information about cellular structures could be obtained. At first, the direct interaction of the tip with

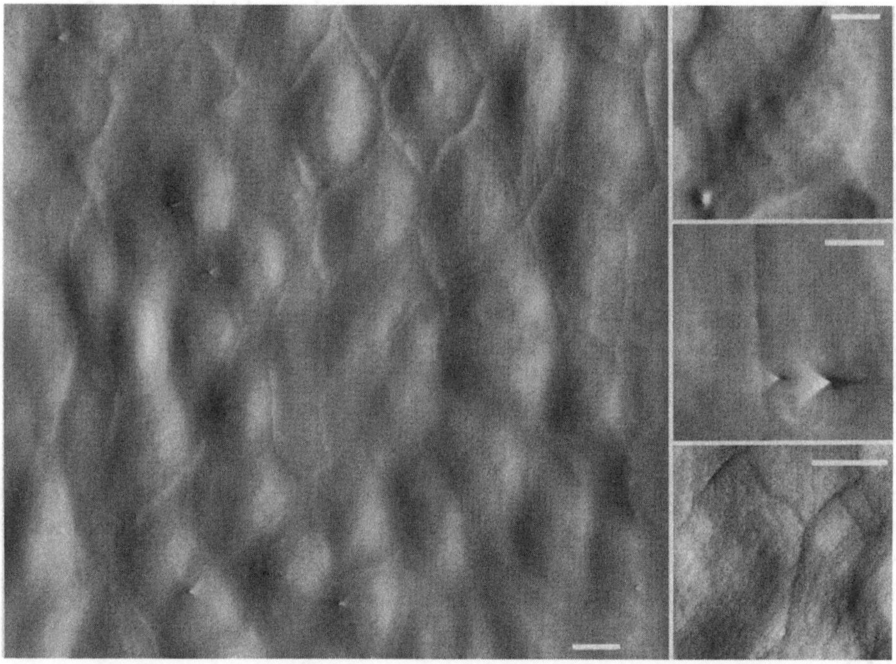

Fig. 1.2. Apical surface of living Madin–Darby Canine Kidney (MDCK) monolayers. MDCK cells were grown to confluence on glass coverslips and subsequently imaged in Hanks balanced salt solution. All imaging was carried out with a Nanoscope III (Digital Instrumens Inc., Santa Barbara, CA) equipped with a 150 μm (J type) scanner and a liquid cell at room temperature. The AFM was operated in contact mode using microfabricated, V-shaped 200 μm long silicon nitride cantilevers (Digital Instruments Inc.) with 40 μm wide legs. The nominal force constant of these cantilevers was 0.06 N/m. Error signal images were collected by monitoring the deflection signal with the gains set as high as possible without the feedback loop oscillating. The overview shows the apical surface of a confluent monolayer. The *large structure* in the center of most cells represents the nucleus. The cell boundaries that delineate neighboring cells are also very prominent (*scale bar*, 10 μm). At higher magnification, different types of surface protrusions are seen: the top *inset* shows a smooth bulge that is typically 2–10 μm wide and 0.1–1 μm tall (*scale bar*, 5 μm), whereas in the middle *inset* sharp spikes are seen (*scale bar*, 5 μm). Several spikes can occur on an individual cell. They are generally 0.1–1 μm tall. In the bottom *inset*, the surface shows a fibrous appearance (*scale bar*, 10 μm). (From Hoh and Schoenenberger, 1994, with permission)

the sample prompted concern that cells might be damaged during imaging. In the meantime, a number of different cell types have been imaged and have remained perfectly viable throughout the process (for a review see Henderson 1994). As discussed in the previous section, immobilization is a critical prerequisite. Thus, cells types with intrinsic adherent growth are most suitable for AFM studies. Epithelial cells that form continuous sheets of cells seem particularly suited. Besides specific cell–substrate interactions that anchor them to an appropriate support (glass or mica), the cell–cell interactions provide additional lateral stability. As shown in Fig. 1.2, cultured Madin-Darby canine kidney (MDCK) cells adhere well to glass coverslips and are not displaced by imaging forces of 1 to 4 nN (Hoh and Schoenenberger 1994). However, only the apical plasma membrane is accessible to the tip. The predominant structures are the underlying nuclei and the cell boundaries that appear as a ridge about 10–100 nm high. At higher magnification, several surface protrusions with different morphologies are seen. The dynamic behavior of these structures (Schoenenberger and Hoh 1994) provides evidence that the cells have not relinquished their viability. In addition, intracellular fibrous structures that run parallel to the apical surface can be discerned. In contrast, the numerous microvilli on the apical surface that are revealed by scanning electron microscopy (Hoh and Schoenenberger 1994) are too delicate to be resolved by AFM. With unsupported membranes, such as whole cells and large vesicles, the resolution achieved by AFM is only about 30–50 nm (Häberle et al. 1991), far below what is required to resolve surface receptors, ion channels, and other individual membrane proteins. In MDCK cells, the apical plasma membrane deforms by approximately 500 nm per nN applied force (Hoh and Schoenenberger 1994). Most likely, the softness of the plasma membrane obstructs further resolution. The deformation may be reduced by applying significantly lower imaging forces. Alternatively, it has been suggested, that cells imaged in tapping mode “harden” under the tapping motion at high frequencies which leads to an improvement of resolution (Putman et al. 1994b). For MDCK cells, however, the images acquired in contact mode or in tapping mode were equivalent (C.-A. Schoenenberger, unpubl. observ.).

1.4.2
Surface and Subsurface of Bovine Articular Cartilage

With the AFM, not only individual cells and cell layers, but also the organization and structure of native tissues can be observed without dehydration of the sample (Jurvelin et al. 1996). This is demonstrated in Fig. 1.3, which displays the surface and subsurface of articular cartilage from a bovine humeral head imaged in phosphate-buffered saline. Articular cartilage is a fiber-composite tissue characterized by a thin, resilient and almost frictionless surface which is important for bone articulation in synovial joints. In addition, the composition, structure, and mechanical properties of the superficial zone are of great significance for the load-bearing function of the underlying cartilage.

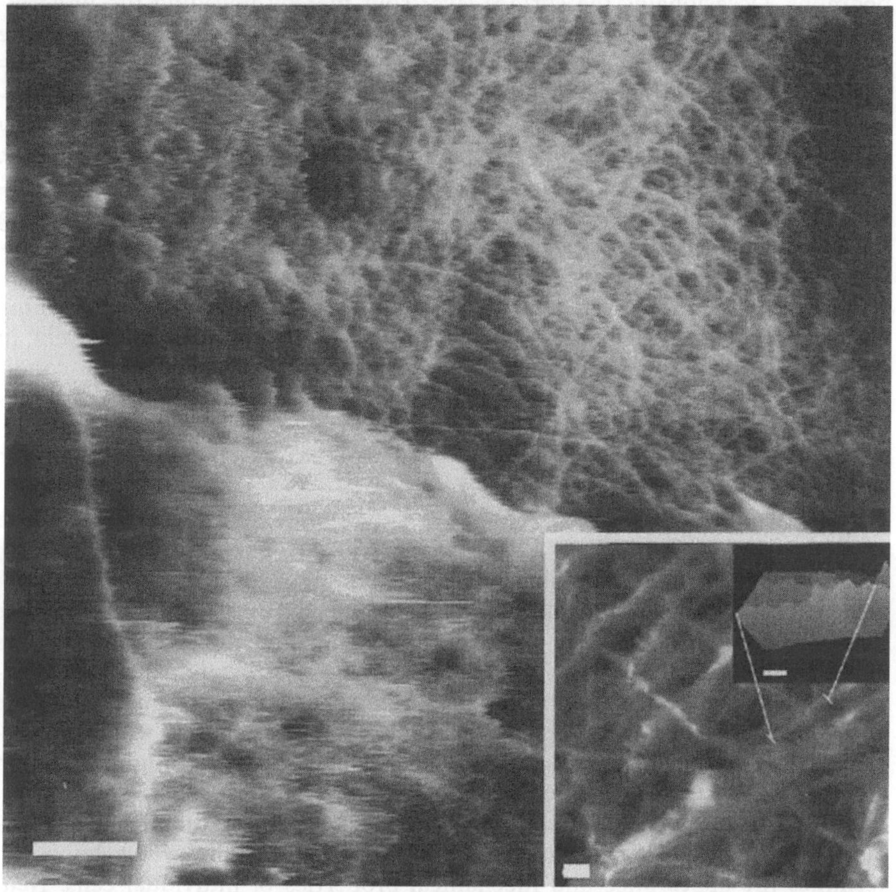

Fig. 1.3. Layered structure of articular cartilage surface. Cylindrical osteochondral plugs were drilled perpendicular to the articular surface of adult bovine humerus and trimmed down to cartilage discs on the bone side which was then glued to a glass substrate. The free articular surface was immersed in phosphate-buffered saline for the entire preparation procedure. Cantilevers (100 µm) with a nominal force constant of $k=0.1$ N/m and oxide-sharpened Si_3N_4 tips (Olympus Ltd., Tokyo, Japan) were used. The tip was positioned on a suitable sample area with a stereomicroscope. Height signal images were collected in contact mode. The articular surface shown in the *bottom left* corner is featureless. The blurred appearance is typical for a high surface viscosity. In areas where the amorphous top layer was removed (*upper half* of the image), the underlying fibrous substructure is revealed (*scale bar*, 2 µm). The *inset* shows a higher magnification of the intricate network of collagen fibers. Individual fibers run mostly parallel to the surface at random orientation. The fibers vary in diameter (mean value ~ 35 nm) and crossovers occur (*scale bar*, 100 nm). A section along the axis of a fiber is shown in the small *inset* (*scale bar*, 50 nm). Irrespective of their diameter, the fibers exhibited a periodic banding of 60±5 nm that is characteristic for collagen fibers. (From Jurvelin et al. 1996, with permission)

As shown in the bottom left corner of Fig. 1.3, the articular cartilage surface is typically covered with an amorphous layer of viscous, nonfibrous material which in the joint acts as a lubricant. The AFM image of this surface suggests that the material adhered to the scanning stylus and that the high viscosity of the surface layer prevented the acquisition of high-resolution topographs. Occasionally, the surface layer split thereby unveiling the underlying collagen meshwork. Height measurements at the edge showed the surface layer to be approximately 500 nm thick. The collagen fibrils are oriented parallel to the surface but the meshwork does not display a distinct organization. At higher magnification (inset), individual fibers could be traced. They are randomly oriented within the plane and cross-overs occur irregularly. The different types of collagen fibers present in the superficial cartilage matrix can not be discerned. However, in the height profile (small inset) along the fiber stretch indicated, a 60 nm periodicity that is characteristic for collagen fibers is evident.

Intriguingly, surface irregularities that have been observed by scanning electron microscopy (Ghadially 1983; Kirk et al. 1994) are not apparent in AFM images of native articular cartilage recorded under physiological conditions. The smooth surface observed by AFM indicates that the irregularities represent artifacts of EM sample preparation.

1.4.3 Intermediate Filaments

Imaging of the articular cartilage subsurface tissue revealed a complex three-dimensional tissue architecture of collagen fibers. Because of this intricate network, not much detail in the structure of the individual collagen fibers could be resolved by AFM. Moreover, mechanical stress or chemical treatment applied to the tissue in order to expose specific features might interfere with the native structure. Isolated supramolecular assemblies which are arranged in a defined geometry seem more likely to convey high-resolution structural information.

Similar to the scaffold of collagen fibers that constitute the articular cartilage subsurface, a distinct network of fibrous structures, the cytoskeleton, determines the appearance and mechanical properties of individual cells. Neurofilaments, members of the intermediate filament family, are the major constituents of the neuronal cytoskeleton (for review see, Heins and Aebi 1994). They are assembled from three polypeptides of which only NF-L self-assembles into 10 nm wide filaments in vitro. All three NF polypeptides consist of a central, α-helical rod domain flanked by an N-terminal head and a C-terminal tail domain.

As discussed in the previous section, many filamentous structures are not effectively immobilized by adsorption and are prone to be displaced by the lateral forces exerted by the stylus when imaged in solution. Therefore, neurofilaments were covalently coupled to an aminopropylated glass surface via a photoreactive cross-linker (Karrasch et al. 1993). As shown in Fig. 1.4, covalently bound filaments assembled from recombinant NF-L polypeptides produced stable AFM images. At low magnification, filamentous structures of variable length are seen. As in negatively stained electron micrographs (data not shown), the majority of the filaments appeared featureless at a resolution of 5 nm. Even at higher magnifications (inset), no structural information was gained. Occasionally, unravelled filaments could be observed. The height of the native filaments measured by AFM (9.5 ± 1.2 nm) was consistent with the

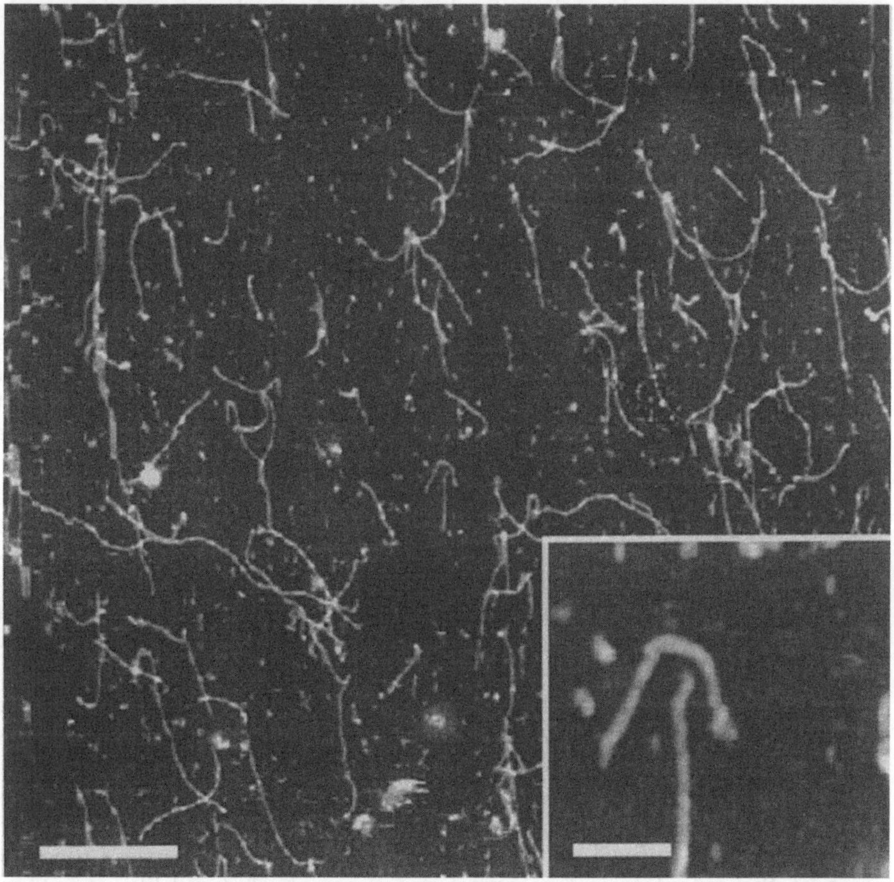

Fig. 1.4. Covalently attached intermediate filament structures imaged in buffer solution. Recombinant full-length mouse NF-L was purified from *E.coli* and reconstituted into filaments by dialysis against filament forming buffer (50 mM MES, 170 mM NaCl, 1 mM DTT, pH 6.25 at 37 °C; Heins et al. 1993). Neurofilaments were covalently immobilized on chemically functionalized glass coverslips via the photoactivatable cross-linker N-5-azido-2-nitrobenzyoloxysuccinimide (ANB-NOS; Karrasch et al. 1993). Contact mode imaging was performed in filament buffer with nominal forces of approximately 1 nN. Cantilevers with a spring constant $k = 0.38$ N/m and a pyramid-shaped tip (Digital Instruments Inc.) were used. At low magnification, the distribution of the filamentous structures is seen. The smaller nonfilamentous features represent either tetramers of the NF-L polypeptide or polymerization products from the chemical modification. (*scale bar*, 2 µm). Structural details are not revealed even at higher magnification (*inset*), but the filaments remained stable during scanning (*scale bar*, 200 nm). (From Karrasch et al. 1993, with permission)

filament dimensions derived from transmission electron microscopy (TEM) data. In contrast, the width of the native filaments appeared to be five times increased over TEM measurements. This discrepancy, as well as the marginal resolution achieved are primarily effects of the tip geometry. With the development of finer tips and a further reduction of imaging forces, an improvement of resolution may be expected. The highest resolution of native filamentous structures achieved so far has been obtained with tapping mode AFM of DNA (Hansma et al. 1994, 1996; Bezanilla et al. 1994).

1.4.4 Nuclear Pore Complex

Bidirectional molecular trafficking between the nucleus and the cytoplasm of eukaryotic cells occurs through the nuclear pore complexes (NPC) which are embedded in the nuclear envelope membrane (NE). The three-dimensional structure of these large supramolecular assemblies ($\sim 1.2 \times 10^8$ kDa) has been extensively investigated by electron microscopy. In electron micrographs of negatively stained nuclear membranes from frog oocytes, NPCs appear as round particles with a diameter of 125 nm and a distinct eightfold rotational symmetry (for review see Panté and Aebi 1995). A basic framework built of eight multi-domain spokes embracing a central pore anchors the NPCs in the membrane. A central plug or channel complex which has been implied to play a role in mediated transport may be present (Unwin and Milligan 1982; Akey and Radermacher 1993). Facing the cytoplasm, the spoke complex is decorated with a massive cytoplasmic ring with eight projecting filaments. On the nuclear side, a corresponding nuclear ring is topped with a basket-like filamentous assembly. Conceivably, the asymmetry along the transport axis reflects the structural basis for the directional nuclear import of proteins, as well as the export of RNAs and ribonucleoprotein particles.

Approximately 100 different polypeptides are believed to constitute the NPC and so far, about two dozen nucleoporins have been biochemically characterized. Very little is known about the mechanisms involved in nucleocytoplasmic transport. AFM offers the exciting possibility of correlating structural changes with the transport mechanism.

The images of unfixed, native NE are shown in Fig. 1.5. Surface topographies of the cytoplasmic face reveal dense arrays of donut-like structures with a diameter of ~120 nm. In most instances, the center appears to be empty. These AFM images exhibit a striking similarity to electron micrographs recorded from quick-frozen/freeze-dried/rotary metal-shadowed samples (Panté and Aebi 1993).

Similar to the plasma membrane of intact cells, nuclear pore complexes are rather soft and spongy structures that are prone to be deformed by the scanning probe. Likewise, the cytoplasmic filaments are too fragile to be resolved and instead are obscured by scanning. The nuclear baskets that most likely give rise to the dome-like appearance of NPCs on the nuclear face are highly susceptible to mechanical damage and can easily be scraped off by the scanning stylus when the forces are higher than 1 nN (Goldie et al. 1994).

1.4.5 Bacteriophage T4 Polyheads

Polyheads are tubular structures folded from a hexagonal lattice (a = b = 13 nm) of capsomeres, each of which contains six major T-even bac-

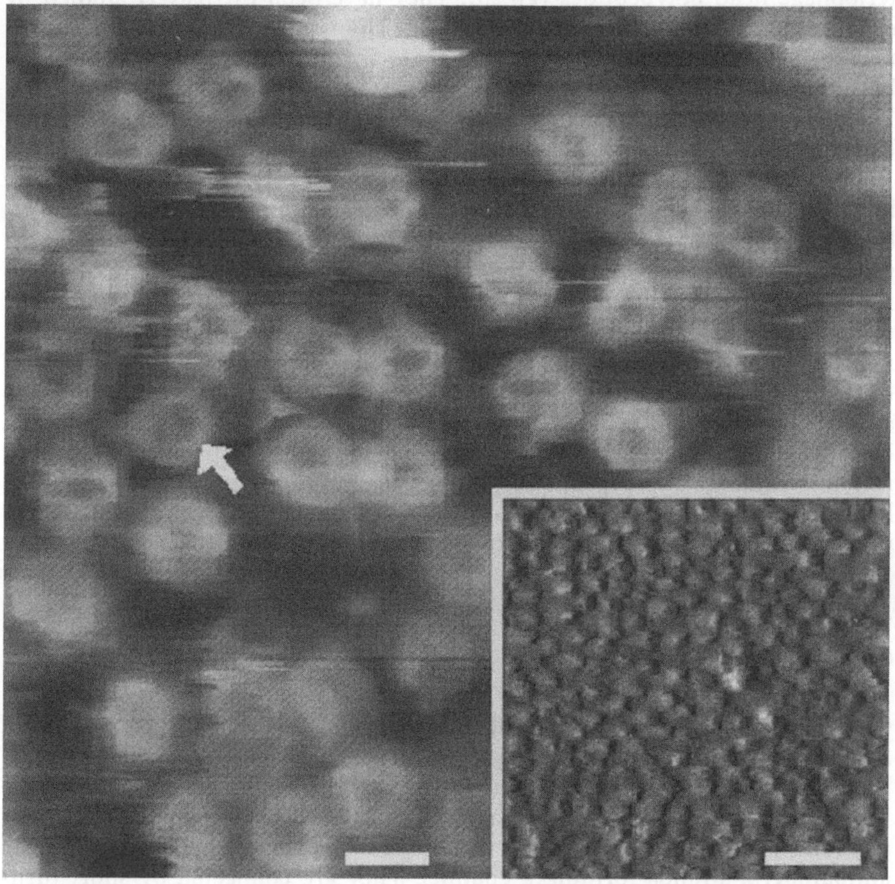

Fig. 1.5. Nuclear pore complexes viewed from the cytoplasmic and nuclear side. Nuclei from *Xenopus laevis* oocytes were manually isolated and placed on glass coverslips coated with muscle adhesive protein. Individual nuclei were then physically ruptured in an oriented manner, i.e., with either the cytoplasmic surface or the nuclear side contacting the substrate. The nuclear envelope preparations were adsorbed in 10 mM Hepes, pH 7.5, 1 mM KCl, 0.5 mM $MgCl_2$ buffer and inspected by light microscopy prior to AFM. Constant-force and error-signal images were recorded in buffer solution with a Nanoscope III using a J-scanner and oxide-sharpened Si_3N_4 tips (Digital Instruments Inc.). The figure shows the cytoplasmic face of an unfixed nuclear envelope. Irregular arrays of donut-shaped nuclear pore complexes (NPC) with a massive annulus surrounding the membrane pore are seen. The apparent diameter of the NPC on the cytoplasmic surface is approximately 120 nm. Occasionally, the center of a donut seems to be filled with a structure (*arrow*) that might correspond to the central plug (*scale bar*, 100 nm). In the error-signal image of the nuclear face (*inset*), NPCs display a dome-like appearance. The apparent dome diameter of ~170 nm is significantly larger and differs from the values based on electron microscopy (~100 nm). This discrepancy is most likely due to a "tip convolution" effect (Keller et al., 1991; *scale bar*, 500 nm). Image courtesy of Daniel Stoffler, M.E. Müller Institute, Biozentrum, Basel

teriophage head polypeptides (gp 23). They have been extensively studied by electron microscopy and image processing. The regular array of a single protein species make them well-suited for AFM investigations provided they can be sufficiently immobilized.

Isolated polyheads were covalently immobilized on a glass surface that had been modified with a photoactivatable cross-linker (Karrasch et al. 1993). Under these conditions, they could be repeatedly imaged in phosphate buffer at low (Fig. 1.6) and high (inset) magnification. In addition to the tubular structures of diverse length, there are many significantly smaller background particles visible at low magnification. Electron microscopy of negatively stained preparations revealed that these particles represent fragments of viral structures. In many instances, preparation techniques for electron microscopy result in a collapse of the tubes into planar double layers. In contrast, height measurements of polyheads imaged in phosphate buffer demonstrated that their cylindrical shape was preserved during AFM in solution. High magnification images (inset) showing individual capsomeres corroborate this finding. The capsomeres exhibit the characteristic pattern of gp 23 hexamers which are clearly seen at the top of the cylinder whereas they appear distorted at the periphery because of their tilt with respect to the stylus.

1.4.6
E. coli OmpF Porin Surfaces

OmpF porin is a major channel-forming protein in the outer membrane of *Escherichia coli*. It functions as a molecular sieve, allowing passage of hydrophilic molecules of up to 600 daltons in both directions. The atomic structure of the trimeric channel protein has been solved by X-ray crystallography (Cowan et al. 1992). Each 37 kD monomer comprises a barrel made of 16 antiparallel β-strands forming a channel. They are connected by short turns on the periplasmic surface and by loops of variable length on the extracellular surface. Likewise, two distinct surface topographies have been observed by AFM of two-dimensional OmpF crystals in solution (Schabert et al. 1995).

For AFM imaging, double-layered rectangular OmpF crystals were adsorbed to freshly cleaved mica. Under these conditions, the periplasmic side of double-layered crystals is exposed to the scanning stylus and can be reproducibly imaged at high resolution at forces below 300 pN. Its unprocessed surface topography is shown in Fig. 1.7. Densely packed triplet channels which protrude less than 0.5 nm from the lipid bilayer confer a relatively smooth appearance to the periplasmic surface. At higher magnification (inset), correlation averaged images reveal structural details at a lateral resolution of 0.8 nm as determined by Fourier ring correlation function (Schabert and Engel 1994). The channels are divided by a 1.2 nm thick tripartite septum. Three protrusions close to the threefold axis encircle a small depression. The channels have an elliptical cross-section of a = 3.4 nm, b = 2.0 nm. Assuming a lateral resolution of 1 nm and a vertical resolution of 0.1 nm, the periplasmic surface topography is in excellent agreement with the atomic structure (Schabert et al. 1995).

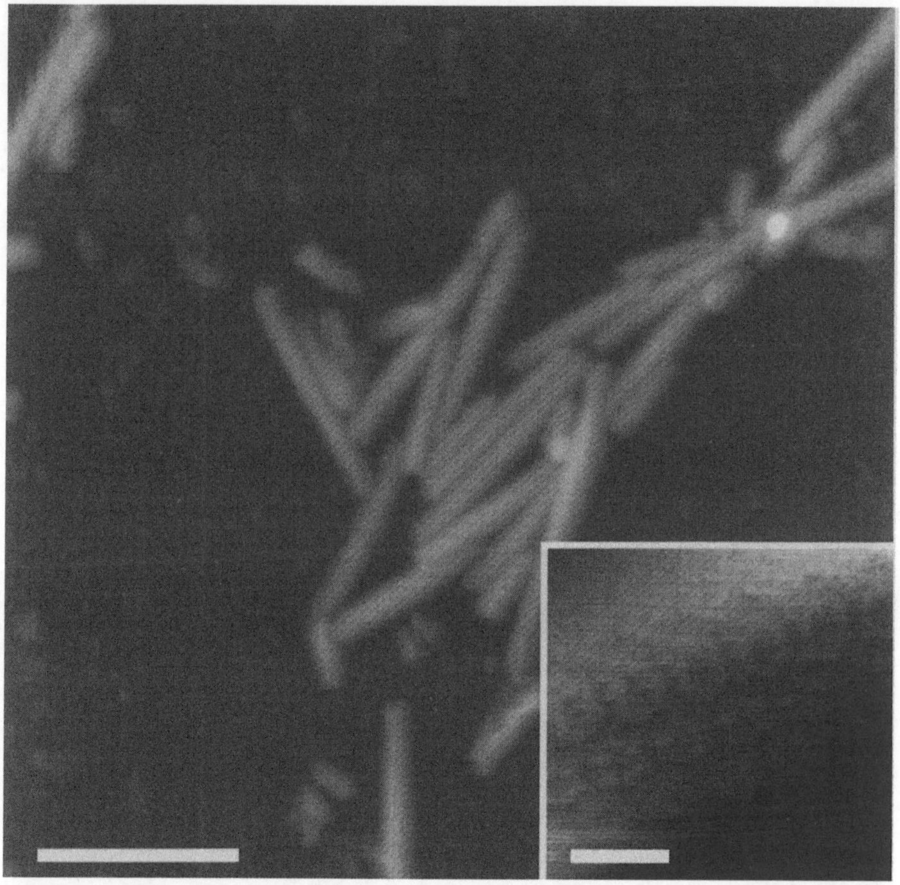

Fig. 1.6. AFM image of bacteriophage T4 polyheads on derivatized glass in buffer solution. A-type polyheads were produced by infecting the nonpermissive *E.coli* B^e with a 20^- mutant of bacteriophage T4. Isolated polyheads were covalently immobilized on a glass surface that had been modified with the photoactivatable cross-linker *N*-5-azido-2-nitrobenzoyloxysuccinimide (ANB-NOS; Karrasch et al. 1993) and imaged in phosphate buffer. At low magnification, the distribution of the polyheads is seen. They exhibited a width of 151±31 nm ($n=30$), a height of 57±14 nm ($n=30$ nm), and a variable length. While the width is significantly enlarged by tip effects, the height values suggest that the cylindrical polyheads which are approximately 60 nm in diameter (Steven et al. 1976), are not compressed by the stylus. Background particles are viral structures (*scale bar*, 1 μm). The *inset* shows the capsomeres at higher magnification. The characteristic pattern of gp 23 hexamers is well resolved (*scale bar*, 40 nm). (From Karrasch et al. 1993, with permission)

1.4.7 Bacteriorhodopsin

Bacteriorhodopsin is a light driven proton pump spanning the inner cell membrane of the archaebacterium *Halobacterium salinarium*. The protein forms a 1:1 complex with the retinal chromophore, which gives the protein its characteristic purple color. The purple membrane (PM) is a naturally oc-

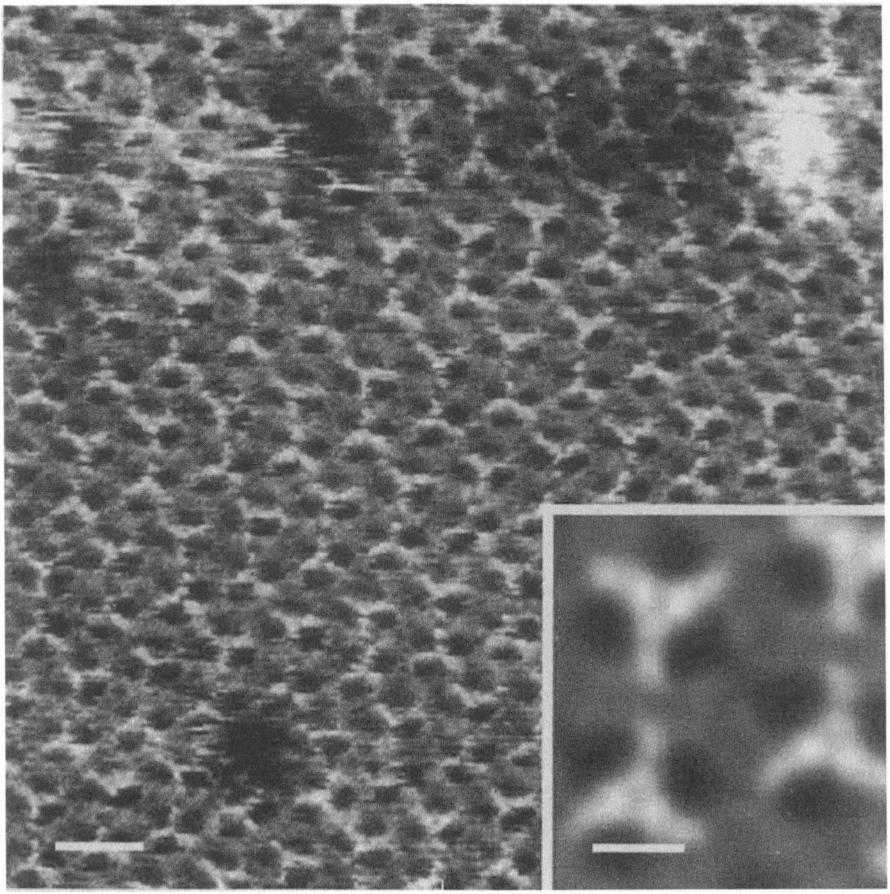

Fig. 1.7. Periplasmic surface of 2D OmpF porin-phospholipid crystals. Double-layered rectangular OmpF crystals that were reconstituted from *E. coli* OmpF porin and phospholipids according to Hoenger et al. (1993), expose the periplasmic surface to the aqueous environment. The topograph was recorded in 20 mM Hepes, 100 mM NaCl, 2 mM $MgCl_2$, pH 7.0 with an oxide-sharpened Si_3N_4 tip on a 100 µm-long cantilever with a spring constant $k = 0.1$ N/m (Olympus Ltd., Tokyo, Japan). Porin trimers are arranged in a rectangular lattice with unit cell dimensions a = 8 nm and b = 14nm. Three channels can be seen per trimer (*scale bar*, 15 nm). The *inset* shows the average of 25 translationally aligned subframes that reveal the fine structure of the periplasmic surface at a lateral resolution of 8 to 15 Å, depending on the criteria used (Schabert and Engel 1994). Three protrusions are clustered about the threefold symmetry axis with a small depression in the middle (*scale bar*, 4 nm). (From Schabert et al. 1995, with permission)

curring two-dimensional array of proteins that consists of 75% (w/w) bacteriorhodopsin and 25% lipids. High-resolution electron cryo-microscopy has led to an atomic model which predicts seven α-helical membrane-spanning segments (Henderson et al. 1990; Grigorieff et al. 1996). Little was known about the loops that connect the individual α-helices at the respective membrane surface until recently. The topography of native purple membrane recorded by AFM at subnanometer resolution (Müller et al. 1995 a, b) provides

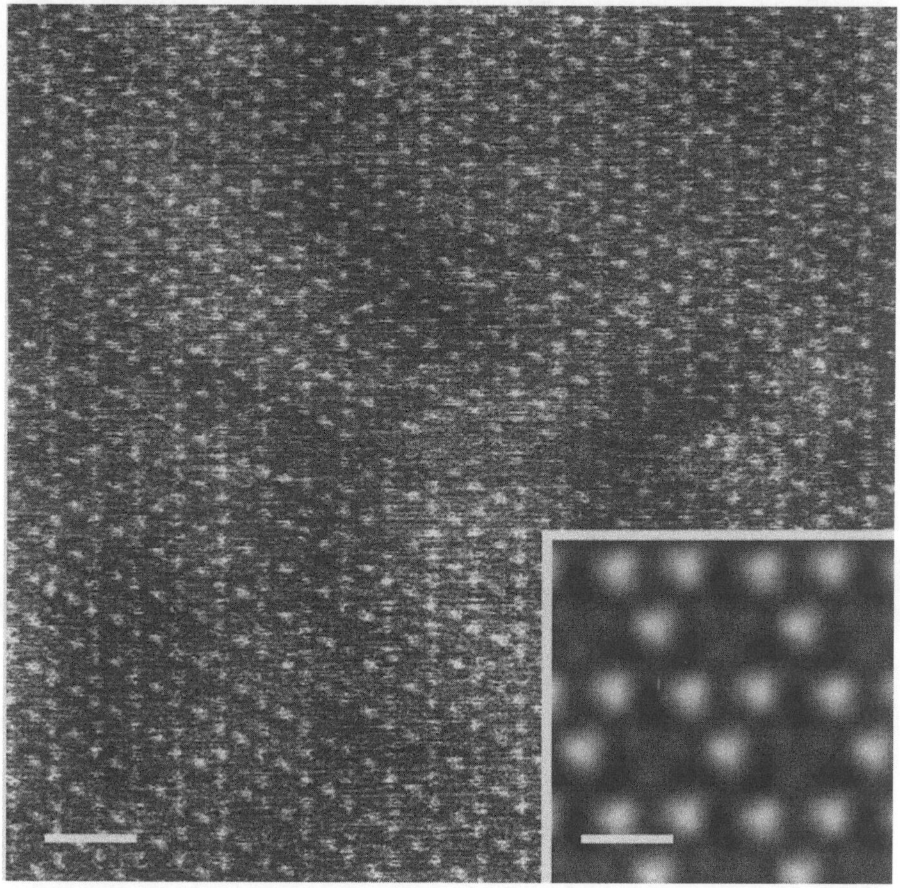

Fig. 1.8. High-resolution topography of the extracellular surface of native purple membrane. Isolated purple membranes were adsorbed at room temperature to freshly cleaved mica from a solution containing 10 μg protein/ml in 10 mM Tris–HCl, pH 8, 150 mM KCl. After 10 min, membranes that had not attached were gently washed away with the same buffer solution and imaging was performed in contact mode AFM with the same type of cantilever as in Fig. 1.3. In the overview, the arrangement of trimeric protrusion on a trigonal lattice (a = b = 6.2±0.2 nm) is recognized (*scale bar*, 12 nm). Enhanced by threefold symmetrization, the correlation averaged images recorded in trace direction (*inset*) reveal three major domains arranged on an equilateral triangle of 2.9 nm side length, and three minor domains separated by 1.9 nm. (From Müller et al. 1996a, with permission)

valuable information that complements the atomic model derived from EM data. Indeed, there is an excellent correlation of the AFM data with the electron-crystallographic refinement of the bacteriorhodopsin structure (Grigorieff et al. 1996).

The surface topography of the extracellular surface is shown in Fig. 1.8. The trigonal lattice is apparent at low magnification. The highest features are trimeric protrusions that project about 0.2 nm from the supporting lipid bilayer. It is conceivable that the carbohydrate moieties of the glycolipids pre-

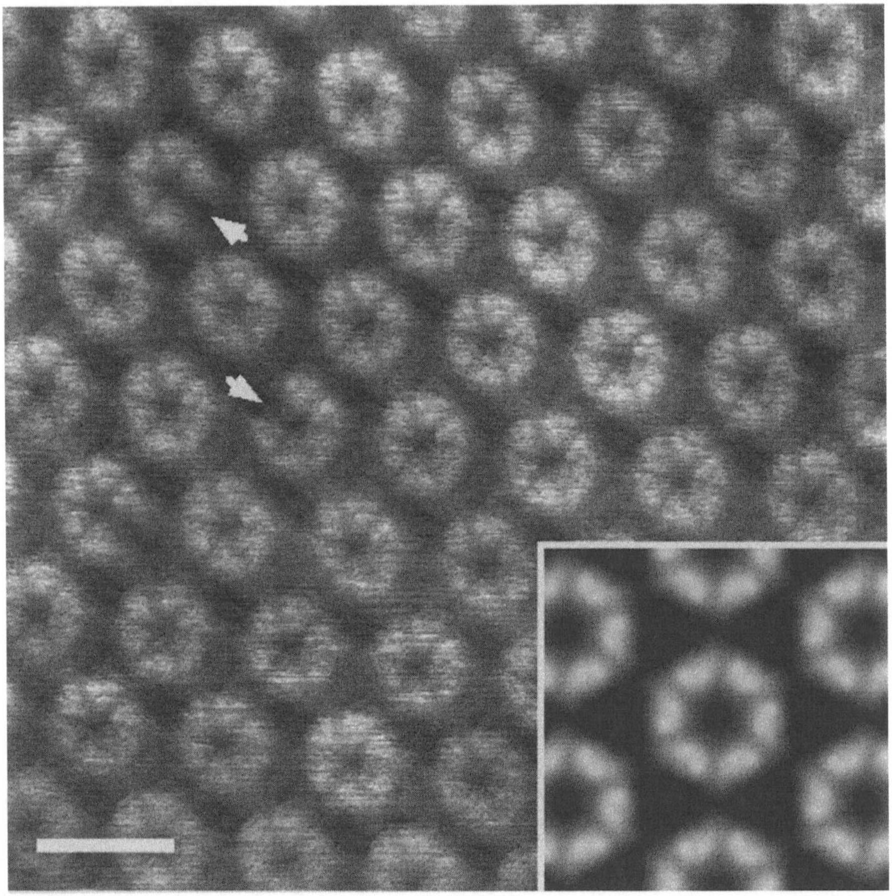

Fig. 1.9. High-resolution AFM topograph of the outer surface of HPI. HPI layers were extracted from *Deinococcus radiodurans* with lithium dodecyl sulfate and deposited on freshly cleaved mica at 20 μg/ml in 10 mM Tris–HCl, pH 8.2, 100 mM KCl, 20 mM $MgCl_2$. Imaging was performed in the same buffer with an oxide-sharpened Si_3N_4 tip (100 μm long cantilever with a spring constant of $k = 0.38$ N/m; Digital Instruments Inc.) and imaging forces of approximately 100 pN. The outer surface exhibits donut-like structures with a core consisting of six V-shaped protrusions and a central pore. Occasionally, a subunit (*arrowheads*) is missing. The six protomers are arranged on a hexagon with a side length of 4.9±0.2 nm (*scale bar*, 20 nm). Six spokes that extend from the corners of the hexagon are more clearly revealed in the averaged and sixfold symmertrized topograph shown in the *inset* (a = b = 18 nm). Image courtesy of Simon Scheuring, M.E. Müller Institute, Biozentrum, Basel

sent in the bilayer prevent the direct contact of the probe with the lipid head groups. After threefold symmetrization of correlation-averaged images recorded at 0.7 nm lateral resolution (inset), one major and one minor trimeric protrusion per bacteriorhodopsin was revealed. The minor trimer had a corrugation amplitude of 0.08 nm. Both tripartite domains are arranged on equilateral triangles. The distance between the major domains was 2.9 nm, while the three minor domains were 1.9 nm apart.

The topography of the cytoplasmic surface is shown in Fig. 1.11. A comparison of this surface with the atomic model derived from EM data will be discussed below.

1.4.8
Hexagonally Packed Intermediate Layer

The crystalline surface layer (S layer) of protein or glycoprotein subunits is an essential consituent of cell envelopes in most archaebacteria and in eubacteria (Baumeister and Lembcke 1992). A typical S layer, the hexagonally packed intermediate layer (HPI) of the radiotolerant eubacterium *Deinococcus radiodurans* has been extensively studied by electron microscopy (Baumeister et al. 1986), scanning tunneling microscopy (Guckenberger et al. 1989), and most recently by atomic force microscopy (Karrasch et al. 1993, 1994; Müller et al. 1996b).

Single and multiple HPI layers extracted from *Deinococcus radiodurans* adsorb with the hydrophilic outer surface to the freshly cleaved mica. If, as shown in Fig. 1.9, double-layered stacks adsorb with one hydrophilic surface, then the hydrophobic surfaces of the two layers will interact with each other, thereby exposing the opposite hydrophilic surface. The high-resolution topograph of the outer hydrophilic surface reveals a hexagonal lattice ($a = b = 18 \pm 0.4$ nm) of donut-like units with six V-shaped protomers arranged around a central pore. To enhance the signal-to-noise ratio, correlation averages ($n = 10$) were calculated and sixfold symmetrized (inset). The V-shaped protrusions are arranged on an equilateral hexagon with a side length of 4.9 ± 0.2 nm, and a spoke-like protrusion emanates from each corner of the core. The height difference between the V-shaped protrusion and the spokes was 1.2 ± 0.2 nm, while the maximum height of the protomers was 2.2 ± 0.7 nm.

The hydrophobic inner surface is exposed to the stylus in single HPI layers that have attached to the mica with their hydrophilic surface. As illustrated in Fig. 1.13, the inner surface displays regular arrays of donut-like structures. The three-dimensional map of negatively stained HPI layers indicates the presence of a small plug about the sixfold axis (Baumeister et al. 1986). Correspondingly, a central plug is observed by AFM in many of the hexameric core structures. The inset shows a montage of sixfold-symmetrized correlation averages of the "closed" and "open" conformation found on the inner surface of HPI.

1.5
Identification of Structures Observed by AFM

One of the main challenges of AFM is the unambiguous identification of the structures imaged and their quantitative interpretation. In most instances, those samples that produced high resolution images were biochemically well characterized and rather homogeneous. Immobilization and imaging conditions can be optimized to improve the image quality. Highest resolution has been obtained with geometrically defined regular arrays of densely packed proteins, specifically with naturally occurring two-dimensional crystals (Hoh et al. 1993; Karrasch et al. 1994; Müller et al. 1995a,b) and with proteins reconstituted into lipid bilayers (Yang et al. 1993; Mou et al. 1995, 1996; Schabert et al. 1995). A tentative assignment of the surface topography to the extra- or intracellular side was possible by comparing AFM images with structural data obtained by electron microscopy and X-ray crystallography provided this information was available. For example, it has been shown for the HPI layer, that topographs differ from those determined by electron microscopy by less than 0.5 nm (Karrasch et al. 1994). However, if the sample is inhomogeneous or poorly characterized by other methods, a direct identification is essential.

Here, we would like to discuss two approaches for structural identification, one depending on a technical development of the instrumentation, and the second taking advantage of the specificity of antigen–antibody interactions. Alone or in combination, these two approaches provide a powerful tool to identify specific structures in the image.

In AFM, the image is generated by probe–sample interactions without direct optical access. With the recent combination of a light microscope and an atomic force microscope (Putman et al. 1993a; Schabert et al. 1994; Hillner et al. 1995; Stemmer 1995) this limitation could be overcome. Simultaneous AFM and optical images allow a direct correlation of information recorded with different imaging modes.

The plasma membrane of living cells is very soft and deforms considerably in response to the imaging forces of ~1–2 nN (Hoh and Schoenenberger 1994). Cells that spread out are often less than 100 nm thick in the periphery. In these areas, relatively rigid intracellular components of the cytoskeleton can be visualized by AFM (Henderson et al. 1992; Chang et al. 1993; Hoh and Schoenenberger 1994). The question whether the stylus penetrates the plasma membrane and directly interacts with the cytoskeleton or whether the soft membrane deforms around the more rigid cytoskeleton has not yet been answered conclusively. The definite identification of cytoskeletal structures is shown in Fig. 1.10. Polymerized actin filaments in human fibroblasts were labeled with fluorescent phalloidin. Coverslips were first examined by fluorescence microscopy (inset) and a selected area was subsequently scanned by AFM. Based on the F-actin-specific labeling in the fluorescent image, the corresponding cytoskeletal structures monitored by AFM can be clearly identified as actin stress fibers. An additional advantage of the combined light/atomic force microscope is that sparse samples can be readily detected and the scan area specifically selected.

Immunolabeling techniques applied in light and electron microscopy have provided a wealth of structural information at the cellular and molecular level. These techniques have recently found their way into combined light/atomic force microscopy. However, the cells were fixed for immunolabeling and the option of imaging native structures by AFM is lost. We have recently reported an alternative approach where the native bacteriorhodopsin surfaces were identified by specific antibodies in a physiological environment (Müller et al. 1996a). At low magnification, individual membrane patches adsorbed to the atomically flat mica appeared indistinguishable, whereas two distinct

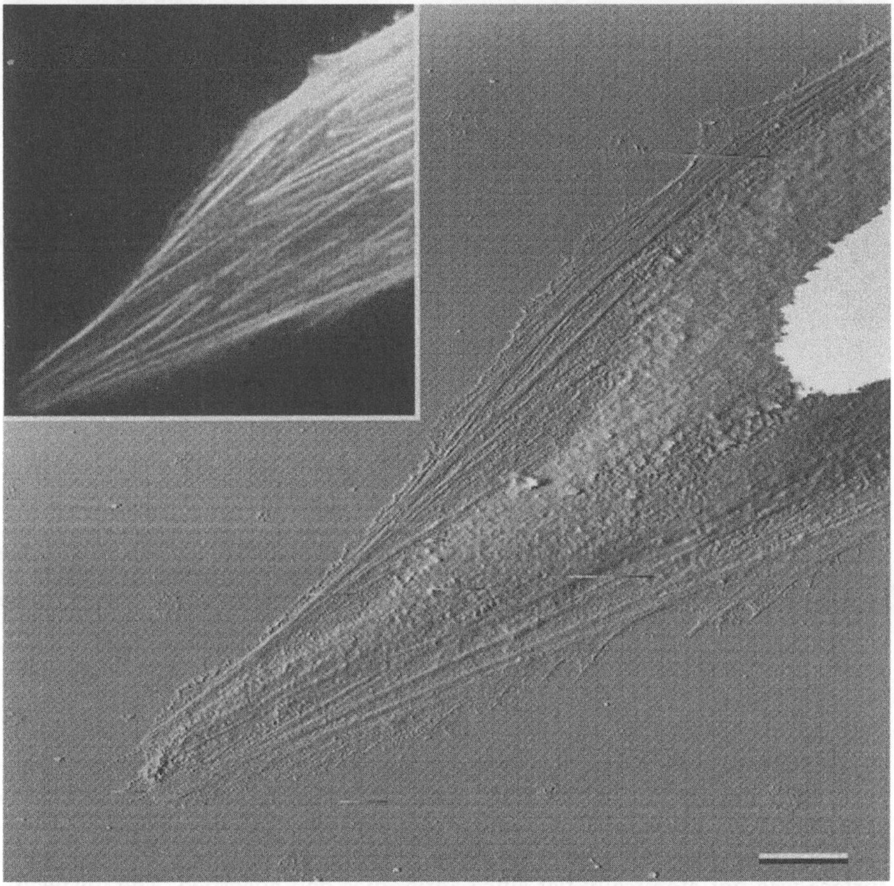

Fig. 1.10. Actin stress fibers visualized by fluorescence and atomic force microscopy. Human fibroblast cells were plated on a glass coverslip glued in a mounting support and cultured for 24 h at 37 °C in a 95% air/5% CO_2 humidified incubator. Cells were permeabilized with 2% octyl-polyoxyethylene, 0.125% glutaraldehyde in a physiological buffer and labeled with fluorescent (TRITC-labeled) phalloidin. Prior to imaging, the buffer was substituted with glucose oxidase, glucose, and catalase to reduce bleaching of the fluorochrome. Images were obtained with a Bioscope (Digital Instruments Inc.) mounted on a Zeiss Axiovert 35 (Zeiss, Oberkochen, Germany). Labeled cells were examined with a 63× oil immersion objective (NA 1.4) and fluorescence images were recorded with a MRXi video camera (Photonic Science, Zeiss, Oberkochen). Simultaneous AFM images were acquired in contact mode using Si_3N_4 Nanoprobe tips (Digital Instruments, Inc.). The phallotoxin produced by *Amanita phalloides* specifically binds to polymerized actin. Fibroblasts display an extensive network of labeled actin stress fibers (*inset*). The intracellular fibrous structures seen in the AFM error signal image correspond to the actin stress fibers (*scale bar*, 10 μm). Image courtesy of Rainer Pansky, M.E. Müller Institute, Biozentrum, Basel

surface topographies were observed at constant imaging forces at high magnification (Müller et al. 1995b). However, within one membrane patch only one type of bacteriorhodopsin structure occurred. This finding suggests that purple membranes attach with either side to freshly cleaved mica. To identify the orientation, purple membrane patches adsorbed to freshly cleaved mica

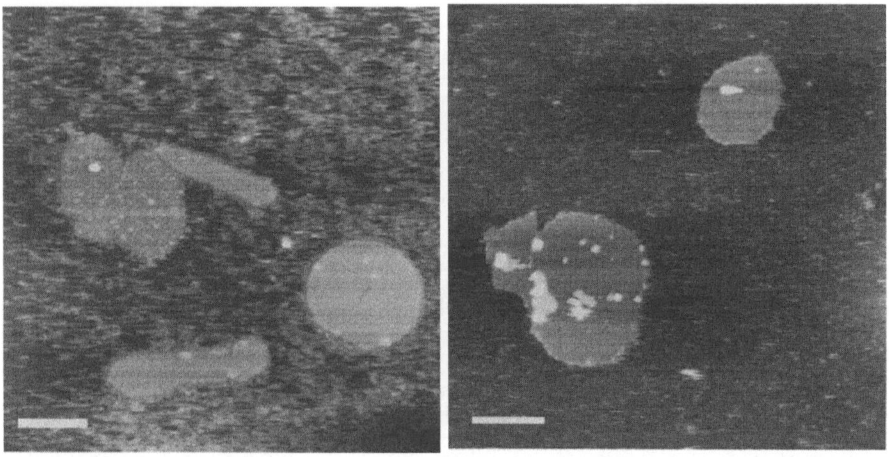

Fig. 1.11. Immuno-AFM of purple membrane. Native purple membranes were adsorbed to freshly cleaved mica in 150 mM KCl, 10 mM Tris, pH 9.2. Antibodies recognizing the carboxy-terminus of bacteriorhodopsin were added to the buffer solution to a final concentration of 2 µg/ml. After a 2 h incubation at room temperature, the samples were directly imaged in contact mode using the same type of cantilever as in Fig. 1.3. *Left panel* Some membrane patches are densely labeled, whereas others display an untextured surface. At loading forces between 100 and 200 pN, antibodies that are unspecifically bound to the mica surface, are also observed. *Right panel* After removal of the epitope by papain digestion, all purple membranes remain untextured in the presence of antibodies. Loading forces above ~250 pN are required to sweep away the antibodies that are bound to the mica surface (data not shown; *scale bars*, 500 nm). (From Müller et al. 1996a, with permission)

were incubated with antibodies directed against the carboxy-terminus of bacteriorhodopsin (Wölfer et al. 1988), and subsequently imaged in their presence (Fig. 1.11, left panel). A number of membrane patches displayed a textured surface whereas others appeared smooth. In addition, the free mica surface was not atomically flat since antibodies had adsorbed. The unspecifically bound antibodies could be displaced at forces above 0.25 nN. The antibodies that had specifically bound to the cytoplasmic membrane surface required forces above 0.8 nN to be removed. The specificity of the antibody recognition is further demonstrated in Fig. 1.11 (right panel). Limited papain digestion of purple membranes removes the carboxy-terminus of bacteriorhodopsin. After papain proteolysis (Fig. 1.11, right panel), all membrane fragments displayed a smooth topography because the exposed cytoplasmic surfaces are no longer recognized by the antibodies.

The specific antigen–antibody interaction can be disrupted by increasing the force. Transient antibody binding allows identification of biomolecules in supramolecular assemblies without perturbing the native conformation. Subsequently, these structures can be imaged at high resolution.

1.6 Conformational Changes Monitored by AFM

The examples discussed in the previous section clearly illustrate that the AFM has the potential to map the surface topography of biological macro-

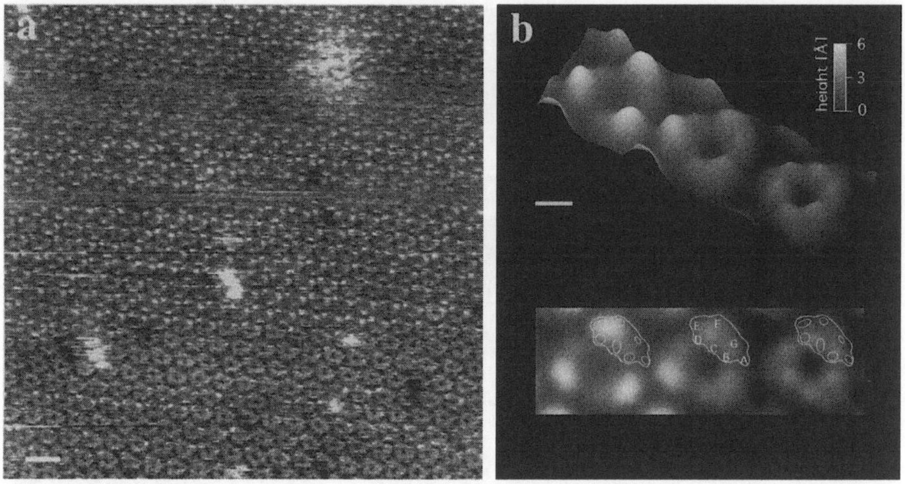

Fig. 1.12. Force-induced conformational change of bacteriorhodopsin. *Panel* **a**, the effect of force variations on the cytoplasmic surface of purple membrane. The initial force of 300 pN (*bottom* of the image) was manually reduced during the scan to 100 pN (*top* of the image). The cantilevers used are described in Fig. 1.3. With the decrease in force, the donut-shaped bacteriorhodopsin trimers transform into units with three pronounced protrusions at their periphery (*scale bar,* 10 nm). The force dependent conformational changes are interpreted in *panel* **b**. A perspective view (*top*) of the transition from native (*left*) to donut-shaped bacteriorhodopsin trimers (*right*) is calculated from several hundred unit cells from images recorded at 50 to 150 pN (*left*) and 300 pN (*right*) that were averaged, 3-fold symmetrized, and scaled according to the microscope parameters. The *central* trimer is a composition of the *left* trimer recorded at 100 pN and the *right* trimer recorded at 300 pN. (*scale bar,* 2 nm). In the corresponding topographs *below,* the helices derived from the atomic model (Henderson et al. 1990) are superimposed. (From Müller et al. 1995b, with permission)

molecules in their native environment at subnanometer resolution. Comparison of topographs with data from both electron microscopy (Karrasch et al. 1994) and X-ray crystallography (Schabert et al. 1995) confirm the high accuracy of the structural information acquired with the AFM. It has been demonstrated that the AFM can directly monitor dynamic processes (Drake et al. 1989; Radmacher et al. 1994; Schoenenberger and Hoh 1994). Ultimately, it should be possible to correlate structural changes with the function of a particular molecule. First steps in this direction are discussed in the following.

Intermolecular forces in the range of a few hundred pN have recently been measured by AFM (Florin et al. 1994; Lee et al. 1994; Dammer et al. 1995, 1996). It is likely that delicate structures contoured by the AFM tip at comparable loading forces change their conformation during the imaging process. Three questions arise: (1) what are the smallest structures that can be reproducibly imaged without perturbing their authentic conformation? (2) Can this limit be extended by reducing the loading forces, and (3) are the conformational changes reversible? Recent experiments with the purple membrane may provide some answers (Müller et al. 1995b). Depending on the loading force, the cytoplasmic surface of PM (Fig. 1.12a) exhibits two confor-

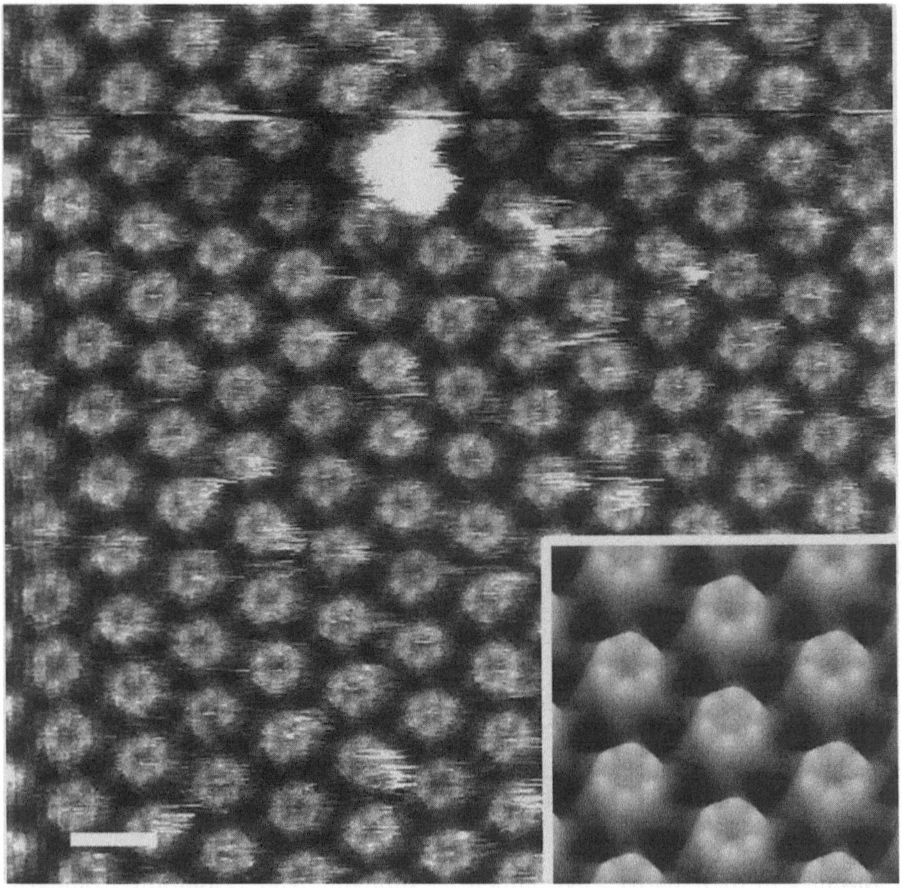

Fig. 1.13. Two conformations of the hydrophobic surface of HPI layers. Single HPI layers are attached with their hydrophilic surface to freshly cleaved mica and expose the inner, hydrophobic surface. The image was acquired in buffer solution (100 mM KCl, 20 mM $MgCl_2$, 10 mM Tris, pH 8.2) with an electron-beam deposited carbon superstylus (Keller and Chih-Chung, 1992) at a vertical force of less than 150 pN. Densely packed donut-like structures are seen. Some pores of the protruding hexameric cores display an open conformation and others an obstructed conformation. It appears that the conformation randomly switches over time (Müller et al. 1996b; *scale bar*, 20 nm). Sixfold-symmetrized correlation averages of both open and close conformations were assembled in a montage (*inset*). The distinct arms emanating from the cores of the hexamers exhibit an anticlockwise rotation. The full grey level range of the average corresponds to 3 nm vertical distance and the distance between the pores corresponds to 18 nm. (From Müller et al. 1996b, with permission)

mations of bacteriorhodopsin. At the bottom of the image, a loading force of ~300 pN results in donut-shaped bacteriorhodopsin trimers which upon reducing the force to ~100 pN (top) transform to trimeric structures with three pronounced protrusions at their periphery. The transition is fully reproducible in the same scan area provided the force does not exceed 500 pN. A perspective view of the transition from native (left) to donut-shaped (right) bacteriorhodopsin trimers is shown in Fig. 1.12b with averaged, threefold symmetrized bacteriorhodopsin contours recorded at 100 pN and 300 pN, respectively. From the position of the seven transmembrane helices which were mapped on the surface topography of an individual bacteriorhodopsin molecule we can tentatively assign surface loops to the protruding structures (Fig. 1.12b, bottom panel). It is noteworthy that the most prominent surface loop which connects helices E and F, projects 0.6 nm from the bilayer surface at low forces, but is not discernible at 300 pN. It is tempting to speculate that the floppy loop bends with higher forces.

As shown in Fig. 1.13, the inner surface of the HPI also exhibited two conformations which is emphasized by the montage assembled from correlation averages of the open and the closed state. Sometimes, individual pores that displayed an open conformation in one scan were found to be closed several scans later and vice versa. It is not known what induces the switching and it can not be absolutely ruled out that the conformational changes are tip-induced. However, the switching of conformations was entirely random and reversible with different types of tips and at forces below 200 pN. For these reasons, the switching may be considered a specific property of the HPI layer and a biological significance of the two conformations seems likely.

1.7 Conclusions and Perspectives

In atomic force microscopy, images are obtained by raster-scanning a sharp stylus over the sample. The surface is contoured by measuring the force between atoms at the surface and atoms at the apex of the stylus in air, vacuo, or, what is most exciting for biologists, in liquid. Within 10 years AFM has developed into an important tool for structural biology. It is the only technique that allows biomolecules to be imaged in their native state at submolecular resolution. The resolution to be achieved, however, depends largely on the type of sample. Complex surfaces with an undefined composition and distribution of proteins, such as the plasma membrane of living cells have yielded only limited resolution. Part of the difficulties are inherent in the sample; the dynamic deformation of plasma membranes in response to the imaging forces and chemical interactions between the surface and the tip prevent high resolution. First steps have been made towards overcoming these difficulties. In tapping mode, the cell appears to stiffen under the oscillating cantilever resulting in a slightly improved resolution (Putman et al. 1994b). Other attempts to decrease deformation include the hypotonic swelling of the cell, or its immobilization in the pores of polycarbonate filters (Kasas and Ikai 1995). Moreover, tip surfaces have been modified to decrease adhesive tip–sample interactions. In some instances, tissue and cell surfaces have been enzymatically treated to remove viscous components (Le Grimellec et al. 1994; Jurvelin et al. 1996). The other major difficulty encountered with heterogeneous macromolecular assemblies is the unambiguous identification of surface structures. Encouraging results have recently been obtained by the use of specific antibodies, and by gold labels (Putman et al. 1993b; Eppell et al. 1995; Müller et al. 1996a). Technical developments of the instrumentation allow the combination of light microscopy with AFM (Putman et al. 1993a;

Schabert et al. 1993; Hillner et al. 1995) which will greatly facilitate the location and identification of the surface. Despite these improvements, imaging cell surfaces at molecular resolution has yet to be demonstrated.

To obtain high resolution surface topographs, samples must be immobilized thereby preventing the displacement of the molecules by the scanning tip. To this end, several approaches have been developed. With better understanding of the mechanisms involved in sample–substrate interactions it will be possible to sufficiently immobilize most samples with a minimum of interference with their structure and function. Images of regular protein arrays have demonstrated the possibility of achieving a lateral resolution of 0.5–1 nm. Sharper tips that have a defined geometry and a reduction of the forces applied during imaging appear to be achievable, and will further improve imaging of biomolecules with the AFM. The force-induced conformational changes observed on bacteriorhodopsin trimers in purple membranes (Fig. 1.12) demonstrate that it is possible to directly monitor conformational changes with the AFM. However, imaging forces should not exceed 100 pN to ensure that the topography represents a native conformation. Together with the observation that one and the same HPI molecule reversibly exhibits two distinct conformations at different times (Fig. 1.13), these findings indicate the feasibility of directly recording function-related conformational changes of native biological macromolecules with the AFM. Tapping mode AFM has been used to detect the conformational changes of active lysozyme, but the magnitude of the height fluctuations observed are not as yet fully understood (Radmacher et al. 1994).

Extrapolating from the progress made during the past ten years, we can look forward to an exciting decade of AFM revealing molecules at work.

References

Aebersold RH, Teplow DB, Hood LE, Kent SBH (1986) Electroblotting onto activated glass. J Biol Chem 261:4229–4238

Akey CW, Radermacher M (1993) Architecture of the *Xenopus* nuclear pore complex revealed by three-dimensional cryo-electron microscopy. J Cell Biol 122:1–19

Albrecht TR, Dovek MM, Kuan SWJ, Lang CA, Grütter P, Krank CW, Pease RFW, Quate CF (1988) Imaging and modification of polymers by scanning tunneling and atomic force microscopy. J Appl Phys 64:1178–1182

Amrein M, Müller DJ (1997) Sample preparation techniques in scanning probe microscopy. In: Roberts C (ed) Biological application of SPM. Techniques in biomedical microscopy. Wiley-Liss, London (in press)

Baumeister W, Barth M, Hegerl R, Guckenberger R, Hahn M, Saxton WO (1986) Three-dimensional structure of the regular surface layer (HPI Layer) of *Deinococcus radiodurans*. J Mol Biol 197:241–253

Baumeister W, Lembcke G (1992) Structural features of archaebacterial cell envelopes. J Bioenerg Biomembr 24:567–575

Bezanilla M, Drake B, Nudler E, Kashlev M, Hansma PK, Hansma HG (1994) Motion and enzymatic degradation of DNA in the atomic force microscope. Biophys J 67:2454–2459

Binnig G, Quate CF, Gerber C (1986) Atomic force microscope. Phys Rev Lett 56:930–933

Binnig G, Gerber C, Stoll E, Albrecht RT, Quate CF (1987) Atomic resolution with atomic force microscopy. Europhys Lett 3:1281–1286

Butt H-J, Siedle P, Seifert K, Fendler K, Seeger T, Bamberg E, Weisenhorn AL, Goldie K, Engel A (1993) Scan speed limit in atomic force microscopy. J Microsc 169:75–84

Chang L, Kious T, Yorgancioglu M, Keller D, Pfeiffer J (1993) Cytoskeleton of living unstained cells imaged by scanning force microscopy. Biophys J 64:1282–1286

Cowan SW, Schirmer T, Rummel G, Steiert M, Ghosh R, Pauptit RA, Jansonius JN, Rosenbusch JP (1992) Crystal structures explain functional properties of two *E. coli* porins. Nature 358:727–733

Dammer U, Popescu O, Wagner P, Anselmetti D, Güntherodt H-J, Misevic GM (1995) Binding strength between cell adhesion proteoglycans measured by atomic force microscopy. Science 267:1173–1175

Dammer U, Hegner M, Anselmetti D, Wagner P, Dreier M, Huber W, Güntherodt H-J (1996). Specific antigen/antibody interactions measured by force microscopy. Biophys J 70:2437–2441

D'Costa NP, Hoh JH (1995) Mussel adhesive protein as a novel substrate for atomic force microscopy. In: Bailey GW, Ellisman MH, Hennigar RA, Zaluzec NJ (eds) Proc Micro Microanal, Jones and Begell, New York, pp 726–727

Drake B, Prater CB, Weisenhorn AL, Gould SAC, Albrecht TR, Quate CF, Cannell DS, Hansma HG, Hansma PK (1989) Imaging crystals, polymers and biological processes in water with AFM. Science 243:1586–1589

Eppell SJ, Simmons SR, Albrecht RM, Merchant RE (1995) Cell-surface receptors and proteins on platelet membranes imaged by scanning force microscopy using immunogold contrast enhancement. Biophys J 68:671–680

Florin E-L, Moy VT, Gaub HE (1994) Adhesion forces between individual ligand-receptor pairs. Science 264:415–417

Fritz M, Radmacher M, Cleveland JP, Allersma MW, Stewart RJ, Gieselmann R, Janmey P, Schmidt CF, Hansma PK (1995) Imaging globular and filamentous proteins in physiological buffer solutions with tapping mode atomic force microscopy. Langmuir 11:3529–3535

Ghadially FN (1983) Fine structure of the synovial joints. A text and atlas of the ultrastructure of normal and pathological articular tissues. Butterworth, London

Goldie K, Panté N, Engel A, Aebi U (1994) Exploring native nuclear pore complex structure and conformation by scanning force microscopy in physiological buffers. J Vac Sci Technol 12:1482–1485

Grigorieff N, Ceska TA, Downing KH, Baldwin JM, Henderson R (1996) Electron-crystallographic refinement of the structure of bacteriorhodopsin. J Mol Biol 259:393–421

Guckenberger R, Wiegräbe W, Hillebrand A, Hartman T, Wang Z, Baumeister W (1989) Scanning tunneling microscopy of a hydrated bacterial surface protein. Ultramicroscopy 31:327–332

Häberle W, Hörber JKH, Binnig G (1991) Force microscopy on living cells. J Vac Sci Technol 9:1210–1213

Häberle W, Hörber JK, Ohnesorge F, Smith DP, Binnig G (1992) In situ investigations of single living cells infected by viruses. Ultramicroscopy 42–44:1161–1667

Hansma PK, Cleveland JP, Radmacher M, Walters DA, Hillner P, Bezanilla M, Fritz M, Vie D, Hansma HG (1994) Tapping mode atomic force microscopy in liquids. Appl Phys Lett 64:1738–1740

Hansma HG, Revenko I, Kim K, Laney DE (1996) Atomic force microscopy of long and short double-stranded, single-stranded, and triple-stranded nucleic acids. Nucleic Acids Res 24:713–720

Hegner M, Dreier M, Wagner P, Semenza G, Güntherodt HJ (1996) Modified DNA immobilized on bioreactive self-assembled monolayer on gold for dynamic force microscopy imaging in aqueous buffer solution. J Vac Sci Technol 14:1418–1421

Heins S, Aebi U (1994) Making heads and tails of intermediate filament assembly, dynamics and networks. Curr Opin Cell Biol 6: 25–33

Henderson E (1994) Imaging of living cells by atomic force microscopy. Prog Surf Sci 46:39–60

Henderson E, Haydon PG, Sakaguchi DS (1992) Actin filament dynamics in living glial cells imaged by atomic force microscopy. Science 257:1944–1946

Henderson R, Baldwin JM, Ceska TA, Zemlin F, Beckman E, Downing KH (1990) Model for the structure of bacteriorhodopsin based on high-resolution electron cryo-microscopy. J Mol Biol 213:899–929

Hillner PE, Walters DA, Lal R, Hansma HG, Hansma PK (1995) Combined atomic force and confocal laser scanning microscope. J Microsc Soc Am 1:127–130

Hoh J, Schoenenberger C-A (1994) Surface morphology and mechanical properties of MDCK monolayers by atomic force microscopy. J Cell Sci 107:1105–1114

Hoh JH, Sosinsky GE, Revel J-P, Hansma PK (1993) Structure of the extracellular surface of the gap junction by atomic force microscopy. Biophys J 65:149–163

Holstein TW, Benoit M, v. Herder G, Wanner G, David CN, Gaub HE (1994) Fibrous minicollagens in hydra nematocysts. Science 265:402–404

Hönger A, Gosh R, Schoenenberger C-A, Aebi U, Engel A (1993) Direct in situ structural analysis of recombinant outer membrane porins expressed in an OmpA-deficient mutant *Escherichia coli* strain. J Struct Biol 111:212–221
Hörber JKH, Häberle F, Ohnesorge F, Binnig HG, Liebich C, Cserny P, Mahnel H, Mayr A (1992) Investigation of living cells in the nanometer regime with the scanning force microscope. Scanning Microsc 6:919–930
Jurvelin JS, Müller DJ, Wong M, Studer D, Engel A, Hunziker EB (1996) Surface and subsurface morphology of bovine humeral articular cartilage as assessed by atomic force and transmission electron microscopy. J Struct Biol 117:45–54
Karrasch S, Dolder M, Schabert F, Ramsden J, Engel A (1993) Covalent binding of biological samples to solid supports for scanning probe microscopy in buffer solution. Biophys J 65:2437–2446
Karrasch S, Hegerl H, Hoh JH, Baumeister W, Engel A (1994) Atomic force microscopy produces faithful high-resolution images of protein surfaces in an aqueous environment. Proc Natl Acad Sci USA 91:836–838
Kasas S, Ikai A (1995) A method for anchoring round shaped cells for atomic force microscope imaging. Biophys J 68:1678–1680
Keller D, Chih-Chung C (1991) Reconstruction of STM and AFM images distorted by finite-sized tips. Surf Sci 253:353–364
Keller D, Chih-Chung C (1992) Imaging steep, high structures by scanning force microscopy with electron beam deposited tips. Surf Sci 268:333–339
Kirk TB, Stachowiak GW, Wilson AS (1994) The morphology of the surface of articular cartilage. Proc 2nd World Congr Biomechanics, vol II, p nm208a
Le Grimellec C, Lesniewska E, Cachia C, Schreiber JP, Defornel F, Goudonnet JP (1994) Imaging the membrane surface of MDCK cells by atomic force microscopy. Biophys J 67:36–41
Lee GU, Chris LA, Colton RJ (1994) Direct measurment of the forces between complementary strands of DNA. Science 266:771–773
Lyubchenko Y, Shlyakhtenko L, Harrington R, Oden P, Lindsay S (1993) Atomic force microscopy of DNA: imaging in air and under water. Proc Natl Acad Sci USA 90: 2137–2140
Martin Y, Wickramasinghe HK (1987) Magnetic imaging by "force microscopy" with 1000 Å resolution. Appl Phys Lett 50:1455–1457
Mazia D, Schatten G, Sale W (1975) Adhesion of cells to surfaces coated with polylysine. J Cell Biol 66:198–200
Mazzola LT, Fodor SPA (1995) Imaging biomolecule arrays by atomic force microscopy. Biophys J 68:1653–1660
Meyer G, Amer NM (1988) Novel optical approach to atomic force microscopy. Appl Phys Lett 53:1045–1047
Mou J, Yang J, Shao Z (1995) Atomic force microscopy of cholera toxin B oligomer bound to bilayers of biologically relevant lipids. J Mol Biol 248:507–512
Mou J, Czajkowsky DM, Sheng S, Ho R, Shao ZF (1996) High resolution surface structure of *E. coli* GroES oligomer by atomic force microscopy. FEBS Lett 381:161–164
Moy VT, Florin E-F, Gaub HE (1994) Intermolecular forces and energies between ligands and receptors. Science 266:257–259
Müller DJ, Schabert FA, Büldt G, Engel A (1995a) Imaging purple membranes in aqueous solutions at subnanometer resolution by atomic force microscopy. Biophys J 68:1681–1686
Müller DJ, Büldt G, Engel A (1995b) Force-induced conformational change of bacteriorhodopsin. J Mol Biol 249:239–243
Müller DJ, Schoenenberger C-A, Engel A (1996a) Immuno-atomic force microscopy of purple membrane. Biophys J 70:1796–1802
Müller DJ, Baumeister W, Engel A (1996b) Conformational change of the hexagonally packed intermediate layer of *Deinococcus radiodurans* monitored by atomic force microscopy. J Bacteriol 178:3025–3030
Müller DJ, Amrein M, Engel A (1997) Adsorption of biological molecules to a support for scanning probe microscopy. J Struct Biol 119:172–188
Ohnesorge F, Binnig G (1993) True atomic-resolution by atomic force microscopy through repulsive and attractive forces. Science 260:1451–1456
Panté N, Aebi U (1993) The nuclear pore complex. J Cell Biol 122:977–984
Panté N, Aebi U (1995) Exploring nuclear pore complex structure and function in molecular detail. J Cell Sci 19:1–11
Peachy NM, Eckhardt JC (1994) Structural studies of ordered monolayers using atomic force microscopy. Micron 25:271–292

Putman CAJ, Van der Werf KO, De Grooth BG, Van Hulst NF, Greve J, Hansma PK (1992) A new imaging mode in atomic force microscopy based on the error signal mode. Proc Soc Photo-Opt Instr Eng 1639:198–204
Putman CAJ, Van Leeuwen AM., De Grooth BG, Radosevic K, Van Hulst NF, Greve J (1993a) Atomic force microscopy combined with confocal laser scanning microscopy: a new look at cells. Bioimaging 1:63–70
Putman CAJ, De Grooth BG, Hansma PK, Van Hulst NF, Greve J (1993b) Immunogold labels: cell-surface markers in atomic force microscopy. Ultramicroscopy 48:177–182
Putman CAJ, Van der Werf KO, De Grooth BG, Van Hulst NF, Greve J (1994a) Tapping mode atomic force microscopy in liquid. Appl Phys Lett 64:2454–2456
Putman CAJ, Van der Werf KO, De Grooth BG, Van Hulst NF, Greve J (1994b) Viscoelasticity of living cells allows high resolution imaging by tapping mode atomic force microscopy. Biophys J 67:1749–1753
Radmacher M, Fritz M, Hansma HG, Hansma PK (1994) Direct observation of enzyme activity with the atomic force microscope. Science 265:1577–1579
Schabert FA, Engel A (1994) Reproducible acquisition of *Escherichia coli* porin surfaces probed by the atomic force microscope. Biophys J 67:2394–2403
Schabert FA, Knapp H, Karrasch S, Häring R, Engel A (1993) Confocal scanning laser-scanning probe hybrid microscope for biological applications. Ultramicroscopy 53:147–157
Schabert FA, Henn C, Engel A (1995) Native *Escherichia coli* OmpF porin surfaces probed by atomic force microscopy. Science 268:92–94
Schoenenberger C-A, Hoh JH (1994) Slow cellular dynamics in MDCK and R5 cells monitored by time-lapse atomic force microscopy. Biophys J 67:929–936
Stemmer A (1995) A hybrid scanning force and light microscope for surface imaging and three-dimensional optical sectioning in differential interference contrast. J Microsc 178:28–36
Sugawara Y, Ohta M, Ueyama H, Morita S (1995) Defect motion on an InP(110) surface observed with noncontact atomic force microscopy. Science 270:1646–1648
Unwin PNT, Milligan RA (1982) A large particle associated with the perimeter of the nuclear pore complex. J Cell Biol 93:63–75
Vinckier A, Heyvaert I, D'Hoore A, McKittrick T, Van Haesendonck C, Engelborghs Y, Hellemans L (1995) Immobilizing and imaging microtubules by atomic force microscopy. Ultramicroscopy 57:337–343
Weihs TP, Nawaz Z, Jarvis SP, Pethica JB (1991) Limits of imaging resolution for atomic force microscopy of molecules. Appl Phys Lett 59:3536–3538
Weisenhorn AL, Maivald P, Butt H-J, Hansma PK (1992) Measuring adhesion, attraction and repulsion between surfaces in liquids with an atomic force microscope. Phys Rev B 45:11226–11232
Weisenhorn AL, Khorsandi M, Kasas S, Gotzos V, Butt H-J (1993) Deformation and height anomaly of soft surfaces studied with the AFM. Nanotechnology 4:106–113
Wölfer U, Dencher NA, Büldt G, Wrede P (1988) Bacteriorhodopsin precursor is processed in two steps. Eur J Biochem 174:51–57
Yang J, Tamm LK, Tillack TW, Shao Z (1993) New approach for atomic force microscopy of membrane proteins: the imaging of cholera toxin. J Mol Biol 229:286–290
Yang J, Mou J, Shao Z (1994) Molecular resolution atomic force microscopy of soluble proteins in solution. Biochim Biophys Act 1199:105–114

CHAPTER 2

Basic Principles and Applications of Confocal Laser Scanning Microscopy

Rolf Borlinghaus* · Bernhard Gröbler

Contents

Carl Zeiss, Jena GmbH, D-077440 Jena, Germany
* Corresponding author. Present address: Leica Lasertechnik GmbH, Im Neuenheimer Feld 518, 69120 Heidelberg, Germany,
Tel.: +49-(0)6221-414–849, Fax: +49-(0)6221-414-833,
e-mail: borlinghaus@llt.de

2.1 Introduction

In the beginning of microscopy the technique used was incident light microscopy. But soon the researcher changed the method and produced very thin preparations for observation in transmitted light. Without any doubt, microscopes are the most important tool in cell biology. It was Hooke who introduced the still recent term "cell" in biology when he carried out research with plant material in the year 1667. This discovery laid the foundations for the understanding of the mechanisms of life.

Comparing the instrument which Leuwenhoek used and recent modern microscopes you cannot see too much similarities maybe except that they both do magnify the object. A significant step was the introduction of the compound microscope by Hans and Zacharias Jansen in 1590. Here, in a first step, the object is magnified by an objective lens and than it is again magnified by a magnifier (the eyepiece). The resulting image is observed by the eye.

An important contribution was made by Ernst Abbe, who was the first to find an optical concept for microscopes and describe image formation theoretically. The results were optical components which could be manufactured with predefined properties. Until this time the production of lenses was a trial and error process. Lenses were made and then tested to see whether they could be used or not. For cell biologists and confocal multifluorescence microscopy the apochromatic color correction introduced by Ernst Abbe is still of enormous importance. Beside this, Abbe also was a person of great humanity and established the Carl Zeiss foundation. The Carl Zeiss Company celebrates its 150th anniversary this year.

An other important advance was the concept of Köhler who found out that the illumination beam path is independent of the observation beam path and can be optimized independently (Köhler illumination 1893).

A completely new contrast method was also introduced by Köhler, who used the principle of fluorescence for the first time in microscopy. In this case it turned out that it was technically more advantageous to switch back to incident light microscopy. Today, all fluorescence microscopes are incident light microscopes. In the field of cell biology, fluorescence microscopy had a renaissance when Coons used his immunohistochemical methods in microscopy (ca. 1950). With these procedures, very specific single structures in the cells become visible. Modern genetics has also developed fluorescence techniques which are now used routinely in research laboratories: fluorescence in situ hybridization (FISH).

2.2 Optical Concept of Confocal Microscopy

The basics of confocal microscopy were founded by Minsky in 1957. In contrast to ordinary microscopes, where Köhler illumination is used to illuminate the whole field homogeneously, here only a single spot is illuminated. Because of the wave properties of light, a mathematical point-like illumination is impossible. The spot which is illuminated is as small an area as possi-

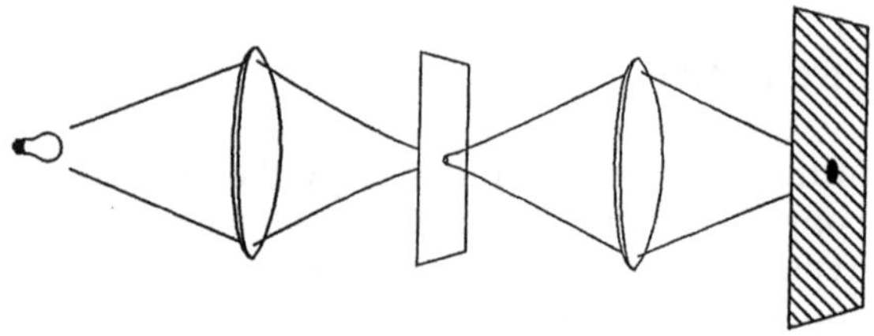

Fig. 2.1. Spot illumination. Light from a light source (*left*) is focused by a first lens to a small aperture. The light which emerges from this pinhole is than focused into the specimen (*right plane*) by a second lens

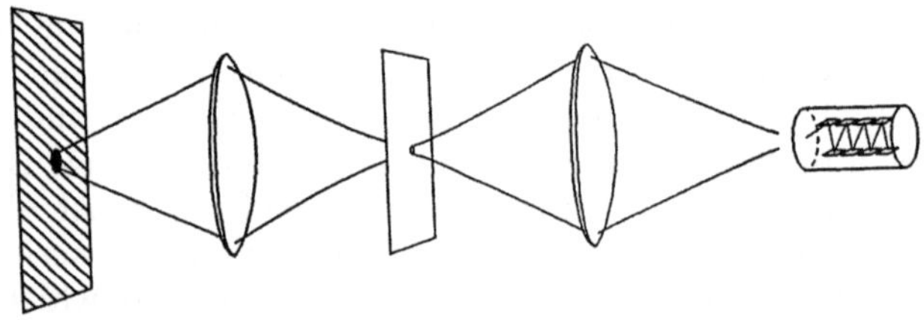

Fig. 2.2. Spot observation. Light emerging from the illuminated spot in the specimen is focused onto a small aperture. Light which passes the aperture is collected by a second lens and finally reaches the sensor (*right*)

ble. This is achieved if the illumination is done with a point-like source. This again is technically impossible. The diffraction limited edge can be achieved if a very small aperture is illuminated and the aperture is then used as a light source (excitation pinhole), or if the light is focused by other means in a diffraction-limited manner onto the focal plane of the specimen (Fig. 2.1). The technical implementation is especially simple if the light source is a laser. Therefore more or less all confocal microscopes today use lasers as light sources.

Similarly, an ideal confocal microscope is equipped with a sensor which has no extension (pointlike). And again this is physically impossible as the sensors which are used in these instruments (mostly photomultiplier tubes) feature extended sensing surfaces. Here again, the trick with the aperture works very well. The spot is imaged in an intermediate image plane (as a diffraction-limited spot). In this position an aperture is inserted which is as small as possible (emission pinhole). Behind this aperture the beam path is not critical (Fig. 2.2). Theoretically a sensor can be placed behind the pinhole without further optics.

If now both concepts are merged, the result is a microscope which is both pointlike illuminating and pointlike observing. This is where a confocal microscope differs from an ordinary microscope (Fig. 2.3).

Apart from yielding some measurements of rather theoretical interest, an instrument like this is not yet ready for useful applications. In order to generate an image, a field of the object has to be scanned point by point and the light has to be recorded. Therefore a means for generating a scanned image

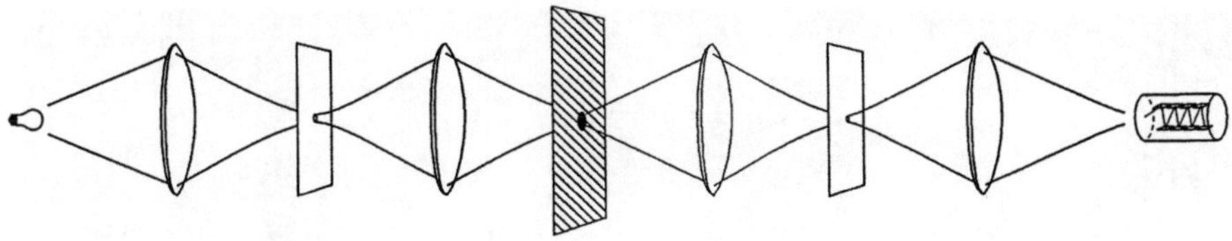

Fig. 2.3. Complete beam path of a confocal device

is necessary. This is very similar to the situation in a scanning electron microscope or, more popularly, in a TV tube.

At least at this point the question may arise as to what the advantage of a confocal microscope might be, as a significant technical outlay is necessary to generate images with such an instrument. Besides some side effects (among them a somewhat improved lateral resolution), confocal microscopy is so strikingly successful because it is possible to optically separate a section from the specimen. On the one hand, this possibility of optical sectioning without deintegrating the specimen gives better images with improved contrast from thick fluorescence specimens (Fig. 2.4) (where sometimes nothing was visible in a normal microscope), and on the other hand, the three-dimensional arrangement of the structure in question can be rendered and analyzed from a sequence of optical sections.

2.3 Beam Scanning System for Fluorescence

Unfortunately the technical implementation of a confocal microscope is not so simple as it might look, at least as far as its principle is concerned. Just the introduction of an aperture in a conventional microscope is not sufficient. In the following the structure of a laser scanning microscope is described. Generally, confocal microscopy works in transmitted light and incident light. Minsky has described a transmitted light confocal microscope. But the effect of optical sectioning is not very strong in transmitted light. The suppression of light from outside the focal plane is much more efficient if incident light methods are used. This was the reason why the confocal revolution in biology took place in fluorescence microscopy.

Although the original way of scanning the specimen by moving the microscope stage in x and y has some advantages which have already been described by Minsky, it also has the disadvantage that the generation of the image is very slow. A big step forward in this respect was the introduction of a beam scanning system. Here the illuminating point is moved over the fixed specimen by means of mirrors, and this kind of principle is now being used for most applications. There are also some other methods which will not be described here. If you are interested in details please see the handbook of confocal microscopy, edited by J. Pawley (1995).

Before we discuss the confocal microscope, we will describe the setup of a non-confocal beam scanning microscope for incident light, as shown in Fig. 2.5. The numbers in parentheses refer to labels in this figure and in Fig. 2.6.

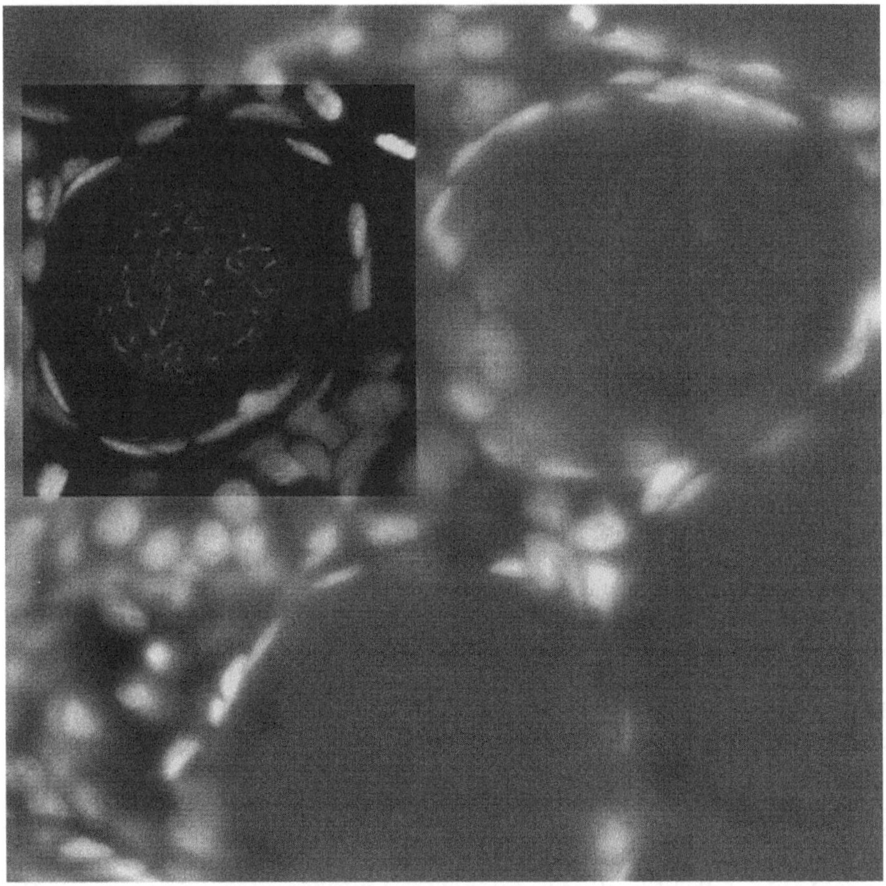

Fig. 2.4. Comparison of normal and confocal imaging. Initially a non-confocal image was recorded. The *insert* shows a region of interest which was recorded confocally. Specimen: Feulgen-stained trophoblast. The DNA in the nuclei of the egg cells is diluted in a large volume. Under normal conditions this fluorescence is hidden by the strong fluorescence of the nourishing cells where the DNA is much denser. The tissue section is some 100 μ thick. The confocal mode reveals the tiny structures of the organized DNA in the egg cells (noodle-shaped fluorescence in the center of the cell.) (Courtesy of Dr. H. Spring, Heidelberg)

The light from the laser (1) is – similarly to an ordinary microscope for fluorescence methods – coupled into the beam path by means of a beam-splitter (2). In the case of fluorescence this in normally a dichroitic beam splitter. For reflected light the most efficient splitter is a grey splitter with a splitting ratio of 1:1. For multichannel fluorescence applications no simple dichroitic splitters are used, but complex systems which have up to four transmitting bands and the same number of reflecting bands.

Behind the splitter, the light is reflected by two mirrors oriented at right angles (3). These scanning mirrors are mounted on the shafts of torsion-motors and move the light spot in the object in the x and y axes respectively.

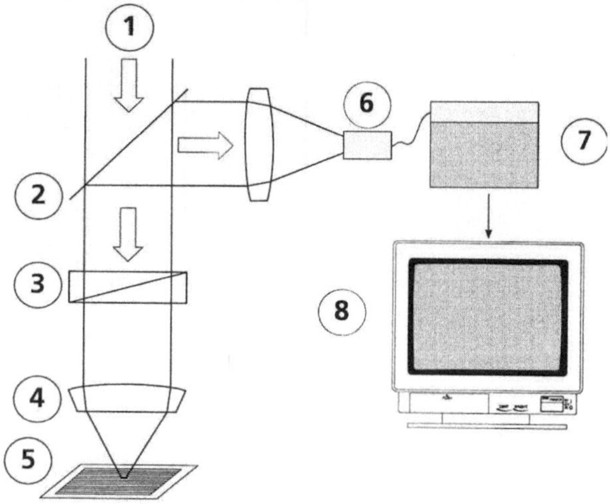

Fig. 2.5. Beam scanning system, for incident light, nonconfocal (without pinhole)

Using the objective lens (4), the light is focused on the specimen (5) into a single spot.

The emitted light is collected by the objective lens and directed by the scanning mirrors back onto the beam splitters. As the light moves at the speed of light from the mirrors into the specimen and back, the mirrors are more or less in the same position for excitation and emission. For this reason, the beam is not moving behind the scanning mirrors. The beam is moved only "below" the scanning mirrors. The emission light is directed by the beam splitter to the sensor (6). Here the intensity change is transformed into an electric signal. Up to now photomultiplier tubes have been the best solution for this task. The electric signal is digitized and stored in a frame store (7) and can than be displayed on a monitor (8).

In order to create a confocal microscope from a laser scanning microscope as described so far, an intermediate image has to be generated behind the beam splitter.

The intermediate image here is again a single point where the light corresponding to the emission at different positions in the specimen is brighter or darker during the scanning procedure. It is in this position of the intermediate image where a confocal aperture (Fig. 2.6, 9) has to be inserted. This aperture is referred to as pinhole in most of the literature.

Figure 2.7 illustrates the effect of the pinhole. Light which is emitted from planes above the focal plane in the specimen is focused a certain distance in front of the intermediate image. The corresponding beam already has a large diameter in the position of the pinhole and therefore is blocked very efficiently by the pinhole. Light which is emitted from the focus fits exactly through the pinhole and can reach the sensor. Light from layers below the focal plane is not yet focused at the pinhole position and thus blocked efficiently. With this arrangement, only light from the focal plane can contribute

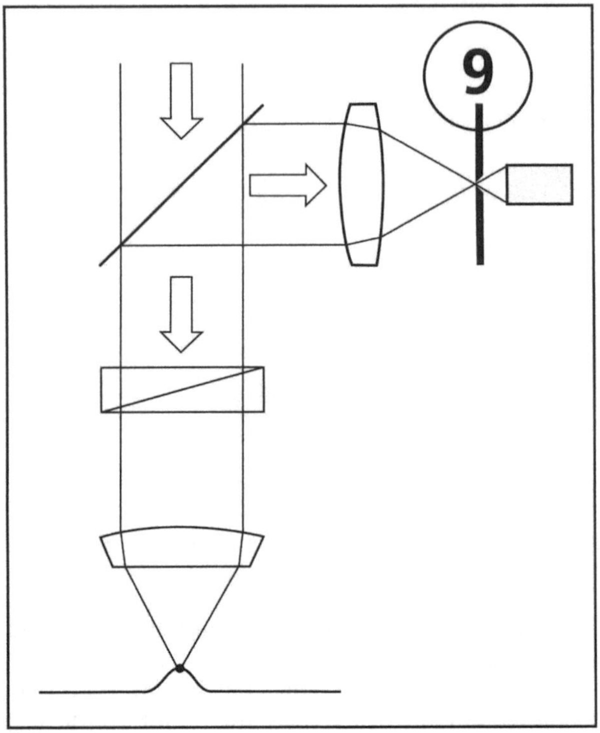

Fig. 2.6. Incident light beam scanning system with detection pinhole

to the recorded image. Contributions from unsharp layers above and below are cut off. The pinhole acts as a spatial filter to generate optical sections.

2.4 Properties of Optical Sections

The great advantage of optical sectioning is the possibility of looking into a cell or tissue specimen without really cutting the specimen into slices by a mechanical microtome. For optical sectioning the preparation is much easier and it is possible to repeat the sectioning later if information is still missing or other reasons made it necessary to reanalyze. But as usual, there is a minor drawback: the shape of the optical section is not as well defined as it is with real sections.

When a thick specimen is viewed in an ordinary fluorescence microscope, there is a small layer which is imaged sharply. The thickness of this layer is called the depth of focus. But this does not mean that information from other areas in the specimen does not reach the eyepieces. On the contrary, the sharp image from the focal plane is disturbed and overlaid by information from all other layers in the specimen. If the fluorescence stain is strong

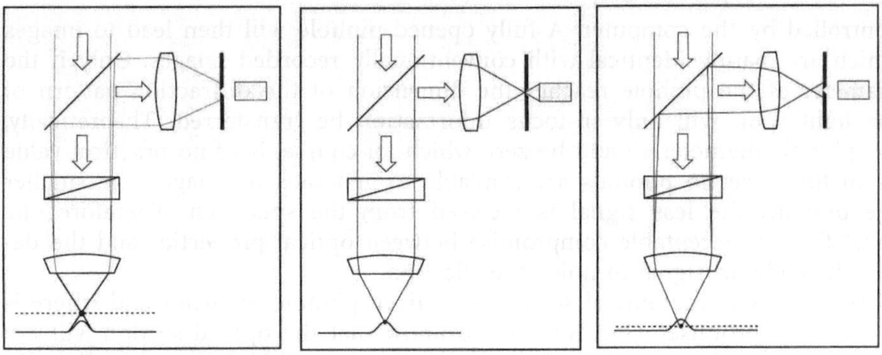

Fig. 2.7. Beam path above, in and below focal plane

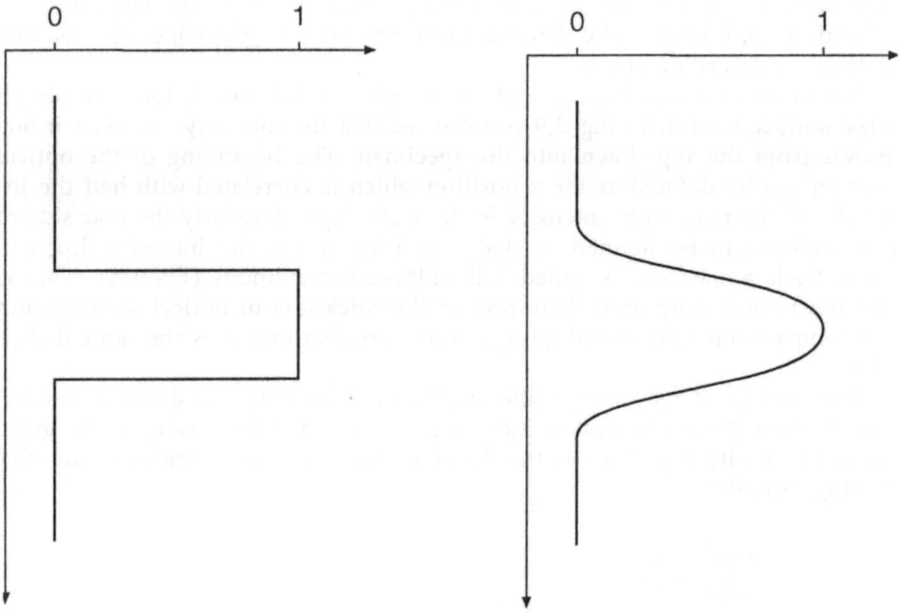

Fig. 2.8. Mechanical, box-shaped profile compared to an optical, bell-shaped profile

enough, this may lead to a situation where the focal plane cannot be seen at all.

Unfortunately the "sharp layer" does not start abruptly at a well defined z-position to end at a similar well defined different z-position but the change from sharp to unsharp is unsharp itself and needs to be described by somewhat complex formulas (Fig. 2.8).

How much of the focal and out-of-focus information is cut out by the confocal microscope depends on the diameter of the detection pinhole. For this reason, in modern instruments the apertures are variable. In the most sophisticated instruments the diameter can be changed continuously and is controlled by the computer. A fully opened pinhole will then lead to images which are mainly identical with conventionally recorded images. Only if the diameter of the pinhole reaches the dimension of the diffraction pattern of the light spot, will only in-focus information be transferred. Theoretically, the pinhole diameter should be zero which, of course, is of no practical value as in this case no photons are available to generate an image. The smaller the pinhole, the less signal is received from the specimen. Therefore one must find an acceptable compromise between optical properties and the demands made on signal to noise specification.

How thick is the optical section at a given pinhole diameter and where is a good compromise? First we have to know that the optical section will not become infinitely thin when the pinhole diameter approaches zero. The reason for this is the wave properties of light, which is also the reason for the well known limitation in lateral resolution present in all classical microscopes (scanning near field microscopes did exceed this limit). The thickness of optical sections must be defined arbitrarily, because the edges are not well defined as described above.

Let us assume one focuses onto a completely flat object, for example an ideal surface mirror. In Fig. 2.9 you can see that the intensity increases if one moves from the top down into the specimen. The beginning of the optical section can be defined as the z-position which is correlated with half the intensity of the maximum intensity in the focus layer. Similarly the rear-side of the section can be defined as that z-position where the intensity drops to 50%. Such a measure is called full-width-half-maximum (FWHM). This is the most commonly used definition of the thickness of optical sections, but for comparisons one should always make sure that one uses the same definitions.

From different reasoning (different theoreticans still find different results) one finds that the thinnest section which one can cut optically at the given parameters with a pinhole of the diameter zero, can be described by the following formula:

$$dz = \frac{0.64 \cdot \lambda_{exc}}{n - \sqrt{n^2 - NA^2}} \, .$$

Unfortunately, the values of *dz* depend on the geometry of the object. A point-like object, e.g., a lysosome, should be described by a formula different from that used for a thin fluorescent surface, e.g., the stained plasma membrane at the edges of epithelial cells grown on a coverslip. As the differences are not that dramatically pronounced, we can stay with the formula given above. This formula is correct for point-like objects. (In the real world this sort of values must be looked at very carefully anyway. The pinhole diameter is more than zero. All the optical parts of the microscope generate errors which will inevitably lead to losses. And the most prominent disturbing factor is the specimen itself, which especially in biology does not represent the simple structure which is assumed by theoreticans. We have rapid changes of the refractive index, dispersion and absorption in all directions, thus in-

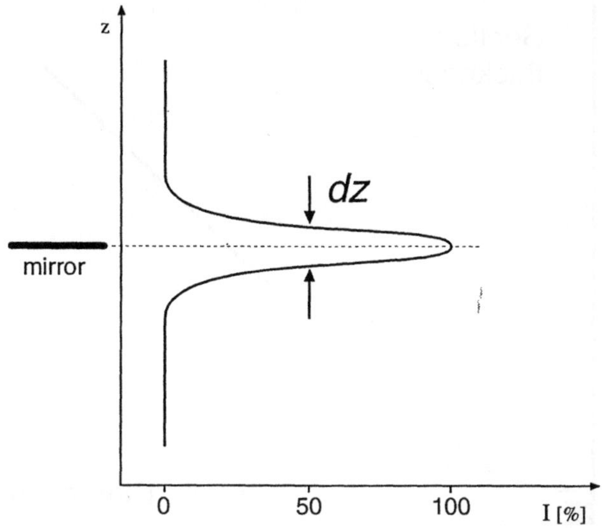

Fig. 2.9. Surface mirror and corresponding intensity profile. The thickness of the optical section is defined as the full-widh (in z-direction) half-maximum (of the light intensity)

troducing the most severe "error" in the system. However, it is these errors which create the image one would like to see.)

In order to get an idea of how thin optical sections in flourescence microscopy could be cut, let us assume blue excitation (e.g., for fluoresceinisothiocyanate, FITC or cyanine 2, Cy2) with 488 nm from an argon laser, a high numerical aperture lens e.g., 1.4 and a resin-embedding with refractive index of 1.5, then:

$$dz = 0.35\ \mu\text{m}\,.$$

In comparison, a dry lens with an aperture of 0.3 cuts slices not thinner than:

$$dz = 7.00\ \mu\text{m}\,.$$

This shows that the aperture should be chosen as large as possible. Especially for vital specimens it is recommended to use high aperture water-immersion objectives (at magnifications 40× or 63×: numerical aperture 1.2).

As already mentioned, the thickness of the sections changes only little if the pinhole is close to zero. How the thickness depends on the aperture diameter can be calculated according to the formula:

$$dz(PH) := \sqrt{\left(\frac{0.88 \cdot \lambda_{em}}{n - \sqrt{n^2 - NA^2}}\right)^2 + \left(\frac{n \cdot \sqrt{2} \cdot PH}{NA}\right)^2}\,,$$

where PH is the pinhole diameter (as projected into the object field) in microns. If the emission wavelength (em was assumed to be 500 nm, the refractive index n was 1.52 and the numerical aperture 1.4; the dependence of the

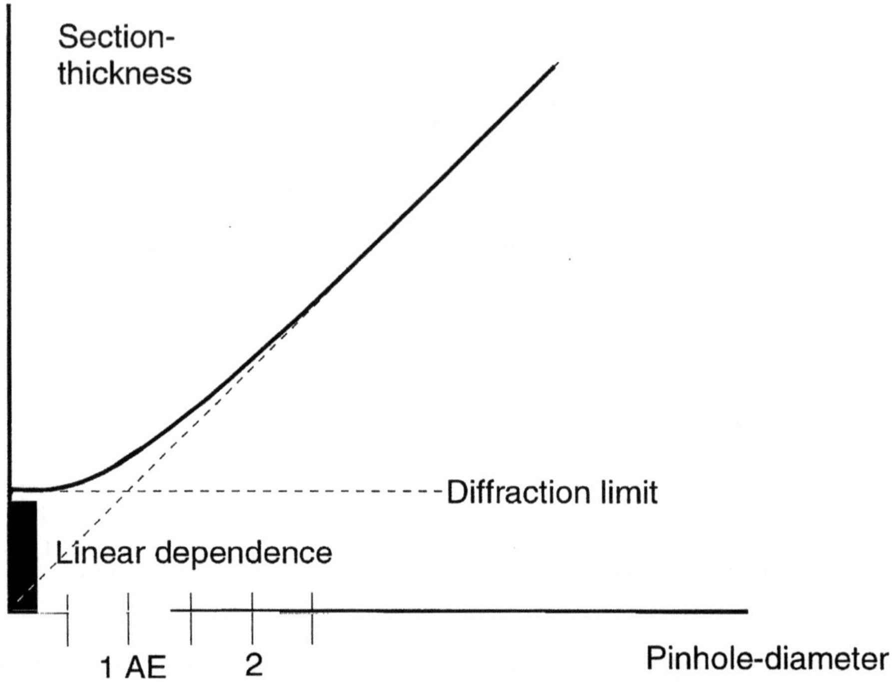

Fig. 2.10. Dependence of section thickness (qualitatively) on pinhole diameter. X-axis: pinhole-diameter, y-axis: thickness of the optical section. Note the change from diameter-independent to linear dependence

thickness on the pinhole diameter is shown in Fig. 2.10. The pinhole diameter corresponds to one Airy-disc diameter under the above conditions at 5 μ.

We can see, that the section thickness at a certain aperture diameter starts to increase suddenly. This diameter coincides with the diameter of the inner diffraction ring, the "Airy disc" of the light spot. If the diameter of the pinhole is just the same as one "Airy unit" most out-of-focus haze is removed but no light is wasted. Good software for confocal microscopes offers automatic setting to this Airy diameter without forcing the user to fiddle around with theoretical backgrounds.

2.5 Multichannel Imaging

Originally, the images which are recorded as described above, do not have a color. The image is generated from digitized current fluctuations which are produced by the photomultiplier tube (PMT) upon photon impacts on the cathode. Such a current is of course colorless. For this reason, the images which are recorded from a single channel are black and white images.

On the other hand, in many cases fluorescence microscopy today means many colors. Different cellular objects are colored by different fluorochromes and the result is an image which shows the interaction of the different com-

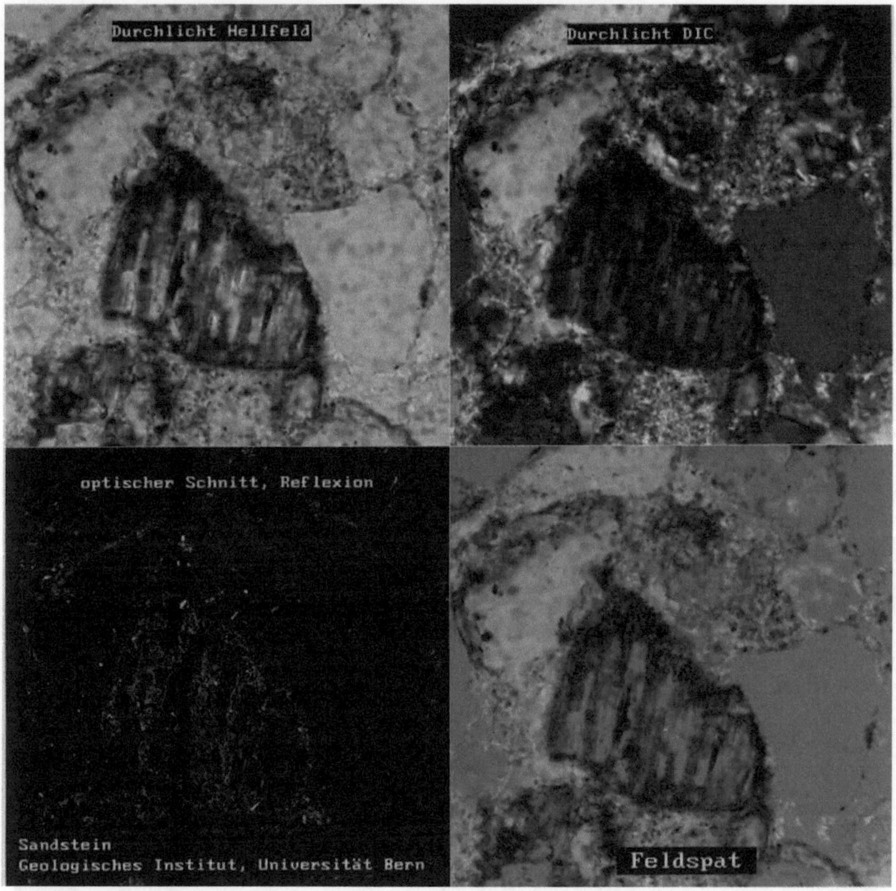

Fig. 2.11. Triple method imaging of a specimen from mineralogy. *Upper left* Transmitted light bright field. *Upper right* Transmitted light differential interference contrast. *Lower left* Reflected light. *Lower right* Color-overlay of the three images

ponents. For this reason, modern confocal microscopes are normally multi-channel instruments (up to four channels at the moment). The different emissions from the individual fluorochromes are optically separated and fed to separate sensors. Several primarily black and white images are generated in parallel. It is also possible to record the single images consecutively. These images can be displayed as a gallery in order to see the interplay of the cellular components or the distribution of the proteins which are analyzed. The advantage of a black and white display of multicolor records in a gallery is the improved perception of black and white images; in addition the distribution of an individual fluorochrome is better visible.

A different way of displaying the interaction of the different objects is to feed the black and white images into different channels of the color video display system. Here, usually three channels are used for red, green and blue. If two or three fluorochromes are used, the system can be used to display

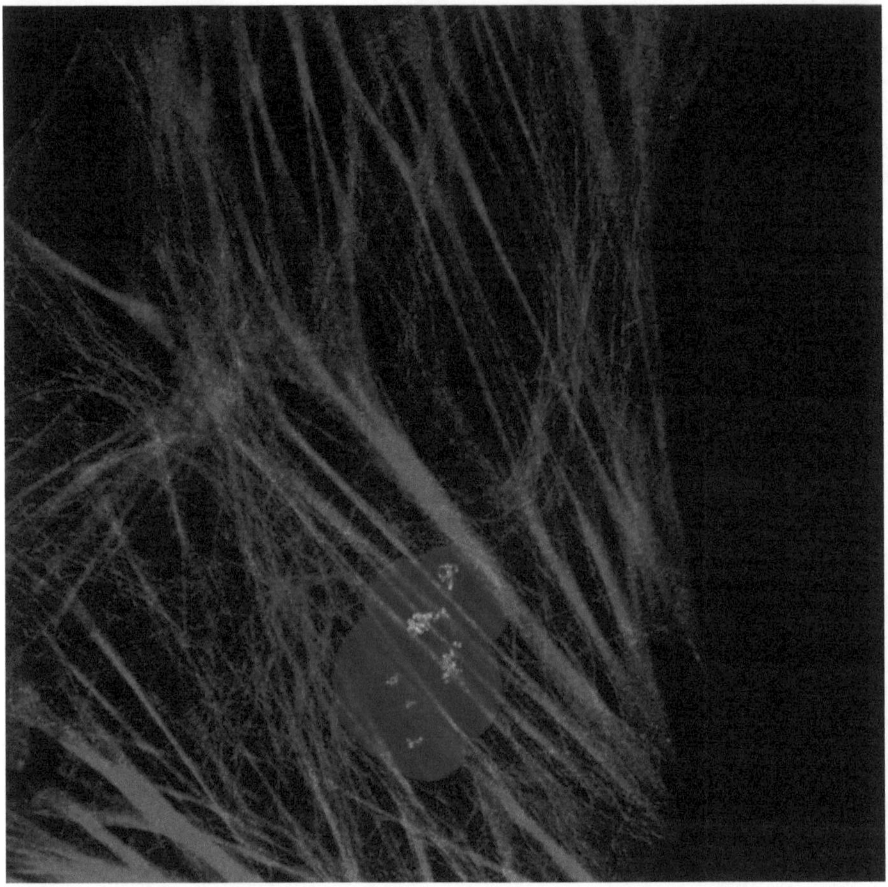

Fig. 2.12. Quadruple fluorescence. Actin-FITC (*green*), Tubulin-Cy5 (*red*), DNA-DAPI (*blue*) and nuclear protein-rhodamine (*white*, synthetically mixed color). Excitation: 364, 488, 543 and 633 nm. Plan Apochromat 63×/1.4 oil lens

these channels in full grey resolution. Of course, the displayed colors and the original colors need not coincide. The advantage is that each channel of the video system displays a single separated fluorescence channel (Fig. 2.11).

The "real" color of the emission can only be displayed if either three PMTs are used with emission filters which represent the sensitivity of our color-receptors in the retina in a way similar to the procedure used in 3CCD video cameras. In this case, the separation is no good because the fluorochrome design and the physiology of the human retina are not correlated. Especially the use of modern near-infrared dyes would be completely impossible, as we cannot see the emission with the eye. A different possibility would be to mix the images from the well separated sensor channels to the three video channels. In this case the video channel signals do not contain specific information from a certain fluorochrome. Furthermore, the grey scale which is used for a single channel is probably reduced (depending on the mixing procedure). Therefore it is not a good idea to display the emission in original color. The often used term "true color" does not describe the function of a real color display, but tells you that the system can display three channels with at least 8 bit grey resolution per channel.

Of course, if one wants to display more than three channels in colors, this is only possible by distributing the multiple signals into the three video channels. [The four-channel video display sometimes requested would not help, as our retina is an RGB system: there are only three different receptors available! (Fig. 2.12)]

2.6 Three-Dimensional Data and Reconstruction

The second main advantage provided by the confocal microscope is the possibility of reconstructing and rendering the three-dimensional organization of the structures in question. In a single optical section the main benefit was that the object of interest all of a sudden pops up out of an ocean of unsharp information. And only the information from the focal plane is transferred. If we now move the focus (using the manual focus knob to begin with), we see different planes, optical sections in different z-positions. So we can cut the specimen at all positions we like, as long as the absorption is not too high and the working distance of the objective lens is not exceeded.

The next step would be to automatically increment the z-position with a motor by a preset distance after each record (Fig. 2.13). These records will be stored as a sequence in the computer's memory. The distances between the single records should not exceed half the thickness of the optical section. Otherwise information which is optically available might be lost during the scanning procedure.

This kind of a sequence contains all the three-dimensional information from the fluorescent object. How can we display this information in the most intriguing way? There are several different solutions:

Gallery Display

If the images are displayed side by side, very much like in a stamp collection, the whole three-dimensional information is available. The two-dimensional information is shown in a single image and the third dimension is covered by the sequence of the individual images. To provide first quantitative information, the z-position of each image can be displayed in the single images.

Extended Depth of Focus

A completely new set of images can be calculated from a sequence, if we compress all optical sections from a stack of images into a single new image. In the simplest case this calculation would be a so-called maximum-point projection. Here, each picture element is traced along the z-axis for the brightest grey value. This brightest value is written in the target image at the position of the picture element. The result is an image which is in focus from top to bottom, regardless of the depth of focus of the objective lens.

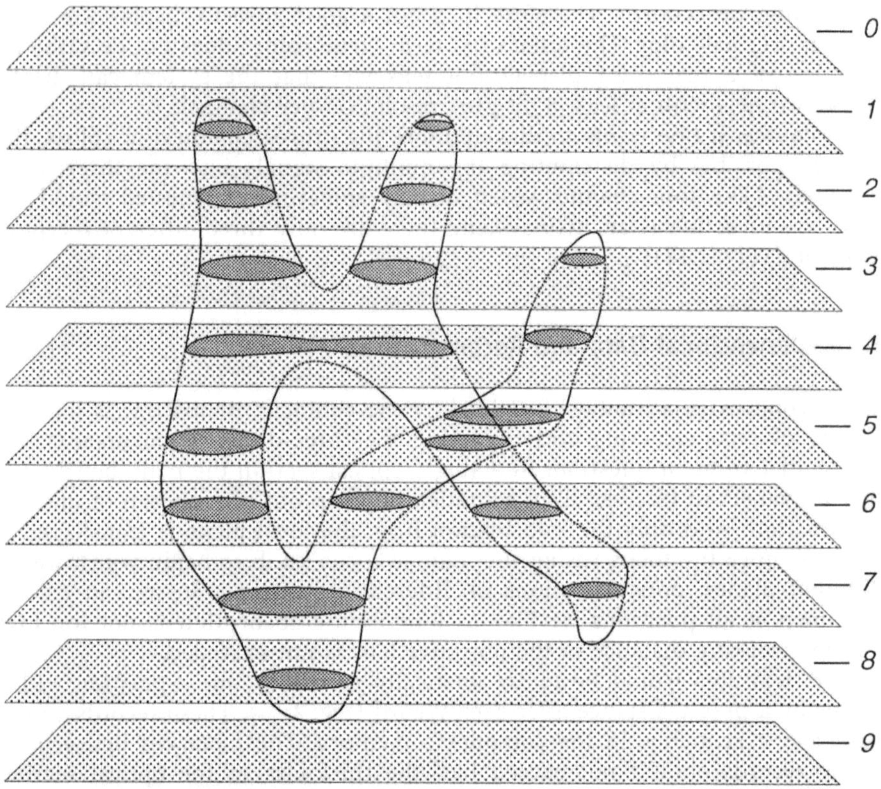

Fig. 2.13. Recoding of a sequence of consecutive z-sections by incrementing the focus position

The depth of focus is virtually infinite. In reality it is the distance which was recorded in the z-direction.

More sophisticated procedures with different weighting and masking of the grey values lead to the definition and rendering of surfaces. Or procedures such as simulated fluorescence process are used to calculate striking 3D-images (Fig. 2.14).

3-D Animation

The projections as described above, can be calculated in different angles of view. So a sequence of images can be generated, where the viewing angle is incremented by a small angle from image to image. The result is a sequence of projections. If this sequence is displayed fast enough on the screen, a three-dimensional impression is generated in our brain, and we can see the structure three-dimensionally.

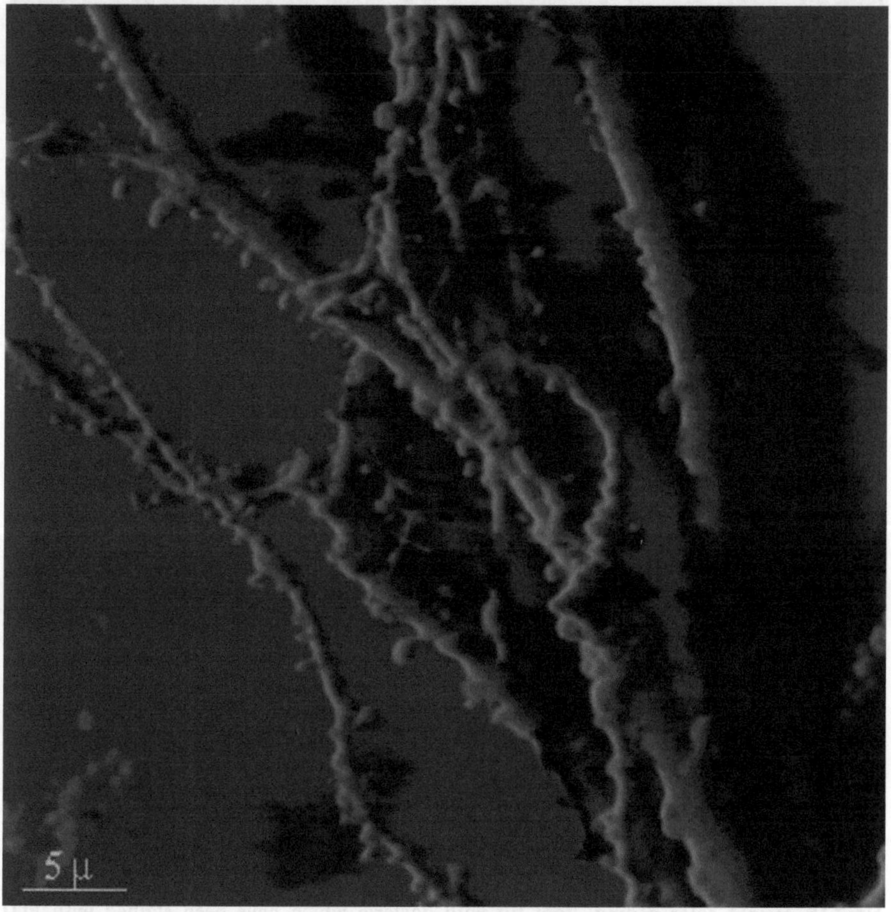

Fig. 2.14. Dendritic spines in a brain section. Visualized by injection of Lucifer yellow. (Courtesy Dr. Gähwiler, Zürich) A sequence of optical sections has been processed with IMARIS software (Bitplane, Zürich) to generate a simulated fluorescence process image

Stereo Images

Two projections which differ only in a small viewing angle make a stereo pair if displayed side by side. The viewer has to fuse the two images and then focus in order to get the 3D impression. It is also possible to display one of the images in red and the other image in green. If one then views the mixed image through goggles which have a red and a green filter instead of lenses, one can see a three-dimensional image (red-green anaglyphs).

Depth-Coding by Color

The third dimension can also be shown in an extended depth of focus image, if the individual sections are displayed in different colors. A color scale bar provides the information on which layers the different parts are located.

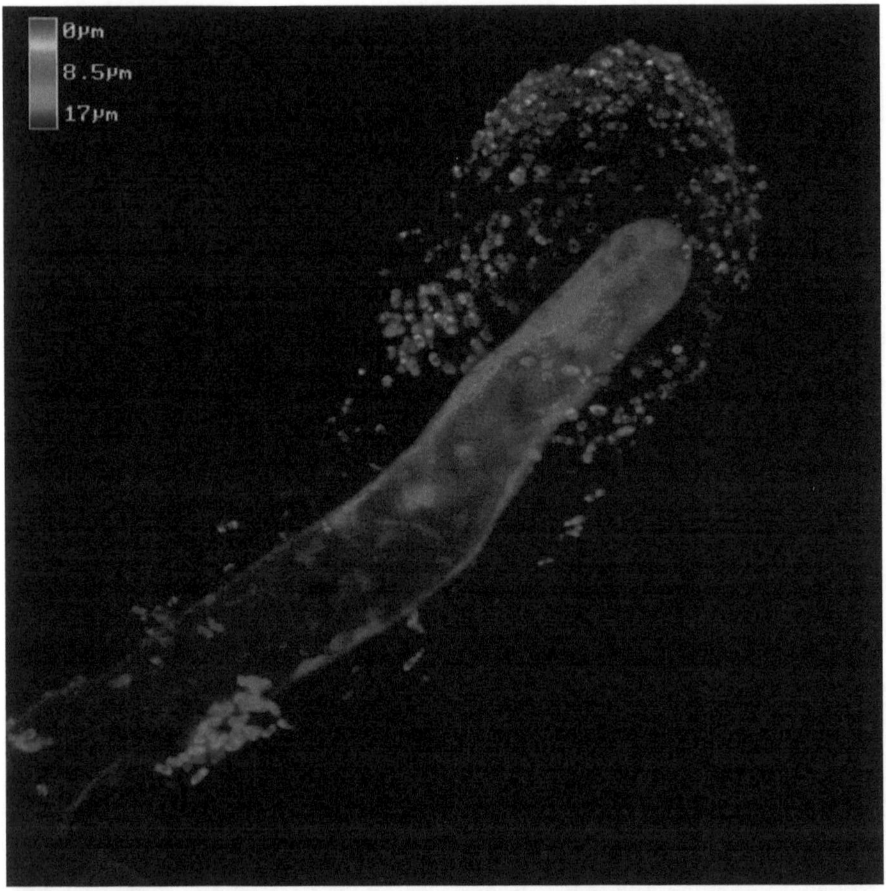

Fig. 2.15. Height color coding. Root tip with bacteria which have been stained with FITC, coupled to probes against ribosomal RNA. (Courtesy Dr. Aßmus, Neuherberg)

Of course, this kind of display is only possible with black and white-sequences, as the color in multichannel images already contains other information (Figs. 2.15–2.16).

Specimen Preparation

Nearly all fluorescence dyes are suited for confocal microscopy. As laser-scanning instruments can be equipped with different lasers simultaneously, one can find laser lines for each dye. The only requirement to be met by the specimen is that the three-dimensional features of the object should be preserved. This is done simply by using spacers which prevent the object from being squeezed between the slide and the cover slip.

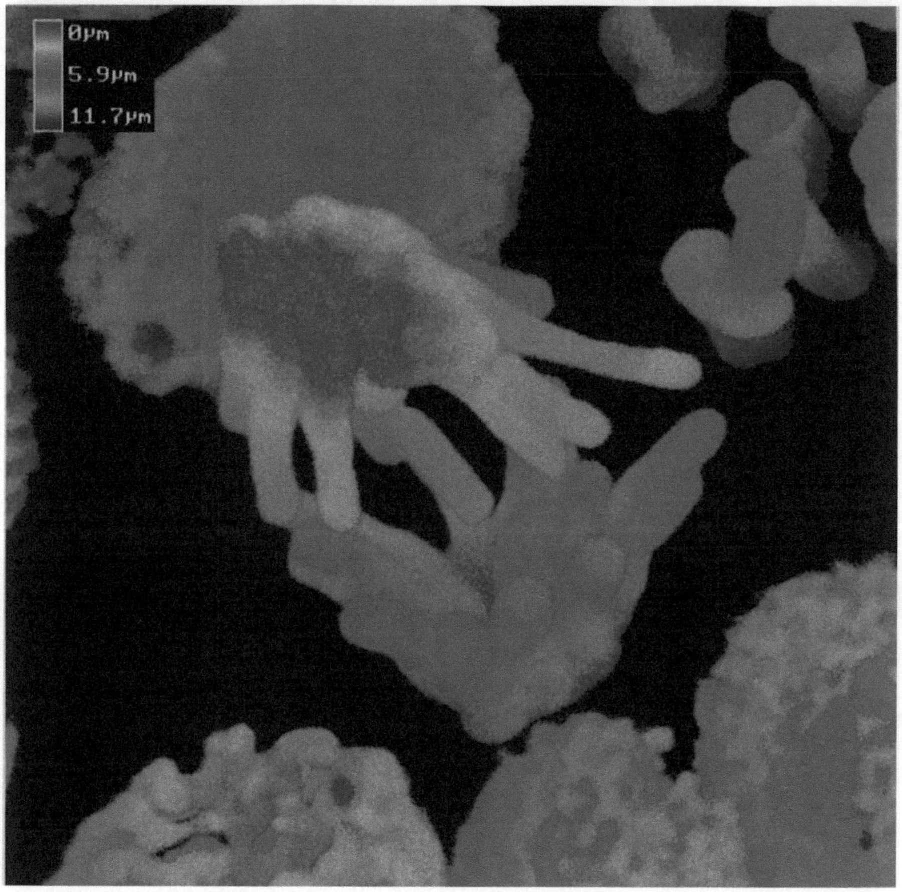

Fig. 2.16. Late anaphase in root tip of *Vicia faba*. DNA stained by mithramycine

2.7
Time Sequences and 4-D Microscopy

An instrument which provides the possibility to record sequences of images, obviously is also capable of collecting a sequence without moving the focus. The result is a time sequence similar to those which are known as time-lapse or slow-motion records (Figs. 2.17, 2.18). The advantage of excellent depth contrast is the significant difference if the sequence was collected with a confocal microscope. Different needs for time resolution make it necessary to adapt the frame format in order to trade off between time resolution and spatial information. At the limit settings, the image consists of only a single line. This line is recorded with a high repetition rate which gives a time resolution in the millisecond range.

These time sequencing method is used mainly in physiology to follow changes of metabolites upon stimulation. Examples are dyes for Calcium,

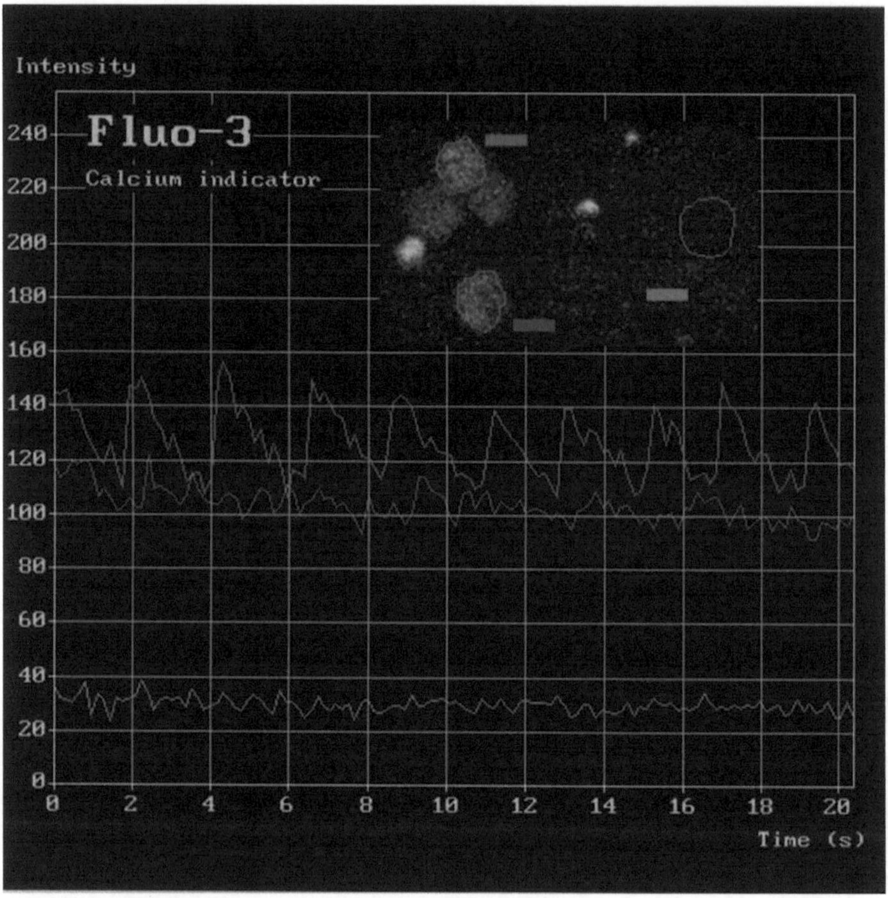

Fig. 2.17. Time sequence and analysis. Recording of Fluo-3 intensity changes in gh-3 cells, indicating autogenous calcium oscillations

pH, and other ions. A very recent developement was the introduction of a UV-excitation ratio method, which allows the use of the very common Ca-dye FURA-2 also for confocal beam scanners. Here the two UV-lines are rapidly switched by means of an acousto-optical tunable filter.

2.8 Conclusion

Confocal microscopy primarily offers the possibility of removing unwanted background in thick specimens. This background is out-of-focus emission which blurs the focus information. Removal of this out-of-focus haze is achieved by optically cutting a section from the specimen. A sequence of these sections collected along the z-axis can be used to render and analyze quantitatively the structure three-dimensionally. The method is useful for both fluorescent and reflecting specimens.

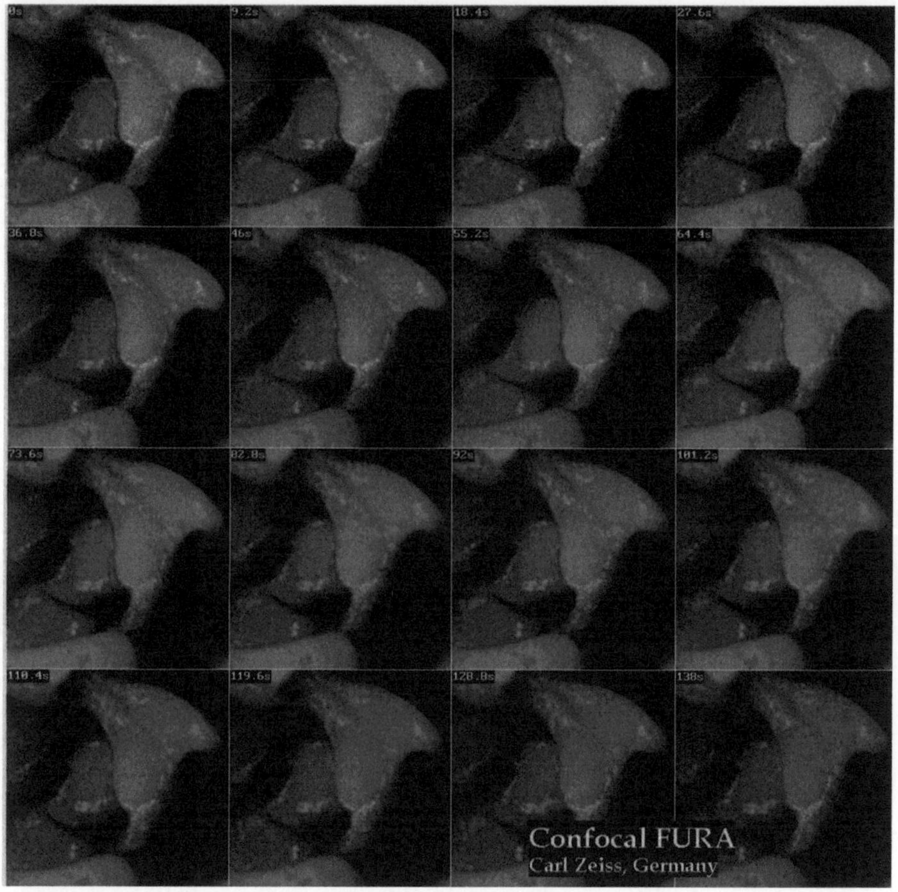

Fig. 2.18. Time sequence of confocal ratio-imaging with FURA-2. Excitation 364 and 351 nm switched by a UV-AOTF. (Courtesy Dr. Nitschke, Freiburg)

Latest techniques are driven by the idea of making better use of the resolution increases attained in confocal microscopy or even of overcoming the classical limits with new approaches. Here we should name multiphoton-excitation, 4-pi-microscopy, theta-microscopy and digital deconvolution techniques.

References

Abbe E (1873) Beiträge zur Theorie des Mikroskops und der mikroskopischen Wahrnehmung. Schultzes Arch Mikr Anat 9:413–468

Coons AH, Kaplan MH (1950) Localization of antigen in tissue cells. II. Improvements in a method for the detection of antigen by means of fluorescent antibody. J Exp Med 91:1–13

Minsky M (1957) US Patent No 3013467, Microscopy Apparatus

Pawley JB (1995) Handbook of biological confocal microscopy, 2nd edn. Plenum Press, New York

Schmitz EH (1981) Handbuch zur Geschichte der Optik, Wagenborgh, Bonn

Wilson T (1990) Confocal microscopy. Academic Press, London

CHAPTER 3

Visualization of Neuronal Form and Function in Brain Slices by Infrared Videomicroscopy

Hans-Ulrich Dodt[1*] · Walter Zieglgänsberger[1]

Contents

[1] Max-Planck-Institute of Psychiatry, Clinical Institute, Clinical Neuropharmacology, Kraepelinstr. 2, 80804 Munich, Germany

* Corresponding address: Max-Planck-Institute of Psychiatry, Clinical Institute, Clinical Neuropharmacology, Kraepelinstr. 2, 80804 Munich, Germany, Tel.: +49-89 30622350, Fax: +49-89-30622402

3.1 Introduction

Brain slices were introduced as a standard preparation for neurophysiological experiments over 20 years ago. Although this technique has greatly advanced our understanding of brain physiology, the utility of this preparation is limited by the difficulty in visualizing individual neurons in standard thick slices. However, the recent introduction of infrared videomicroscopy has solved this problem. With the aid of this technique, it is possible to visualize neurons within brain slices in great detail and neuronal processes can be patch-clamped under visual control. Infrared videomicroscopy has also been applied successfully to other fields of neuroscience such as neuronal development and neurotoxicity. A further development of infrared videomicroscopy allows the visualization of the spread of excitation in slices making the technique a tool for investigating neuronal function and the pharmacology of synaptic transmission.

3.2 Infrared: the Advantages of Light of a Long Wavelength

The fact that infrared radiation penetrates through many kinds of matter more easily than visible light has been extensively exploited in astrophysics (Scheffler and Elsässer 1987). Thus it appeared plausible to utilize this phenomenon to visualize neurons in the depths of brain slices, which are opaque to visible light (Dodt and Zieglgänsberger 1990) but more translucent at longer wavelengths (MacVicar 1984; Eggert and Blazek 1987). Since contrast and not resolution sets the limits for the visibility of small structures such as dendritic spines, the loss in spatial resolution, due to the longer wavelength is only moderate. Thus, for example, highly contrasted stained dendritic spines can be seen through a medium resolution 40× power objective in transparent Golgi preparations. At present, the properties of the optics and detectors available limit the useful wavelength for infrared illumination to l_{max} = 780 nm. However, in preliminary experiments we have found that the introduction of CCD cameras which allow the use of longer wavelengths can further increase the contrast of structures in deeper layers of a slice. Unfortunately, such CCD cameras possessing a sufficiently high dynamic range do not yet provide images at video rate and are still very expensive.

Presently infrared videomicroscopy employs infrared illumination, differential interference contrast (DIC) or similar optics, and contrast enhancement by video. Infrared illumination is obtained by placing an infrared filter in front of the light source. The image is projected on the image tube of an infrared sensitive camera (see below). The optics used have to visualize unstained neurons, which are primarily phase objects, i.e., they only change the phase of a light wave passing through them, unlike coloured objects which mainly change its amplitude. Therefore, the phase gradients of the nervous tissue need to be converted into amplitude gradients by the optics in order to be visualized. The optical system also has to provide good optical sectioning to prevent the image from becoming obscured by objects lying above or below the plane of focus. DIC meets these criteria for thin objects (Allen et al. 1968). Thus, until recently this technique was used most commonly in videomicroscopy (Dodt 1992). At a wavelength of 780 nm, conventional sheet polarizers start to lose the polarizing effect necessary for DIC images and therefore calcite polarizers (Glan-Thompson prisms) have to be used instead. However, a new contrast system (gradient contrast) which has recently been developed eliminates all polarizing elements (Dodt H.-U., patent pending).

Unfortunately, even with appropriate optics, the contrast obtainable with infrared illumination is still weak. A further contrast enhancement of 5–10 times can be achieved with videocameras developed for cell biology (Hamamatsu) (Allen 1985; Inoue 1986; Weiss 1989). The most widely used image tube for videomicroscopy (Newvicon) is also sensitive in the near infrared up to 850 nm. The combination of infrared illumination, the new gradient contrast system and contrast enhancement by video allows the visualization of neuronal structures up to a depth of 80–100 μm within brain slices.

3.3 It Is Possible to Perform Infrared Videomicroscopy with Both Inverted and Upright Microscopes

As water-immersion objectives with high numerical apertures (N.A.) were not initially available, infrared videomicroscopy was developed with an inverted microscope (Dodt and Zieglgänsberger 1990; Dodt 1993). The experimental setup consisted of a brain slice chamber with a coverslip bottom mounted on a patch-clamp tower. The slices were held gently from above with thin threads glued to a perspex or platinum frame to keep them submerged in standard Krebs–Ringer solution. The maximal N.A. of the optics was 1.4 for the objective and 1.2 (in aqueous medium) for the condenser. As patch-clamp studies are facilitated by the use of upright microscopes, the Axioskop FS (Zeiss) is now used with an N.A. of 0.9 for the objective and an N.A. 1.4 for the condenser, e.g. for patch-clamping and applications of substances on dendrites (see below).

3.4 Visualizing Neurons with High Resolution

One of the main advantages of infrared videomicroscopy is that one can study topographical details of neuronal networks without staining. Neuronal structures that are only a few μm above or below the focal plane do not blur the image, due to the optical sectioning capability of this system. Figure 3.1 shows cell bodies and fibers of neocortical neurons of the adult rat in vitro. By optical sectioning, fibers are traceable for considerable distances allowing a great variety of neuronal specializations to be visualized on dendrites of pyramidal and nonpyramidal neurons. Studies using conventional staining methods (Fleischhauer et al. 1972; Peters and Kara 1987) have described that dendrites often form bundles with an interdendritic spacing of about 50 μm. This spatial periodicity of bundles is especially prominent in the visual cortex. The exact role these dendritic bundles play in information processing

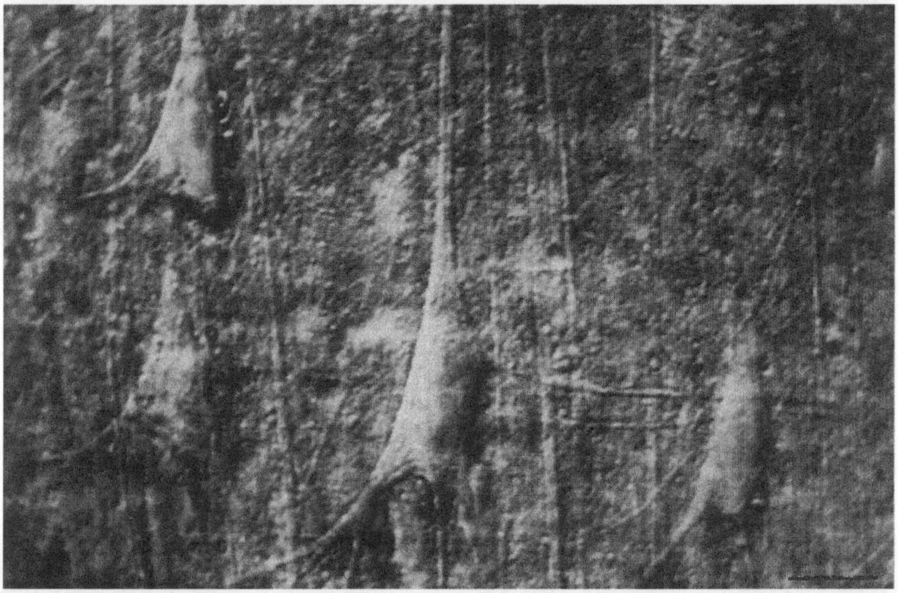

Fig. 3.1. Pyramidal neurons and dendritic bundles in lamina 2/3 of the rat visual cortex as seen with infrared videomicroscopy. Dendritic bundles are a prominent feature of the visual neocortex. By repeated focusing of the microscope through different focal planes, a three-dimensional impression of the neuronal network can be obtained. *Scale bar*: 10 μm

and storage has yet to be elucidated (Swindale 1990). The beaded structures, often found on dendrites, might represent spineheads but the identity of these small structures remains to be clarified in a correlative videomicroscopic and electronmicroscopic study. The fact that one can visualize numerous neuroanatomical details without any staining could also make infrared videomicroscopy a valuable technique when a rapid diagnosis is required, for example, in the case of brain tumors.

3.5 Patch-Clamping in the Infrared: Towards a Physiology of Cell Compartments

Since the description of the relevant techniques (Edwards et al. 1989; Blanton et al. 1989), patch-clamping of neurons in brain slices has become very popular. In the past, standard DIC optics and "thin slices" of 100-150 μm have been required to produce sufficient contrast to visualize neurons. This technique is limited by the restricted contrast of standard DIC optics. Consequently patch-clamp recordings could only be obtained under visual control from the cell soma. The adaptation of infrared videomicroscopy for an upright microscope allows visualization of the dendritic tree in detail and enables the patch-clamping of soma and dendrites (Stuart et al. 1993), even simultaneously (Stuart and Sakmann 1994, 1995). Dendritic currents in voltage-clamp, dendritic action potentials in current-clamp, and dendritic single-

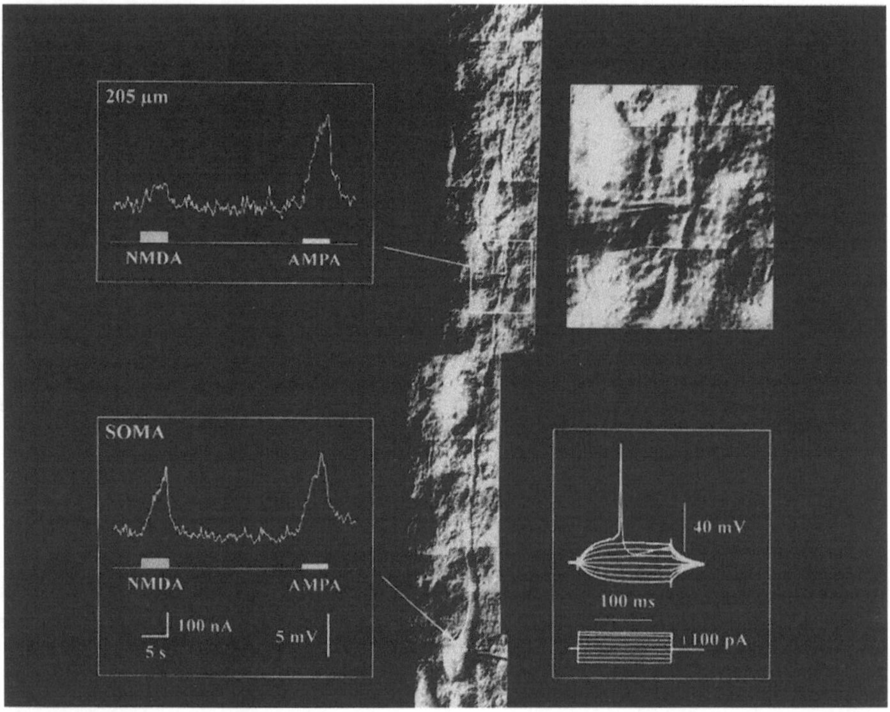

Fig. 3.2. Infrared videomicroscopy image of a layer 5 neuron in a 300 μm thick neocortical slice during whole cell recording and microiontophoresis. The *lower inset* at the *right* shows voltage deflections and an action potential of the neuron elicited by de- and hyperpolarizing current pulses applied to the neuron at resting membrane potential (–65 mV). The *upper inset* at the *right* shows the dendrite in contact with the microiontophoretic pipette. The *insets* at the *left* show the effect of somatic and dendritic application of NMDA and AMPA

channel currents in outside-out patches have all been recorded (for review see Yuste and Tank 1996). The somata of neurons up to a depth of 80 μm can be seen and their dendrites have been traced for more than 700 μm. It is also possible to apply substances repetitively, even to remote sites of dendrites, from a micropipette which can be separately moved by a fully motorized manipulator system (INFRAPATCH, Luigs and Neumann, Germany) after a somatic patch recording has been established (Fig. 3.2, Frick et al. unpubl. observ.). In a recent study we investigated the distribution of L-glutamate receptor subtypes (NMDA- and AMPA-receptors) on the soma and apical dendrite of neocortical layer V pyramidal neurons (Dodt et al. 1996).

3.6 Neuronal Development: Observing Cell Migration in the Slice

Neuronal migration is one of the basic phenomena in postnatal brain development (Rakic 1990). Until now, only a few studies using confocal microscopy have been able to directly visualize the dynamics of these morphologi-

A

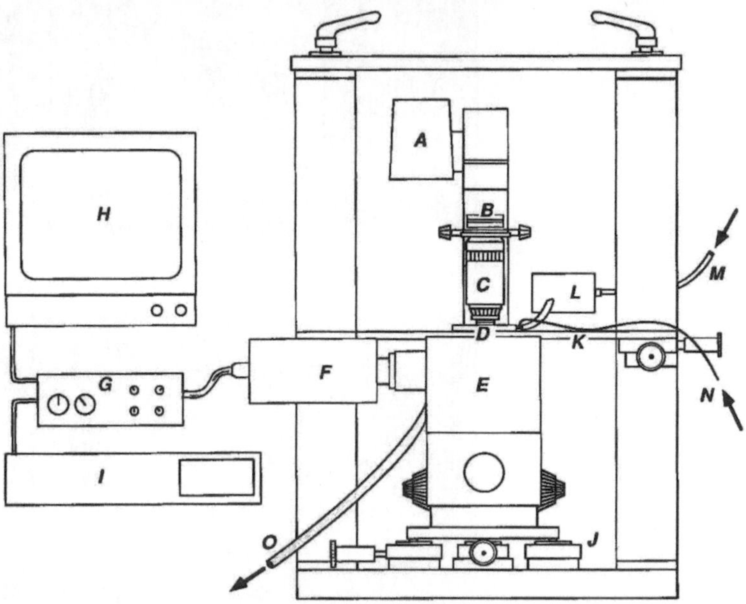

B

Fig. 3.3 A–B. Infrared videomicroscope as used for the experiments on neurotoxicity and neuronal migration. A General view. B Schematic drawing of the infrared video system used for the neurotoxicity experiments. *A*, halogen lamp; *B*, infrared filter; *C*, condensor; *D*, brain slice chamber; *E*, inverted microscope; *F*, Newvicon camera; *G*, camera control unit with variable offset, gain and shading correction; *H*, monitor; *I*, time-lapse recorder; *J*, X-Y translator to move microscope; *K*, stage of the patch-clamp tower; *L*, prechamber for heating of perfusion medium; *M*, inflow; *N*, drugs from syringe pump; *O*, outflow

Fig. 3.4. Neuronal migration in the cerebellum of a young rat (postnatal day 3). The nucleus of the granule neuron indicated by *arrows* is moving towards the pia in the external granular layer. *Scale bar*: 5 μm

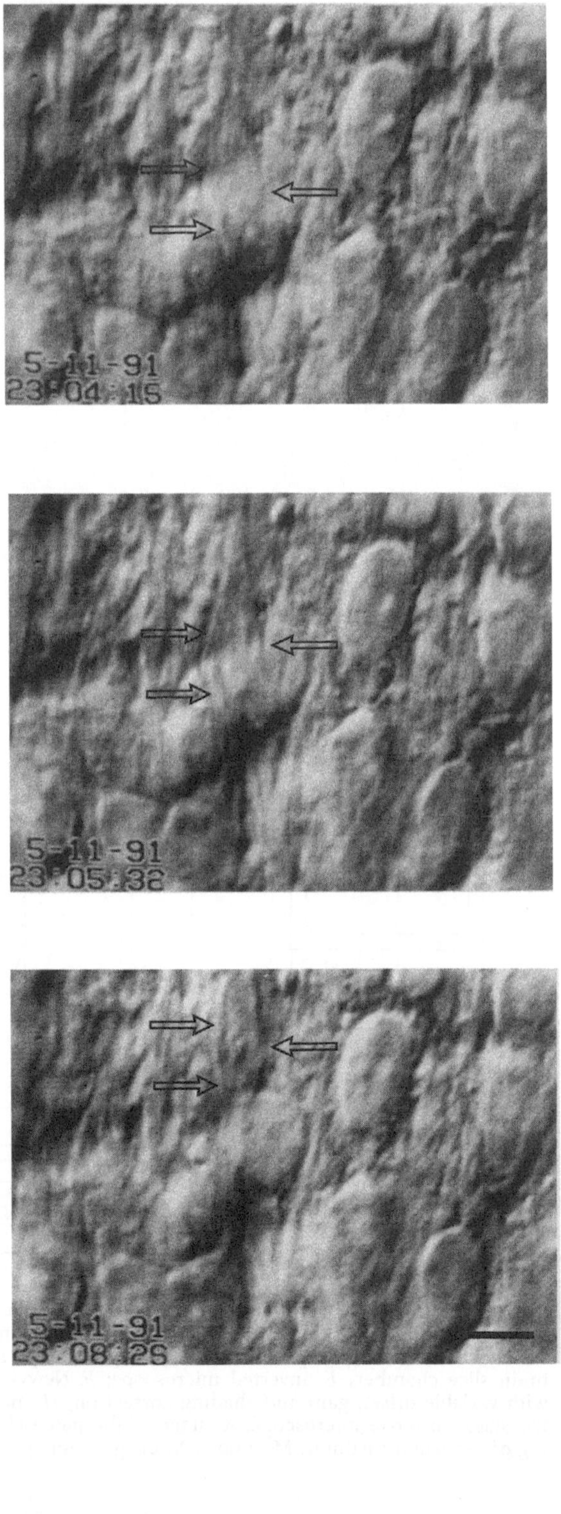

cal changes (O'Rourke et al. 1992; Komuro and Rakic 1993). Consequently, practically all our knowledge about neuronal migration has been deduced from static anatomy of fixed and stained tissue at various stages of brain development (Altman 1980) or from cell culture (Hatten 1990). Infrared videomicroscopy, in combination with time-lapse recording (Fig. 3.3), allows the visualization of neuronal migration in brain slices of animals at different developmental stages. The migration of neurons in their natural environment can be seen and their movement relative to each other becomes apparent. Since infrared videomicroscopy is not phototoxic, neuronal migration has been followed already for up to 24 h (Hager et al. 1995). Factors influencing migration can be included in the standard Krebs–Ringer perfusion medium (Liesi et al. 1995).

By employing this technique it has been possible to observe migratory activity in the external granular layer of the cerebellum between postnatal days 1–14. In the first postnatal week rapid "explorative" movements of cellular processes of granule cells can be seen. The migrating neurons move via translocation of their nuclei inside preformed processes with a speed that ranges between 6–70 μm/h with an average of 28 μm/h (Fig. 3.4). In the first five postnatal days, vertical migration of external granule cells prevails. After this period, the vertical migration of granule cells is accompanied by horizontal neuronal migration. Apart from movement and rotation of nuclei, organelle transport can also be detected in fiber bundles. Furthermore, it was found that neuronal nuclei frequently rotate, turning up to 360° within 15 min (Hager et al. 1995). Such rotations of the chromatin have been previously described only in cell culture (De Boni and Mintz 1986). The role of this energy dependent process remains obscure. In addition to migration it is also possible to follow mitotic processes. Neuronal migration has also been observed in the neural tube of intact young mice embryos where time-lapse recordings reveal the movement of neurons inside the neural tube of living E9 mice embryos kept in standard Krebs–Ringer perfusate.

3.7 Neuronal Death: Looking at What Goes on During Neurotoxicity

Previous neurotoxicity studies employing excitatory amino acid transmitters and other substances have mainly been performed in vivo where the results can be seen only after the animal has been killed (Buzsaki et al. 1989) or in cultured cell systems using mainly embryonic neurons which do not necessarily carry the same set of receptors as neurons of adult animals (Choi 1987). With infrared videomicroscopy and time-lapse recording the dynamics of neurotoxicity can be investigated in neurons of adult animals retaining the three-dimensional relationships of neuronal and non-neuronal cells. In such preparations, the physical constraints of swelling imposed on neurons in an intact tissue are to some extent maintained.

This technique was first employed to study the neurotoxicity of L-glutamate and the effect of anoxia on neurons located about 50 μm deep in rat neocortical slices. Using time-lapse recording the dynamics of cell swelling after bath application of glutamate agonists or during anoxia were revealed

(Dodt et al. 1993). To examine the effect of anoxia, the superfusion medium and the slice chamber were equilibrated with 95% N_2/5% CO_2. The neurotoxic effect of the agonist applied, or the effect of anoxia, were assessed by counting the number of preserved neurons (Rothman 1985; Choi 1987).

Using this approach, the dynamics of the morphological changes induced by excitatory amino acid (EAA) superfusion can be studied simultaneously in about ten neurons per test. The application of N-methyl-D-aspartate (NMDA, 50 μM) for 10 min was found to irreversibly alter the morphological characteristics of the neurons within the visual field (Fig. 3.5 A, B). After about 3 min the somata of some neurons began to swell or became no longer discernible in the tissue. Kainate and *α*-amino-3-hydroxy-5-methyl-4-isoxazolepropionic acid (AMPA) had similar effects in this concentration range, whereas millimolar concentrations of L-glutamate were required to induce comparable morphological changes. The morphological changes induced by EAAs can be completely prevented by the competitive NMDA antagonist D(-)-2-amino-5-phosphonovaleric acid (D-APV) or the non-competitive NMDA antagonist MK-801. Concentrations which protect about 90% of the neurons are 30 μM D-APV and 1 μM MK-801. The slightly higher protective potency of MK-801 compared to APV is in general agreement with findings in cell culture (Rothman 1992). Anoxia induces irreversible morphological changes similar to the EAA agonists 20–30 min after deprivation of the oxygen supply (Fig. 3.5 C and D). This technique may therefore be useful for determining the effects of potentially neuroprotective compounds on identified neurons in situ and may consequently facilitate the screening of drugs for stroke therapy.

3.8 Visualizing the Spread of Neuronal Excitation by the Use of Infrared

The so-called intrinsic optical signal (IOS) is a signal which correlates with neuronal excitation but shows a slower time course. This phenomenon has been used to map neuronal excitation in vivo (Bonhoeffer and Grinvald 1991) and in vitro (Grinvald et al. 1982; MacVicar and Hochman 1991). Use-dependent oxygen comsumption has been suggested as a mechanism underlying the IOS in vivo, however, the nature of the IOS in vitro is still obscure, although changes in light scattering properties have been discussed.

Darkfield microscopy, a method well established to image scattered light, has been adapted to visualize the IOS as an indicator of neuronal excitation in in vitro preparations (Dodt et al. 1993). Unstained slices of the neocortex are placed in the slice chamber of the inverted microscope which is equipped

Fig. 3.5 A–D. Morphological changes induced by NMDA and oxygen deprivation in neocortical slices. **A** Neurons and dendritic bundles in lamina 2/3 of the somatosensory cortex of the rat in control. **B** The same neurons after application of 40 μM NMDA for 10 min. Neurons are swollen or no longer discernible. **C** Another neocortex slice in control. **D** The brain slice after 30 min of oxygen deprivation, only one neuron is still clearly discernible. *Scale bar*: 10 μm

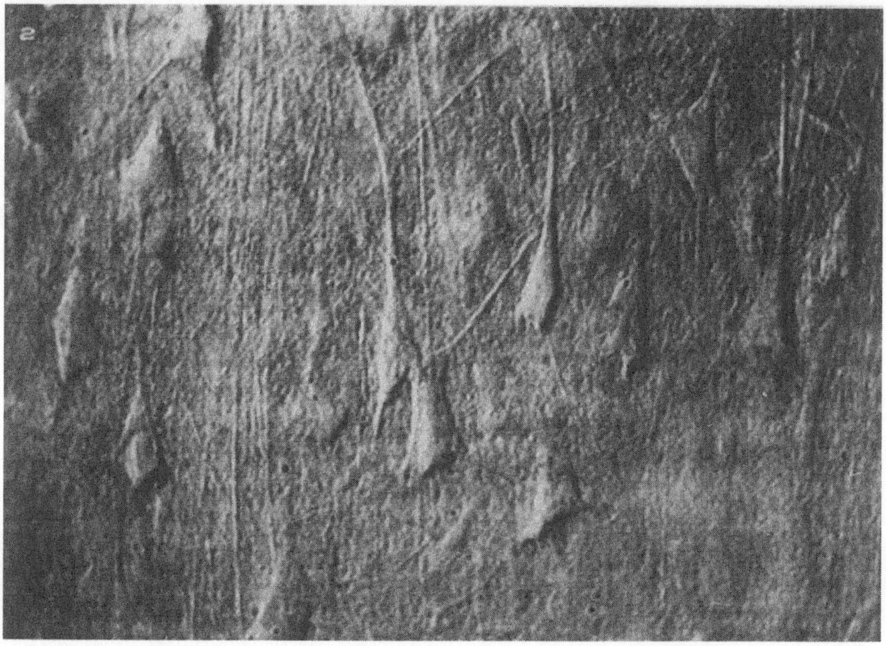

A

B

C

D

with a long-distance condenser modified for darkfield microscopy. They are illuminated with near infrared light and imaged with a low power 2.5× objective projecting onto the target of an image tube of an infrared sensitive camera as used for standard infrared videomicroscopy. An averaged darkfield image is stored in the computer memory and subtracted in real time from the incoming image. The contrast of the resulting difference image is digitally enhanced and the image is displayed in false colours.

Brief tetanic stimulation (50 Hz trains of 2 s with pulses of 0.2 ms duration and 1.5–3.5 V amplitude) of the white matter generally evoked a column-like pattern of the IOS in the adjacent neocortical areas (Fig. 3.6). The optical signal reached its maximum amplitude 3 s after the end of stimulation and decreased to baseline level in 30-60 s. The column-like area of increased brightness often exhibited a clearly discernable waist in the region of lamina IV (>80%). To prove a correlation between the IOS and the electrical neuronal activity of neurons, field potentials were recorded in the region where the IOS was observed previously. The amplitudes of the field potentials and the intensity of the IOS closely matched their spatial distribution. The maximum amplitude of the field potential was recorded in lamina II/III in the center of the column of increased brightness. The amplitude of the field potential decreased towards the border of the IOS (Fig. 3.6B).

To establish that the IOS is driven by synaptic transmission, mediated via the activation of EAAs, we added the L-glutamate antagonists 6-cyano-7-nitroquinoxaline-2,3-dione (CNQX; 10 μM) plus D-APV (100 μM) simultaneously to the perfusion medium. The combination of these compounds reduced the IOS to 13% of its control value on average. The marked depression of the IOS was partially (>50%) reversible after 45 min of washing (Fig. 3.7).

Since there is evidence that inhibitory GABAergic interneurons in lamina IV of the neocortex may delimit the spatial spread of synaptic excitation, we tested the effects of $GABA_A$ agonists and the $GABA_A$ antagonist bicuculline on the IOS. The substances were applied for 15 min to the bath (Fig. 3.8). The $GABA_A$ antagonist bicuculline (10 μM) increased the area covered by the IOS to about 200% of the control, whereas the $GABA_A$ agonist muscimol (10 μM) reduced the spread of excitation to 75%. Both effects were reversible within 45 min. Similarly to muscimol, the neuroactive steroid 5*a*-tetrahydrodeoxy-corticosterone (5*a*-THDOC; 10 μM), which enhances $GABA_A$ receptor-mediated chloride currents (Rupprecht et al. 1993) also reduced the spread of excitation to 70%. The reduction induced by the steroid was not reversible within the observation period.

The disappearance of the IOS after the blockade of synaptic transmission by CNQX and D-APV, demonstrates that intact synaptic transmission, mediated via EAA receptors, is a prerequisite for the generation of the IOS in the neocortex. These findings are in agreement with observations in the hippocampus (MacVicar and Hochman 1991). The importance of GABAergic synaptic inhibition for delimiting the spatial spread of neuronal excitation reflected by the IOS is highlighted by the actions of $GABA_A$ agonists and the antagonist bicuculline. The column-like pattern of the IOS with a waist at the level of lamina IV may result from a massive control by inhibitory interneurons found in this lamina in the neocortex (Eccles 1981; Szentagothai 1978).

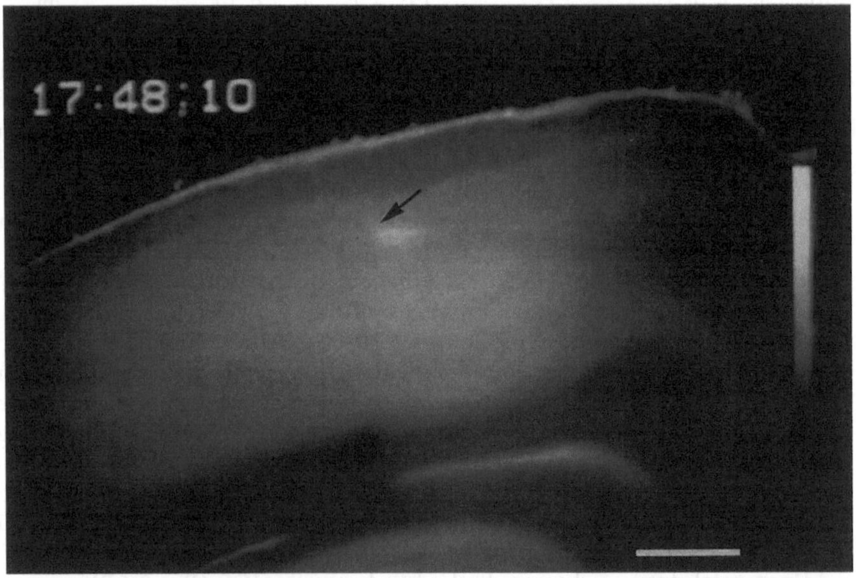

A

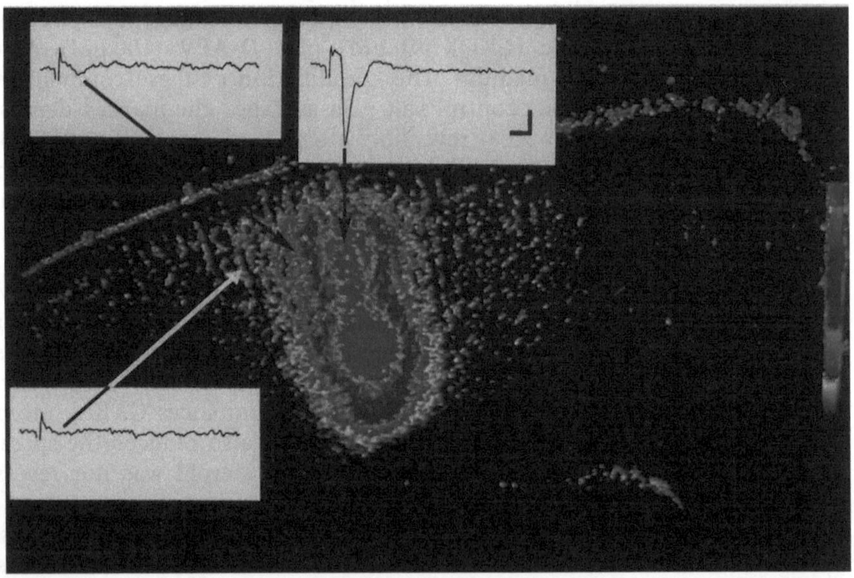

B

Fig. 3.6 A–B. Neuronal excitation in the neocortical slice visualized by infrared–darkfield videomicroscopy. The correlation between IOS and electrical activity is demonstrated by recording of the optical signal and the field potential. **A** Darkfield image of the neocortical slice before image processing. The location of the recording electrode, placed here at the side of the IOS, is indicated by an *arrow*. *Scale bar*, 0.5 mm. **B** Tetanic stimulation (3 V, 50 Hz for 2 s) of white matter induces an area of increased brightness, here colour-coded in *red*. The recording electrode is placed in layer II/III of the neocortex where the maximum of the IOS is observed. The field potential is elicited with single pulses of 5.5 V. The different positions of the recording electrode are indicated by *arrows*. The field potential with the highest amplitude was recorded in the center of the IOS. *Calibration bars*: 5 ms, 1 mV

Fig. 3.7 A–C. The L-glutamate antagonists CNQX and D-APV reduce the IOS. Both antagonists were added simultaneously to the perfusion medium. Control (**A**), IOS 30 min after begin of bath application (**B**) and 30 min of washout (**C**)

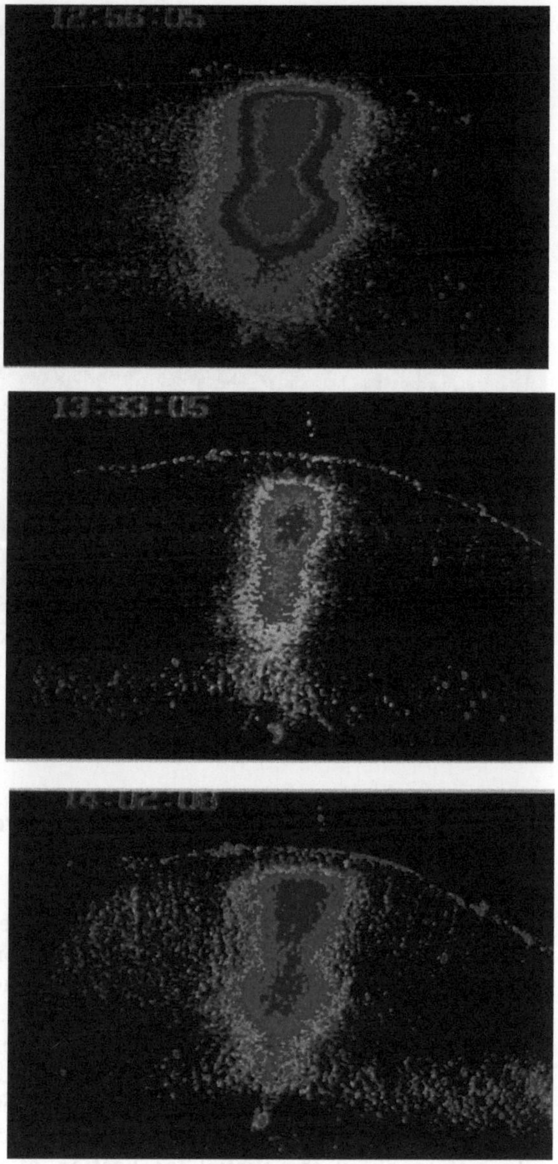

A

B

C

The most simple explanation for the finding that THDOC reduced the IOS is an enhancing action on GABAergic synaptic transmission. This is suggested by studies which showed that this steroid increased $GABA_A$ receptor-mediated synaptic potentials in rat neocortical neurons (Teschemacher et al. 1995).

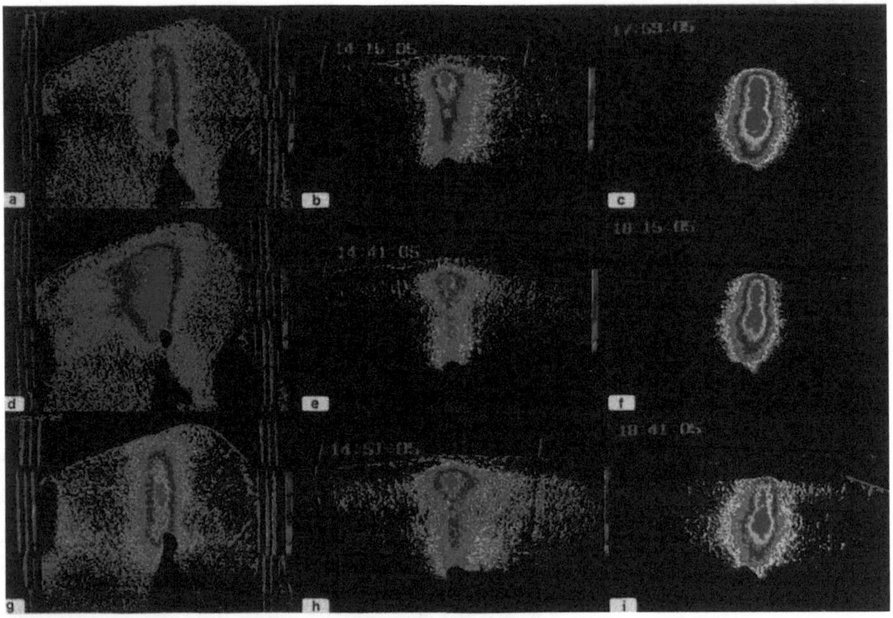

Fig. 3.8 a–i. Effects of bicuculline, muscimol and 5α-THDOC on the spread of excitation. The three substances were added to the perfusion medium to elicit a 10 μM concentration for 15 min. Control of bicuculline (**a**), muscimol (**b**), 5α-THDOC (**c**), after addition of the substances (**d–f**) and after 10–60 min of washout (**g–i**)

3.9 Perspectives: Visualizing the Spread of Neuronal Excitation with Cellular Resolution

Twenty-five years ago Cohen, Keynes and Hille described that the light-scattering properties of a nerve change when action potentials travelled along its axons (Cohen 1968). This fast signal still can not be exploited for imaging in slices without the use of voltage sensitive dyes. However, in a study utilizing cultured neurons of aplysia, which have a dense network of axonal collaterals, it was shown that scattered light can be used to detect action potentials by optical recording techniques (Stepnoski et al. 1991). The investigators used several photodiodes as detectors to record the summed activity without information about the spatial localization of neuronal excitation. These results indicate, that scattered light may be used to monitor action potentials also in more complex preparations.

It has been recently suggested in a study using ion-sensitve electrodes that the generation of the IOS involves an increase in cell volume which reduces extracellular space (Holthoff and Witte 1996). As relative movements of cell boundaries can be monitored with nanometer sensitivity using videomicroscopy, minute displacements of the cell membrane associated with the IOS should be detectable. Since cell volume changes affect many intrinsic optical properties of neurons such as light scattering, birefringence and phase retardation, highly sensitive CCD-detectors should be able to detect such changes.

3.10 New Stimulation and Visualization Techniques to Come

By combining the technologies described above with new fiber optics, small (20 µm diameter) preselected areas, for example on soma or dendrites of central neurons, can be briefly irradiated with UV light to release caged compounds, such as L-glutamate (Adams and Tsien 1993; Callaway and Katz 1993) in their vicinity to probe for differences of neurotransmitter sensitivity (Dodt et al. 1996).

The resolution of this system can be further enhanced to areas as small as 1 µm in diameter by using a UV-Laser (Dodt et al. 1996). By stimulating the discharge of neighboring neurons by local release of L-glutamate, the connectivity in neuronal networks can be analyzed (Dodt et al., unpubl. observ.).

The possibilities of visualizing neuronal elements with conventional optical techniques have certainly not been fully exploited. Classical microscopy was always the microscopy of thin samples. How the wealth of classical techniques (polarization, interference, phase contrast) performs in thick samples has still to be elucidated. Even minor adaptations of the classical techniques might render some of them suitable for the investigation of thick samples. The key feature of any such technique will always be the reduction of stray light. Furthermore, the third dimension waits to be introduced to the analysis of living neuronal tissue, by the adaptation of optical stereoscopic methods.

References

Adams SR, Tsien R (1993) Controlling cell chemistry with caged compounds. Annu Rev Physiol 55:755–784

Allen RD (1985) New observations on cell architecture and dynamics by video-enhanced contrast optical microscopy. Ann Rev Biophys Biochem 14:265–290

Allen RD, David GB, Nomarski GJ (1968) The Zeiss-Nomarski differential interference equipment for transmitted-light microscopy. Z Wiss Mikr Mikrotech 69:193–224

Altman J (1980) Postnatal development of the cerebella cortex in the rat. J Comp Neurol 145:353–398

Blanton MG, Lo Turco JJ, Kriegstein AR (1989) Whole cell recording from neurons in slices of reptilian and mammalian cerebral cortex. J Neurosci 30:203–210

Bonhoeffer T, Grinvald A (1991) Iso-orientation domains in cat visual cortex are arranged in pinwheel-like patterns. Nature 321:579–585

Buzsaki G, Freund TF, Bayardo F, Somogyi P (1989) Ischemia induced changes in the electrical activity of the hippocampus. Exp Brain Res 78:268–278

Callaway EM, Katz LC. (1993) Photostimulation using caged glutamate reveals functional circuitry in living brain slices. Proc Natl Acad Sci USA 90:7661–7665

Choi DW (1987) Ionic dependence of glutamate neurotoxicity. J Neurosci 7:369–379

Cohen LB (1973) Changes in neuron structure during action potential propagation and synaptic transmission. Physiol Rev 53:373–418

De Boni U, Mintz AH (1986) Curvilinear, three-dimensional motion of chromatin domains and nucleoli in neuronal interphase nuclei. Science 234:863–866

Dodt H-U (1992) Infrared videomicroscopy of living brain slices. In: Kettenmann H, Grantyn R (eds) Practical electrophysiological methods. Wiley-Liss, New York, pp 6–10

Dodt H-U (1993) Infrared-interference videomicroscopy of living brain slices. In: Dirnagl U, Villinger A, Einhäupl K (eds) Optical imaging of brain function and metabolism. Plenum Press, New York, pp 245–250
Dodt H-U, Zieglgänsberger W (1990) Visualizing unstained neurons in living brain slices by infrared DIC-videomicroscopy. Brain Res 537:333–336
Dodt HU, Frick A, Rüther T, Zieglgänsberger W (1996) Photostimulation using caged glutamate reveals differential distribution of excitatory amino acid receptors on rat neocortical neurons. Eur J Physiol. (Suppl) 431: R18
Dodt H-U, Hager G, Zieglgänsberger W (1993) Direct observation of neurotoxicity in brain slices with infrared videomicroscopy. J Neurosci Meth 50:165–171
Dodt H-U, Holländer H, Zieglgänsberger W (1993) Infrared DIC-Videomicroscopy of living brain slices. Soc Neurosci (Abstr) 19:272
Eccles J (1981) The modular operation of the cerebral neocortex considered as the material basis of mental events. Neuroscience 6:1839–1856
Edwards FA, Konnerth A, Sakmann B, Takahashi H (1989) A thin slice preparation for patch clamp recordings from neurones of the mammalian central nervous system. Pflügers Arch 414:600–612
Eggert HR, Blazek V (1987) Optical properties of human brain, meninges and brain tumors in the spectral range of 200 to 900 nm. Neurosurgery 21:459–464
Fleischhauer K, Petsche H, Wittkowski W (1972) Vertical bundles of dendrites in the neocortex. Z Anat Entwickl-Gesch 136:213–223
Grinvald A, Manker A, Segal M (1982) Visualization of the spread of electrical activity in rat hippocampal slices by voltage-sensitive optical probes. J Physiol 333:269–291
Hager G, Dodt H-U, Zieglgänsberger W, Liesi P (1995) Novel forms of neuronal migration in the rat cerebellum. J Neurosci Res 40:207–219
Hatten ME, Mason C (1990) Mechanisms of glial-guided neuronal migration in vitro and in vivo. Experientia 46:907–916
Holthoff K, Witte OW (1996) Intrinsic optical signals in rat neocortical slices measured with near-infrared dark-field microscopy reveal changes in extracellular space. J Neurosci 16(8):2740–2749
Inoue S (1986) Videomicroscopy. Plenum Press, New York
Komuro H, Rakic P (1993) Modulation of neuronal migration by NMDA receptors. Science 260:95–97
Liesi P, Hager G, Dodt H-U, Seppälä I, Zieglgänsberger W (1995) Domain specific antibodies against the B2 chain of laminin inhibit neuronal movement in the neonatal rat cerebellum in situ. J Neurosci Res 40:199–206
Macvicar BA (1984) Infrared video microscopy to visualize neurons in the in vitro brain slice preparation. J Neurosci Methods 12:133–139
Macvicar BA, Hochman D (1991) Imaging of synaptically evoked intrinsic optical signals in hippocampal slices. J Neurosci 11:1458–1469
O'Rourke N, Dailey M, Smith S, Mcconnell S (1992) Diverse migratory pathways in the developing cerebral cortex. Science 258:299–302
Peters A, Kara DA (1987) The neuronal composition of area 17 of rat visual cortex. IV. The organization of pyramidal cells. J Comp Neurol 260:573–590
Rakic P (1990) Principles of neural cell migration. Experientia 46:882–891
Rothman SM (1985) The neurotoxicity of excitatory amino acids is produced by passive chloride influx. J Neurosci 5:1483–1489
Rothman SM (1992) Excitotoxins: Possible mechanisms of action. Ann NY Acad Sci. 648:132–139
Rupprecht R, Reul JMHM, Trapp T, Van Steensel B, Wetzel C, Damm K, Zieglgänsberger W, Holsboer F (1993) Progesterone receptor-mediated effects of neuroactive steroids. Neuron 11:523–530
Scheffler H, Elsässer H (1987) Physics of the galaxy and interstellar matter. Springer, Berlin Heidelberg New York, pp 239
Stepnoski RA, Laporta A, Racuccia-Behling F, Blonder GE, Slusher, RE, Kleinfeld D (1991) Noninvasive detection of changes in membrane potential in cultured neurons by light scattering. Proc Natl Acad Sci USA 88:9382–9386
Stuart G, Sakmann B (1994) Active propagation of somatic action potentials into neocortical pyramidal cell dendrites. Nature 367:69–72
Stuart G, Sakmann B (1995) Amplification of EPSPs by axosomatic sodium channels in neocortical pyramidal neurons. Neuron 15:1065–1076
Stuart GJ, Dodt H-U, Sakmann B (1993) Patch-clamp recordings from soma and dendrites of neurons in brain slices using infrared videomicroscopy. Pflügers Arch 423:511–518

Swindale NV (1990) Is the cerebral cortex modular? Trends Neurosci 13:487–492
Szentagothai J (1978) The neuron network of the cerebral cortex: a functional interpretation. Proc R Soc Lond B 201 (1978):219–248
Teschemacher A, Zeise ML, Holsboer F, Zieglgänsberger W (1995) The neuroactive steroid 5α-Tetrahydrodeoxycorticosterone increases GABAergic postsynaptic inhibition in rat neocortical neurons in vitro. J Neuroendocrinol 7:233–240
Weiss DG (1986) Visualization of the living cytoskeleton by video-enhanced microscopy and digital image processing. J Cell Sci (Suppl) 5:1–15
Yuste R, Tank DW (1996) Dendritic integration in mammalian neurons, a century after Cajal. Neuron 16:701–716

CHAPTER 4

Time-Resolved Imaging of Membrane Potentials and Cytoplasmic Ions at the Cellular Level with a 50×50 Fiber Array Photodiode Camera

S. Hosoi[1,4] · H. Tsuchiya[1] · M. Takahashi[1] · M. Kashiwasake-Jibu[2*]
K. Sakatani[3] · T. Hayakawa[1,2]

Contents

[1] Central Research Lab, Hamamatsu Photonics K.K., Hamakita, Shizuoka 434, Japan
[2] Tsukuba Research Lab, Hamamatsu Photonics K.K., Tsukuba, Ibaraki 300–26, Japan
[3] Neuroscience Research Lab., Inst. for Clinical Science, China–Japan Friendship Hospital, Beijing 100029, P.R. China
[4] Present address: Laboratory of Molecular BioPhotonics, 5000 Hirakuchi, Hamakita, Shizuoka, Japan
* Corresponding address: Tsukuba Research Lab, Hamamatsu Photonics K.K., Tsukuba, Ibaraki 300–26, Japan, Tel.: +81-053-584-0250,
Fax: +81-053-584-0260, e-mail: mkjibu@hpk.trc-net.co.jp

4.1
Introduction

The brain, one of the last frontiers for humankind, performs highly sophisticated information processing such as memory, recognition and thinking. Recently, functional imaging techniques, such as positron emission tomography (PET) and functional magnetic resonance imaging (MRI), have been revealing many new insights into such brain activities. However, further improvements in spatial and temporal resolution for these techniques are still needed to study detailed activities of neurons and their networks. To fill this gap, the so-called optical method has been developed by the pioneering groups (see reviews Cohen and Lesher 1986; Grinvald and Frostig 1988), and this method may be considered to be at an experimental stage, mainly because, in spite of special instrumentation and expertise, the data obtained are not easy to interpret as the network activities in the brain. In this sense, even now to many neuroscientists electrophysiology would still be the method of choice for such studies.

Our main purpose is to show that simple upgrading of such optical devices to an imaging system based on the photodiode array detectors is indeed feasible and that obtaining this kind of data as sequential images is almost comparable to taking a movie with a video camera and may become routine for cellular functional studies in neurophysiology. We will also present some of the precautions for setting up such a system and of the experiences for successful imaging.

4.2
Photodiode Array Detector

Since the pioneering works of Cohen and others, scientists have used optical monitoring methods to study electrical events in the nervous system (Chien and Pine 1991; Momose-Sato et al. 1991; Orbach and Van Essen 1993; Elias et al. 1993; Fukunishi et al. 1993; Wellis and Kauer 1994; Sugitani et al. 1994; Sutor et al. 1994; Sawaguchi 1994; Antic and Zecevic 1995; some examples between 1991 and April 1996 in a MEDLINE search result using the words optical and voltage-sensitive dye). Although some authors tried to utilize CCD detectors, most of them constructed recording systems using specially fabricated photodiode-array chips. The photodiode-array chip is easier to install into a system for detecting electrical events faster than the rate at which the conventional TV camera can capture. Data obtained through each photodiode have a temporal profile similar to intracellular electrophysiological recording. Therefore, it is expected that this kind of optical technique can extend neurophysiological research to studies on highly parallel information processing in the brain not normally attainable with glass microelectrode techniques. In this expectation, it would be very important to test the feasibility of constructing such a device capable of capturing submillisecond to millisecond events as sequential images by increasing the number of photodiodes. Acquiring data as images requires more detectors in the same area of the image plane projected from a microscope. However, simple calculation indicates that putting more pixels into the same area of interest means less light gathered by each pixel, which leads to deterioration of the signal-to-noise ratio. So there must be a compromise between image resolution and signal quality. Also, through our project we have attempted to show whether or not this type of camera can capture electrical activities of neurons by single-shot recording, i.e. without signal integration. Further, since using this kind of instrument is supposed to require much expertise and experience, we also have tried to develop a system that can easily be used even by inexperienced researchers with only minimum knowledge of electronic and/or electrophysiological techniques.

4.3 Construction of a Fiber Array Camera

At the time we started this project, photodiode arrays available for us had 10×10 or 12×12 elements (Centronics, USA). Later, Hamamatsu Photonics introduced a photodiode array having 16×16 elements (ARGUS/PDA, Hamamatsu Photonics, Japan), which, in spite of its 16-bit data acquisition enabling off-line subtraction of the fluctuating heart-beat signal for live brain specimens, still cannot be considered as an imaging system. One way to upgrade this type of system was to fabricate a special array chip having more pixels. Instead, we have taken another way of utilizing optical fiber coupling between the image plane and the detector circuits to construct an imaging system with ordinary photodiodes on a circuit board, which permitted us a more flexible layout (Fig. 4.1). We chose an array of 50×50 as a compromise between the number of elements and image resolution, and even this compromise requires 2500 channels of parallel detector circuits. A further benefit of this fiber-coupled construction is its efficient electrostatic shielding of the detector circuits having high input impedance at the initial stage of photocurrent amplification. By using 0.5 mm (o.d.) optical fibers, the camera head has a dimension of 25×25 mm^2. We prefer a fiber made of plastics (Eska, Mitsubishi Rayon, Japan) to that made of quartz glass. The main advantages of plastic fibers are: ease of handling (tough and flexible) and a very thin clad layer. However, it should be noted that the transmission loss (i.e., reflections at the air/fiber interfaces and absorption within the fiber) of optical fiber affects the signal intensity. The overall transmission loss by these factors was roughly estimated experimentally to be 10 to 20% with 1.5 m of plastic fibers.

We have tried to construct photocurrent amplifying circuits to be as simple as possible (Fig. 4.1E). Even though the cost of one channel of circuit is trivial, the sum of 2500 channels could drive the total construction cost very high. Therefore, the overall circuit was designed to minimize the number of

Fig. 4.1 A–E. 50×50 fiber-array photodiode camera. **A** An overview of the system. **B** Side view of the detector circuit box. **C** Fiber-matrix head. **D** Block diagram of the camera system. **E** Simplified detector circuit. After amplification, the signal was transferred to a sample and hold circuit, multiplexers and then to A/D converters

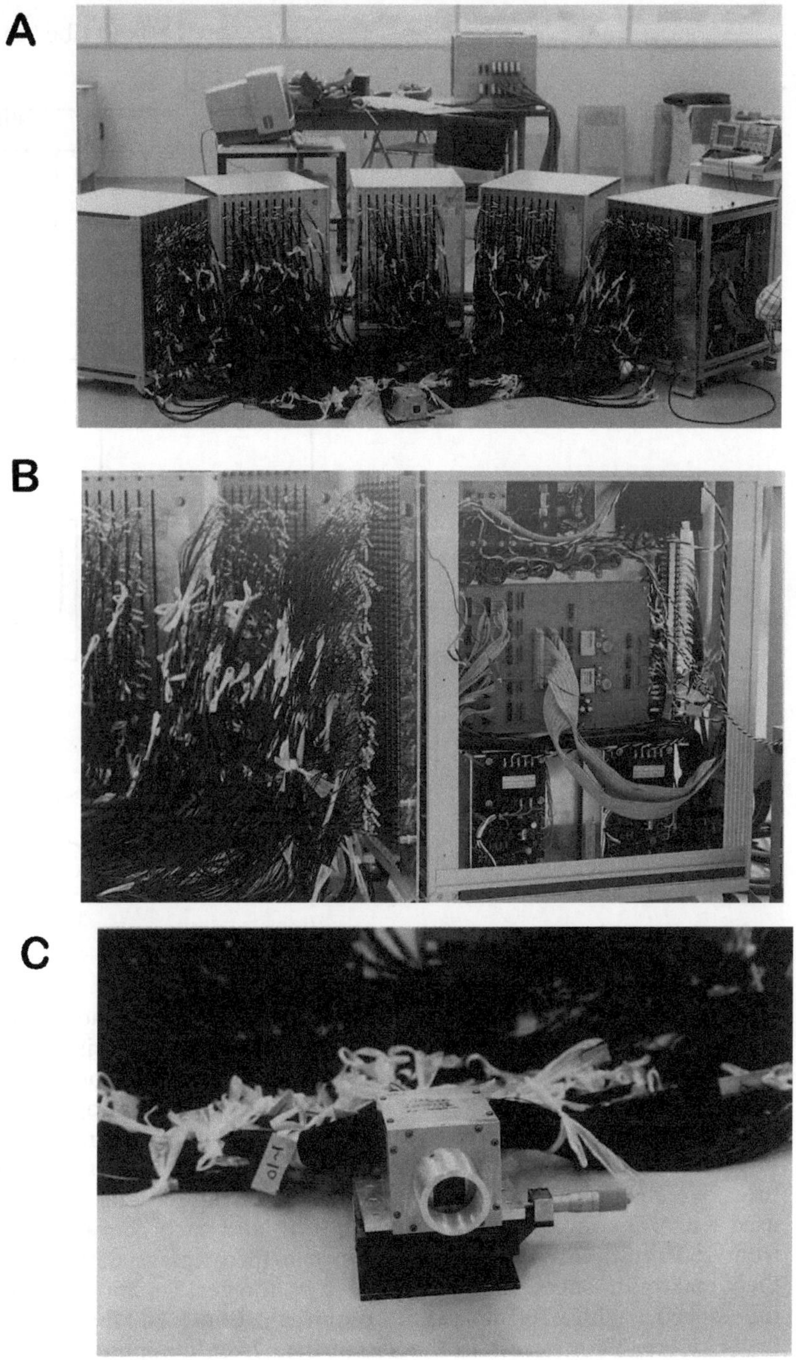
A
B
C

D

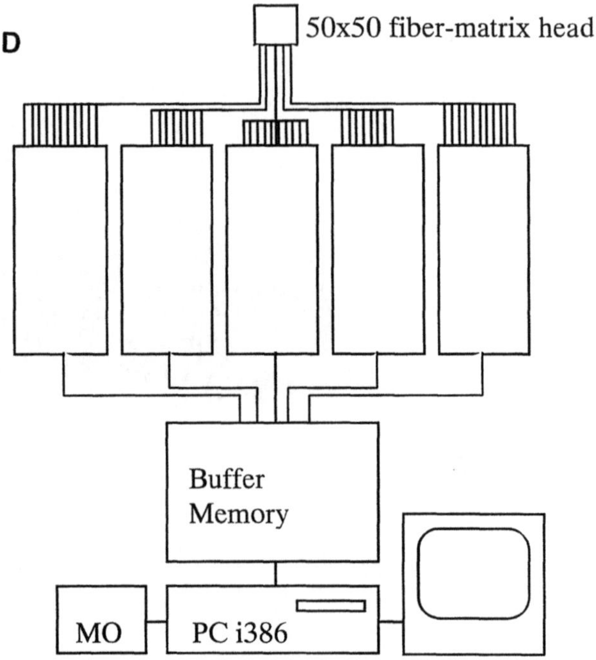

E

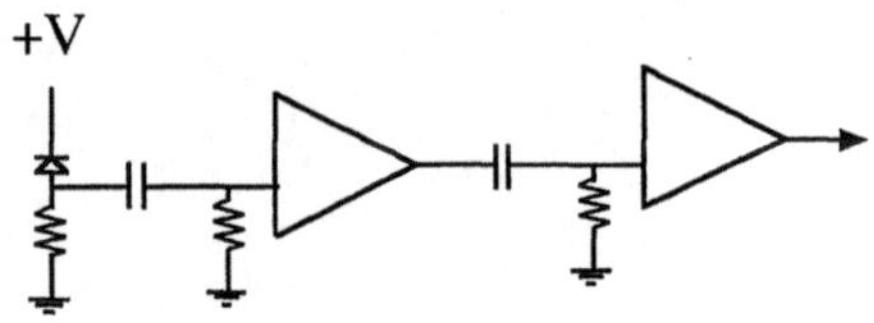

operational amps. To allow the circuit parameters to be modified after construction, we utilized 16-pin dip sockets to mount the resisters and capacitors connecting photodiodes and operational amps. Also, because constructing 2500 channels of A/D conversion and digital data storage with commercially available units is very costly and requires modification and a great deal of programming for the data analysis and display as images and motion pictures, we custom made sample-and-hold, A/D, digital data storage units. We used ten A/D converters (12-bit resolution), and each converter receives data from 250 multiplexed channels, which can form image data every 256 μs at their maximum speed. The storage of 4096 images, i.e. about 1 s recording at full speed, with 16-bit depth requires about 20 Mbytes of memory ($=2\times2500\times4096$). Access to the stored data from computer was obtained by use of a parallel I/O board. The data analysis and display program, which runs on Windows 3.1, was developed using C++ with some parts optimized by writing in assembly language.

The camera thus constructed and equipped with an ordinary camera lens was sensitive enough to capture images of fast moving objects, such as a moving hand or a punctured rubber balloon with just an ordinary room light. It was also possible to reproduce an experiment on imaging of membrane potentials in the voltage-sensitive dye-stained sea urchin egg under external voltage application (Kinoshita et al. 1988; Takahashi et al. 1993). For data recording, after the lamp shutter was first opened, the camera was allowed to wait for about 5 s until the AC-coupled amplifier circuit output leveled off to 'zero', at which point recording was started and the stimulus voltage was triggered 65 ms after the start of recording. All sequences of events were controlled by a parallel output of the computer.

4.4 Conditions to Consider for Single-Shot Imaging

By comparing acquired signal data from the same microscope image plane with this camera (50×50) and those of its predecessors, such as ARGUS/PDA (16×16), it is evident that one photodiode in the 50×50 camera only gets about ten times less light, which means that getting the signal with the same signal-to-noise ratio requires at least ten times more signal integration. We have tried to optimize circuit parameters, optics, dye-staining and sample preparations. Realization of this attempt will prove the true usefulness of this type of camera, in the sense that the camera can capture not only a time-averaged picture but also a single-shot event in the nervous system.

Fluorescence vs. Absorption. We confine our consideration to the fluorescence detection method since even a small brain from a frog is optically opaque under a microscope and absorption measurement obscures the site of nerve excitation within a thick brain. Besides, recent reports show that fluorescence change of some newer dyes is expected to be greater than 1% of the total fluorescence emitted. Therefore, fluorescence measurement will give a better chance of capturing single-shot events.

Important Factors to Consider for Optimization. Probably the best method for successful imaging is to synthesize a dye with much larger fluorescence change on electrical excitation. Another more practical thing to try is to make fluorescence bright enough for the camera. The main factors affecting the fluorescence brightness are: illumination, and the light collection efficiency of the microscope.

Illumination. We can use either a tungsten/halogen lamp, a xenon-arc lamp, high-pressure mercury-arc lamp or a laser. In terms of brightness, a high-power laser would be most advantageous. It requires, however, more careful setup for even illumination over a wide area with low temporal luminance fluctuation. Among the three kinds of lamps, a tungsten/halogen lamp is the best suited for constant and even illumination over a wide area. In the present attempt, the priority requirement for brightness leads to the choice of a high-pressure mercury-arc lamp, which has a spectral peak output at about

546 nm suitable for excitation of voltage-sensitive dyes. One drawback of manufacturer-supplied Hg lamps is that the luminance intensity fluctuates (less than 1%) at the AC frequency, due to the ripple noise of the DC power supply.

Magnification and Other Optical Factors. Microscope objective lenses usually have larger numerical apertures (N.A.; index for light collecting power) with larger magnification. In epifluorescence microscopes, however, the overall fluorescence brightness depends on both N.A. and efficiency of excitation light focusing. As it turned out, the intensity of fluorescence at each photodiode was stronger for the 20×, 40× and 10× Nikon Fluor lenses, in this order. With these lenses and one of the standard relay lens before the fiber matrix, each photodiode covers an area of 12.5×12.5, 6.75×6.75, 25×25 μm^2, respectively. With a 20× or 40× objective lens, each photodiode can monitor signal at the cellular level. It would be of some help to eliminate unnecessary glass windows and filters in the optical paths. We simply used a green band-pass filter and a green dichroic mirror unit provided by a manufacturer (Olympus, Japan). Also, preventing back scattering in the dichroic mirror units reduces background fluorescence.

4.5 Single-Shot Imaging of Electrical Activity in the Brain at the Cellular Level

Frog Brain Preparation. *Xenopus* frog brains with a part of the olfactory nerve retained were isolated after anesthesia. This kind of brain preparation can respond to electrical stimulation in a glucose-supplemented saline solution for a few to several hours. This brain preparation is still covered with dura membrane, which prevents the brain from dye staining. After removal of the dura with surgical forceps dye staining was greatly facilitated.

Dye Staining. Our choice of dye was either RH-292 or RH-414, which makes 1 mM phosphate-buffered solution without precipitation. Although, according to literature and data sheets (Molecular Probes, USA), di-4-ANNEPS and di-8-ANNEPS give more fluorescence change per mV membrane potential, RH-292 and -414 were more water-soluble and more resistant to photobleaching. Staining was simply done by immersing the brain preparations into 10 to 100 µM dye solution (diluted with 0.7× Hanks–Wallace solution; 500 ml of 1× Hanks–Wallace solution contains: NaCl 4.0 g, KCl 0.2 g, $CaCl_2$ 0.07 g, $MgSO_4 \cdot 7H_2O$ 0.1 g, Na_2HPO_4 $2H_2O$ 0.03 g, KH_2PO_4 0.03 g, $NaHCO_3$ 0.175 g and glucose 1.0 g) for 1 h prior to the experiment. We then transferred the stained brain into a recording chamber filled with the 0.7× Hanks–Wallace solution at room temperature. Extensive dye washing was not necessary since free dyes were only weakly fluorescent.

Recording Chamber. The recording chamber consisted of 2 mm-thick Lucite with a hole in the center and a large cover glass glued at the bottom. For

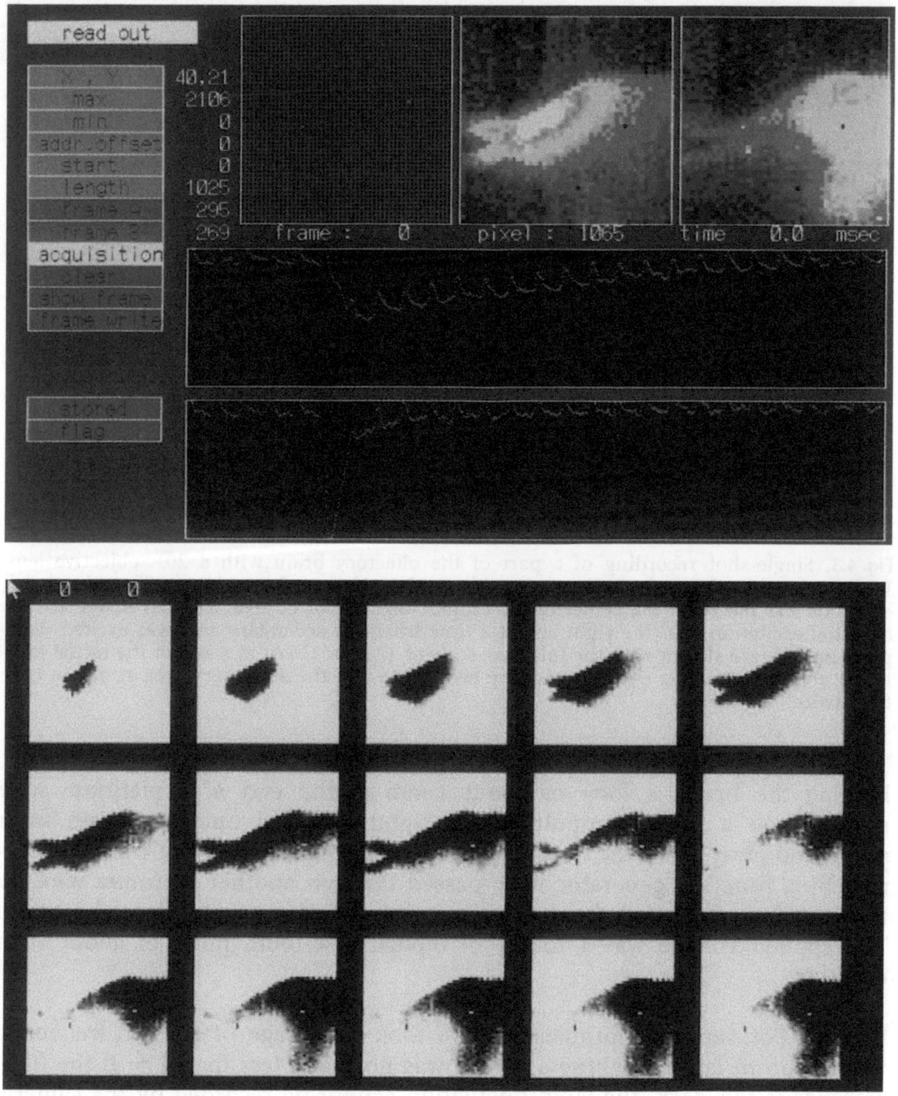

Fig. 4.2. Single-shot recording of electrical excitation spreading in a frog olfactory brain after olfactory nerve stimulation. The objective lens used was 10×, the area of view being 1.25×1.25 mm. The brain was stained with RH-414. With membrane depolarization fluorescence of this dye decreases. *Top* Two representative color-coded images of a part of the olfactory nerve and the olfactory brain. The color was coded so that fluorescence decrease corresponding nerve excitation is shown in *yellow* to *red.* The stimulating electrode was touched at the base of the olfactory nerve from above of these images. The *left* one is an image at time after the initial excitation spread, and the *right* one at time after the secondary area was excited. Temporal profies are shown with the full time scale of 262.4 ms. *Red* at a site of the initial excitation area; *green* at a site of the secondary area. Note that the ripple noise (50×2 Hz) of the lamp is seen in the opposite direction to the nerve excitation signal. *Bottom* Images were shown every 1.024 ms

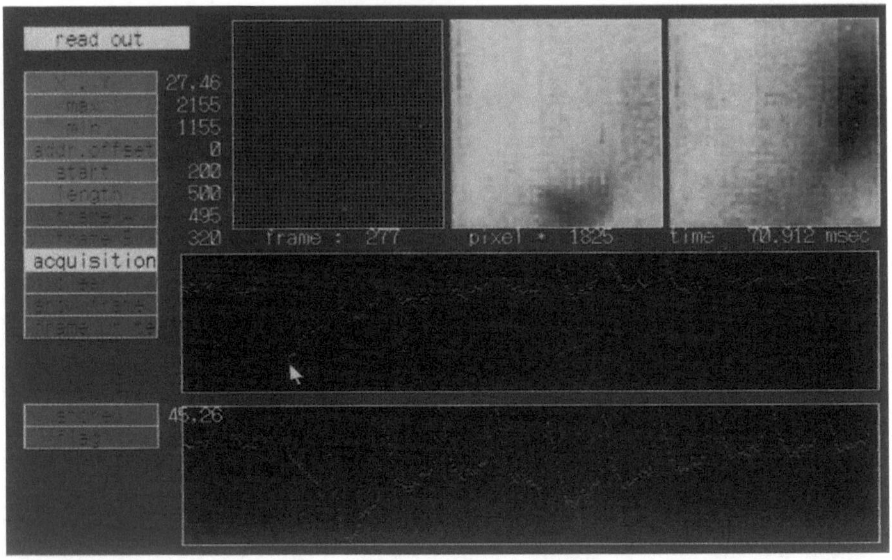

Fig. 4.3. Single-shot recording of a part of the olfactory brain with a 20× objective lens. With a 20× objective each pixel corresponds to 12.5×12.5 μm in the image. The stimulating electrode was placed at the center of the bottom slightly out of area. The left image at time of initial excitation, and the right one at a time when the secondary area was excited. Temporal profiles are shown with the full time scale of 128 ms. *Green* at a site of the initial excitation area; *red* at a site of the secondary area. Note that the secondary area seems to have fired twice

holding the brain, a loop of about 1 mm at the end of a platinum wire mounted on a micromanipulator was lightly pressed onto the brain from above and away from the observation area. Stimulating current pulses generated by a function generator were passed through another platinum wire on the stimulus point and the grounding wire placed far away from the brain. The applied voltage was 1 to 5 V of bipolar saw-tooth pulse of about 1 ms duration.

Results. For staining optimization, we took advantage of the AC frequency fluctuation of lamp brightness, which was normally less than 1%. If the fluorescence is too dark, the lamp fluctuation cannot be recorded by the camera. When the fluorescence was bright enough and the brain was electrically active, transient fluorescence decrease due to nerve excitation could be recorded as a signal several times larger than the lamp fluctuation (Fig. 4.2). The shape of the lamp fluctuation of this particular setup and this particular lamp allowed us to produce images from the noise-free time regions in the recording (Figs. 4.2 and 4.3). For more detailed studies, it is desirable to eliminate the AC noise by adding a feed-back circuit to the DC power supply.

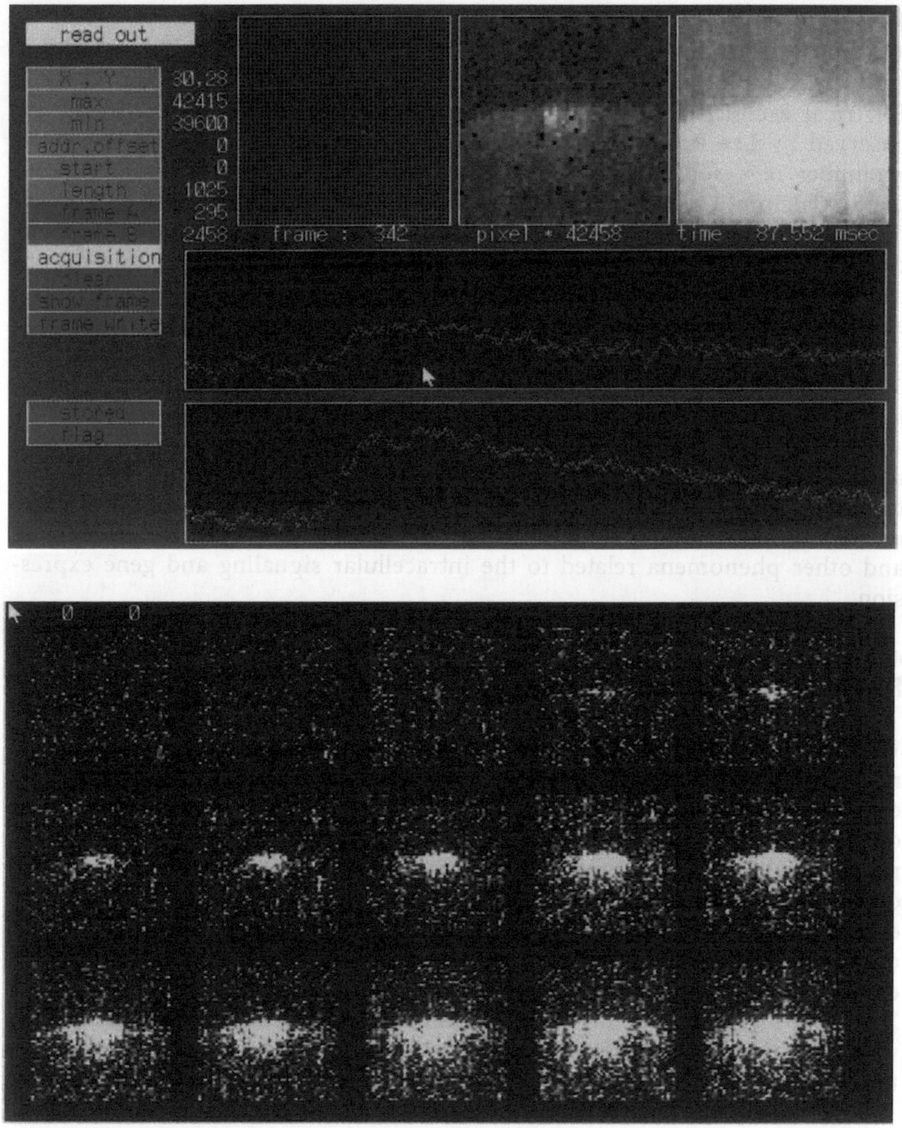

Fig. 4.4. Time-resolved cytoplasmic Ca^{2+} imaging. The microscope setup was the same as in Fig. 4.2 with a 10× objective lens. The calcium indicator dye used was Calcium Orange-AM. Fluorescence of this dye increases with increasing cytoplasmic Ca^{2+}. The signal was averaged over 20 times of repeated recording. *Top* Two representative images at an area of the brain cortex. The stimulating electrode was placed on the cortex from above in the center. Temporal profies are shown with the full time scale of 262.4 ms. *Red* At a site of the initial excitation area; *green* at a site of the surrounding area. *Bottom* Images were shown every 1.024 ms. Note that the onset of cytoplasmic Ca^{2+} increase was delayed about 3 ms as compared with that of electrical excitation

4.6 Imaging of Cytoplasmic Ca^{2+} Transients Preceded by Electrical Activity

Electrical excitation-driven Ca^{2+} transients in the nervous system were first observed by Lev-Ram and Grinvald (Lev-Ram and Grinvald 1987). Recently, a number of cytoplasmic Ca^{2+} ion indicator dyes were introduced. Among them, Calcium Orange has similar excitation and fluorescence spectra to those of the RH dyes so that this dye can be used for experiment without changing the experimental setups. We have preliminarily tried this dye to test whether or not fast cytoplasmic transients of Ca^{2+} can be seen following electrical stimulation. Calcium Orange-AM (Molecular Probes) was loaded by immersing a dura-removed frog brain into 100 μM Calcium Orange-AM in 0.7× Hanks–Wallace solution for about 2 h. The brain was then washed once in the solution and mounted into the chamber as described above. The Ca^{2+} signal was weak and could only be seen after signal integration by repeated stimulation (Fig. 4.4). Further experimental optimization would be needed for single-shot recording. This kind of approach would be useful for studies on involvement of the cytoplasmic Ca^{2+} transients in the memory processes and other phenomena related to the intracellular signaling and gene expression.

4.7 Future Directions and Possibilities

We could show that this type of camera can capture electrical activities of neurons by single-shot recording at the cellular level and that the camera can easily be used by inexperienced researchers with only minimum knowledge of electronic and/or electrophysiological techniques. Once the camera was properly set up, experiments were easily operated by click of a mouse without assistance. And, although vibration should be carefully attenuated, no electrical shielding around the recording chamber was needed. The major drawback of this prototype is its bulky size. It would be most desirable to reduce its size to a more manageable dimension. Also, at present, each of the 2500 fibers must be connected individually. Elucidating a more convenient way of connection would greatly reduce the time needed for the experimental setup. Despite its drawbacks, it should be emphasized that a number of studies could be done to reveal cellular mechanisms of the brain functions such as functional cellular architecture of the brain, relationships of electrical activities and the intracellular signaling, or systematic anatomical neurotransmitter identification and its overall effects on brain functioning.

References

Antic S, Zecevic D (1995) Optical signals from neurons with internally applied voltage-sensitive dyes. J Neurosci 15:1392–1405

Chien CB, Pine J (1991) Voltage-sensitive dye recording of action potentials and synaptic potentials from sympathetic microcultures. Biophys J 60:697–711

Cohen LB, Lesher S (1986) Optical monitoring of membrane potential: methods of multisite optical measurement. Soc Gen Physiol Ser (USA) 40:71–99

Elias SA, Yae H, Ebner TJ (1993) Optical imaging of parallel fiber activation in the rat cerebellar cortex: spatial effects of excitatory amino acids. Neuroscience 52:771–786
Fukunishi K, Murai N, Uno H (1993) Spatiotemporal observation of guinea pig auditory cortex with optical recording. Jpn J Physiol (Suppl 1) 43:S61–66
Grinvald A, Frostig RD, Lieke E, Hildesheim R (1988) Optical imaging of neuronal activity. Physiol Rev 68:1285–1366
Kinoshita K Jr, Ashikawa I, Saita N, Yoshimura H, Itoh H, Nagayama K, Ikegami A (1988) Electroporation of cell membrane visualized under a pulsed-laser fluorescence microscope. Biophys J 53:1015–1019
Lev-Ram V, Grinvald A(1987) Activity-dependent calcium transients in central nervous system myelinated axons revealed by the calcium indicator Fura-2. Biophys J 52:571–576
Momose-Sato Y, Sakai T, Komuro H, Hirota A, Kamino K (1991) Optical mapping of the early development of the response pattern to vagal stimulation in embryonic chick brain stem. J Physiol (Lond) 442:649–668
Orbach HS, Van Essen DC (1993) In vivo tracing of pathways and spatio-temporal activity patterns in rat visual cortex using voltage sensitive dyes. Exp Brain Res 94:371–392
Sawaguchi T (1994) Modular activation and suppression of neocortical activity in the monkey revealed by optical imaging. Neuroreport 6:185–189
Sinha SR, Patel SS, Saggau P (1995) Simultaneous optical recording of evoked and spontaneous transients of membrane potential and intracellular calcium concentration with high spatio-temporal resolution. J Neurosci Methods 60:49–60
Sugitani M, Sugai T, Tanifuji M, Onoda N (1994) Signal propagation from piriform cortex to the endopiriform nucleus in vitro revealed by optical imaging. Neurosci Lett 171:175–178
Sutor B, Hablitz JJ, Rucker F, ten Bruggencate G (1994) Spread of epileptiform activity in the immature rat neocortex studied with voltage-sensitive dyes and laser scanning microscopy. J Neurophysiol (USA) 72:1756–1768
Takahashi M, Tsuchiya H, Hosoi S, Hayakawa T (1993) Development of 50×50 fiber-array photodiode camera. Proc SPIE (USA) 1757:111–114
Wellis DP, Kauer JS (1994) GABAergic and glutamatergic syanaptic input to identified granule cells in salamander olfactory bulb. J Physiol (Lond) 475:419–430

CHAPTER 5

Micromanipulation of Macromolecules: How to Measure the Stiffness of Single Microtubules

Harald Felgner[1,2*] · Rainer Frank[1] · Manfred Schliwa[1]

Contents

[1] Adolf-Butenandt-Institut (Zellbiologie), Ludwig-Maximilians-Universität, 80336 München, Germany
[2] Lehrstuhl für Biophysik E22, Technische Universität München, 85748 Garching, Germany
* Address for correspondence: Adolf-Butenandt-Institut (Zellbiologie), Schillerstrasse 42, 80336 München, Germany, Tel.: +49–89-5996-875, Fax: +49-89-5996-882, e-mail: Harald.Felgner@physik.tu-muenchen.de

5.1 Optical Tweezers and Microtubules

Optical tweezers are a combination of an intense light source and a standard light microscope, making it possible to grab and manipulate optically refracting particles in solution with the momentum of light. Ashkin and Dziedzic first used optical tweezers to manipulate biological objects such as viruses and bacteria in 1987. Since then, single-beam laser traps have frequently been employed to hold, move, and deform cells and subcellular particles. Examples include, but are not restricted to, the holding of yeast cells within a trap for more than one cell cycle (Ashkin et al. 1987), the manipulation of nuclei and organelles in plant cells (Ashkin and Dziedzic 1989; Leitz et al. 1994) and protozoa (Aufderheide et al. 1992), the displacement of chromosomes or chromosome fragments in cultured cells (Berns et al. 1989; Seeger et al. 1991), the blockage of axonal transport (Martenson et al. 1993), and cell sorting (Buican 1991; for reviews, see Block 1990; Kuo and Sheetz 1992; Weber and Greulich 1992). These studies have demonstrated convincingly that intracellular organelles up to the size of nuclei as well as whole cells can be displaced or deformed without damaging effects. Apart from these in vivo experiments, optical tweezers-based techniques are widely used to measure physical parameters of single biological molecules in vitro, for example the forces of single molecular motors and their step sizes (Finer et al. 1994; Kuo and Sheetz 1993; Svoboda et al. 1993), or the elastic parameters of DNA (Perkins et al. 1995) or microtubules (Kurachi et al. 1995).

In this chapter we describe yet another application of the use of optical tweezers, namely, the direct micromanipulation of cytoskeletal polymers and its exploitation to determine the elastic properties of single microtubules.

Microtubules are polymers consisting of α- and β-tubulin subunits which form a hollow cylinder with a diameter of 24 nm (Mandelkow and Mandelkow 1994). They are part of the eukaryotic cytoskeleton and are involved in many cellular functions such as mitosis, cell motility, and organelle transport. As extended cytoskeletal polymers they contribute to the shape and polarity of cells. To serve as cytoskeletal stabilizers, microtubules must resist tensile and compressive forces, a property reflected in their flexural rigidity. Different classes of proteins are known to interact with microtubules. One class, the microtubule-associated proteins (MAPs), are known to stabilize microtubules and to promote their assembly (Mandelkow and Mandelkow 1995). The anti-tumor drug taxol modifies microtubule morphology in mammalian cells, and stabilizes the polymer against depolymerizing agents (Schiff et al. 1979). Whereas taxol-treated microtubules are more stable dynamically, they were found to be less rigid mechanically than untreated microtubules in two studies (Dye et al. 1993; Venier et al. 1994). Gittes et al. (1993), on the other hand, found the rigidity of taxol-stabilized microtubules to be higher than that of untreated microtubules (Kurz and Williams 1995). Knowing the flexural rigidity and its modification by different agents is essential to an understanding of the role of microtubules as cytoskeletal polymers.

Single-beam gradient force optical traps consist of a laser beam that is highly focused by the objective lens of a standard light microscope which, at the same time, is used for observing the microscopic object. Cellular organelles or whole cells can be directly trapped, while polystyrene or glass beads can be used as handles to manipulate single molecules. The application described here is unique in that a macromolecular complex (a microtubule) is handled directly under microscopic control. Microtubules attached at one end to the coverslip are captured at the other end, deflected, and released. The shapes of the microtubules are observed and recorded by video microscopy as they return to their resting position. During this movement, the hydrodynamic forces of viscous flow counteract the elastic restoring forces which actively drive the microtubules relative to the surrounding buffer medium. To determine the flexural rigidity on which this movement is based, the bending of a microtubule is analyzed under a given velocity distribution along its length.

5.2 Optical Tweezers Setup

The optical tweezers are formed by a cw plane-polarized Nd:YAG laser beam (wavelength 1064 nm, Spectron, Rugby, England), expanded by a telescope to a diameter approximately filling the back aperture of a microscope objective (Plan-NEOFLUAR 100×/1.30) and coupled into an upright light microscope (Zeiss Axioskop, Oberkochen, Germany) via the fluorescence illumination path as shown in Fig. 5.1. The laser beam is fixed relative to the image plane, but objects near the coverglass may be manipulated in all three directions by moving the microscope stage via stepping motors (Märzhäuser EK32 and MCL, Wetzlar, Germany) under computer control (Apple Macintosh IIfx,

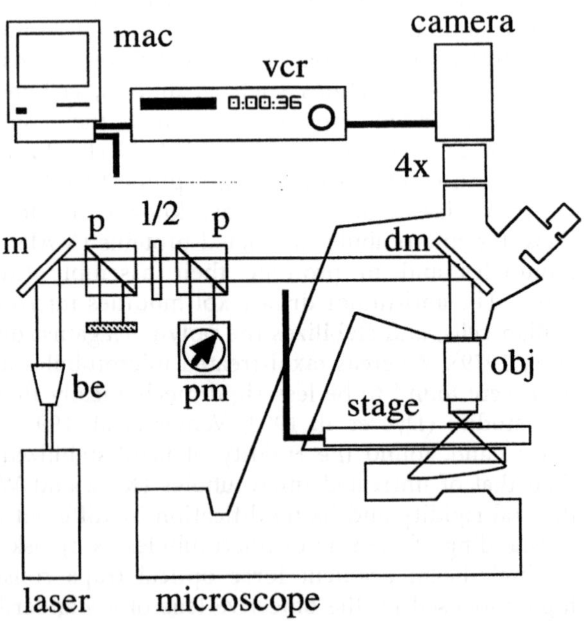

Fig. 5.1. Schematic representation of the laser trap setup. The beam of a Nd:YAG laser is expanded by a beam expander (*be*) and coupled into the objective (*obj*) of a light microscope via mirrors (*m*) and a dichroic mirror (*dm*) after passing to polarizing beam splitter cubes (*p*) and a $\lambda/2$-wave plate (*l/2*). The laser power is monitored on a pyroelectric power meter (*pm*). The movement of the microscope stage is controlled from the computer, as is the acquisition and processing of microscope images from the Nevicon camera

Cupertino, USA). Cells are monitored by phase contrast optics (4× magnified), and images are recorded with a Newvicon camera (Hamamatsu C2400-07, Herrsching, Germany) onto a SVHS video recorder (Panasonic AG-7350, Düsseldorf, Germany). During all experiments a laser output power of approximately 1.5 W is coupled into the epifluorescence illumination path of the microscope, which results in a power of nearly 700 mW at the specimen due to losses in the microscope optics.

5.3 Digital Video Processing

Processing of the resulting images is performed on the same Macintosh IIfx computer used to control the microscope stage with the public domain NIH Image software (written by Wayne Rasband at the US National Institutes of Health and available from ftp://zippy.nimh.nih.gov). The image processing board (PixelPipeline, Perceptics, Knoxville, USA) is used to store the video sequences recorded with the video recorder during the experiment one field at a time into the computer memory. Overall magnification of the whole optical system is such that 43 nm in the object plane correspond to one pixel in the computer image files. Field-by-field digitization allows us to obtain a time resolution of 50 Hz. The positions of the microtubule ends and of the laser beam are tracked interactively by the experimenter. The fixed microtubule end serves as a point of reference. By calculating the other positions as differences to this point in each field, all coordinates are given in the frame of reference of the unbent microtubule. Furthermore, the position differences correct for time-base errors during still frame capture. The movements of the microscope stage are controlled via user-written algorithms integrated into the image processing software. A joystick enables fine control of the stage to bend single microtubules.

5.4 The Bending of a Stiff Rod

We consider the bending of the microtubule to be described as the linear elastic bending of a circularly symmetric rod. If the rod is considered isotropic, its flexural rigidity is the product of the Young's modulus E and the geometrical moment of inertia I. E a measure of the elasticity of the material of the rod, whereas I describes the geometry of the cross-section of the rod.

Let x be the coordinate along the unbent rod and y the coordinate orthogonal to that in the plane of bending (Fig. 5.2). When the resulting deflection of the rod may be considered small compared to its total length L, the deformation by a force f per unit length is described by the following equation (Feynman et al. 1964; Landau and Lifschitz 1986):

$$f = EI\frac{d^4y}{dx^4}\,. \tag{1}$$

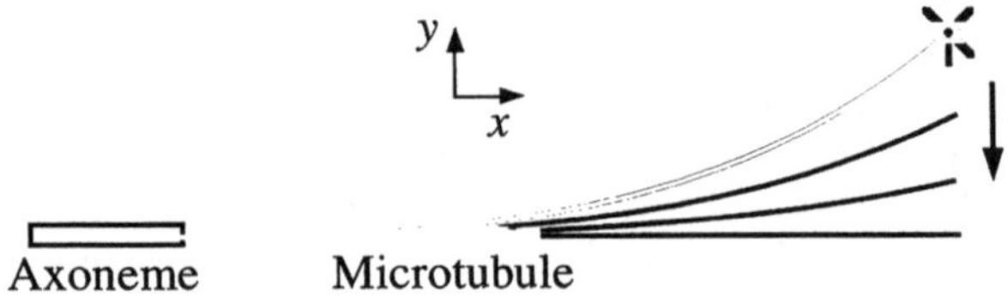

Fig. 5.2. Relaxation movement. As soon as the laser power is switched off, the free end of the microtubule relaxes to its straight resting position. x and y positions of the free end are tracked over time

The shape $y(x)$ of a bent microtubule is derived from this differential equation by succesive integration. The boundary conditions in the chosen frame of reference are for a rod fixed at one end ($x=y=0$):

$$y|_{x=0} = 0 \tag{2}$$

$$\left.\frac{dy}{dx}\right|_{x=0} = 0 \tag{3}$$

and for small deflections

$$\left.\frac{d^2y}{dx^2}\right|_{x=L} = 0 \tag{4}$$

$$\left.\frac{d^3y}{dx^3}\right|_{x=L} = 0\,. \tag{5}$$

When a single force F_{single} is pulling at the end of the rod in the direction y, the shape of the microtubule is obtained as

$$y(x) = \frac{F_{single}\,L^3}{6EI}\left[-\left(\frac{x}{L}\right)^3 + 3\left(\frac{x}{L}\right)^2\right]. \tag{6}$$

In an analogous way the shape of the rod may be calculated when it is bent by a constant force density f_{const} along its length. That is the case for a microtubule in a hydrodynamic flow of constant velocity. The resulting bending shape is:

$$y(x) = \frac{f_{const}L^4}{24EI}\left[\left(\frac{x}{L}\right)^4 - 4\left(\frac{x}{L}\right)^3\right] + 6\left(\frac{x}{L}\right)^2\Bigg]. \tag{7}$$

Last consider a linearly increasing load along the rod, $f=f_{\max}x/L$, with the maximum force density $f_{\max}$ at the free end of the rod. This applies to the situation when, for example, a fluid with zero velocity at the fixed microtubule end and a maximum velocity at the free end bends a microtubule. The corresponding result is:

$$y(x) = \frac{f_{\max}\,L^4}{120EI}\left[\left(\frac{x}{L}\right)^5 - 10\left(\frac{x}{L}\right)^3 + 20\left(\frac{x}{L}\right)^2\right]. \tag{8}$$

The rod is maximally deflected at its free end ($x=L$) to

$$y(L) = \frac{11\,f_{\max}\,L^4}{120EI}\,. \tag{9}$$

5.5 Hydrodynamic Forces on a Cylinder

The measurements to determine the elastic properties of single microtubules rely on hydrodynamic forces counterbalancing the elastic restoring forces. When a cylinder is moving perpendicular to its long axis through a Newtonian fluid with the velocity v_{const}, the force per unit length may be written as (Doi and Edwards 1986)

$$f = \frac{4\pi\eta\, v_{const}}{\ln(L/2d)} \,. \tag{10}$$

η denotes the (dynamic) viscosity of the fluid and d is the diameter of the cylinder.

When the fluid velocity increases linearly along the microtubule length as $v = v_{max}x/L$, the force per unit length is given as (Venier et al. 1994)

$$f = \frac{2\pi\eta\, v_{max}}{\ln(L/2d)} \,. \tag{11}$$

Inserted into Eq. (9), a microtubule acted upon by a hydrodynamic flow with zero velocity at the fixed end and a maximum velocity v_{max} at the free end has a maximum deflection of the free end of

$$y(L) = \frac{11\pi\eta\, v_{max} L^4}{60EI\ln(L/2d)} \,. \tag{12}$$

5.6 Microtubule Manipulation by Optical Tweezers

The microtubules used in these studies were prepared from phosphocellulose-purified porcine brain tubulin isolated according to standard procedures (Shelanski et al. 1973; Mandelkow et al. 1985). Axonemes of *Chlamydomonas reinhardtii* (Witman 1986) were attached to polylysine-coated coverslips and incubated with purified tubulin. The microtubules polymerized from the plus-ends in this way have a length of typically 5–20 μm. Taxol (10 μM final concentration) or microtubule-associated proteins (MAPs) can then be added after polymerization as desired. The buffer used was a standard microtubule polymerization buffer (Felgner et al. 1996) supplemented with proteolysis inhibitors. The coverslip was placed on a slide and sealed, forming a specimen chamber with a height of approximately 15–25 μm. Observations were made at a temperature of 25–28 °C in the specimen chamber.

The single beam optical gradient trap allows us to grab objects of a refractive index higher than the refractive index of the surrounding medium at a single point. Microtubules may be directly trapped and manipulated in a direction perpendicular to their long axis. To bend a microtubule directly with the laser tweezers, the microtubule was fixed at one end. In all our measurements microtubules polymerized from axonemes parallel to the coverslip were used. They were bent in the viewing plane of the microscope by apply-

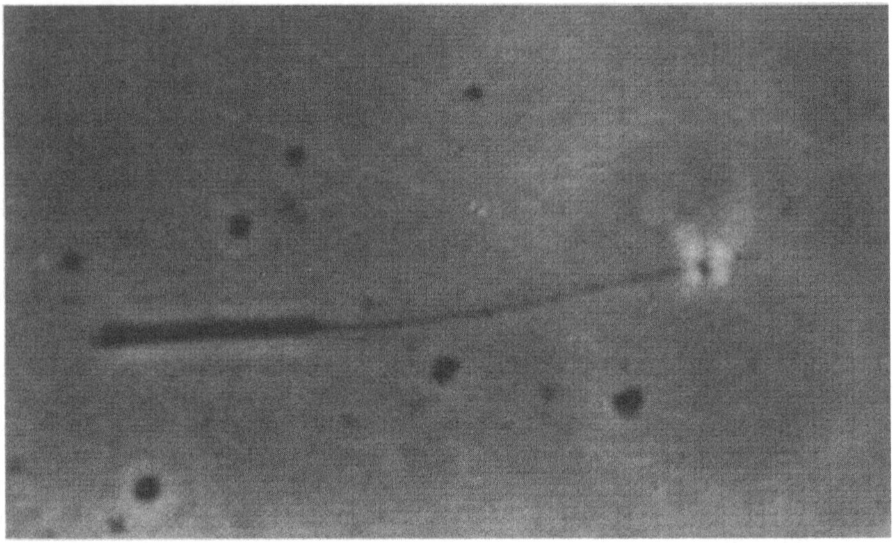

Fig. 5.3. Digitally enhanced phase-contrast image of a microtubule. The microtubule is shown immediately before its release from the optical tweezers (*four bright spots*)

ing a force with a component perpendicular to the microtubule near its free end.

When the laser beam is focused onto the microtubule, the microtubule is trapped in the focal point of the laser. By moving the stage and with it the axoneme and the fixed microtubule end in a direction approximately perpendicular to the long axis of the microtubule, the microtubule is bent. The laser power is switched off after bending the microtubule a certain distance (Fig. 5.3). The velocity of the relaxation movement is determined by the relation between the elastic force driving back the microtubule and the hydrodynamic force of the buffer acting on the moving microtubule and resisting this movement. The movement of the microtubule through buffer is a case of low Reynolds number hydrodynamics, where inertia effects are small relative to frictional forces. The movement of the free end of the microtubule is therefore described by a first-order differential equation which can be solved by a simple exponential decay. If $y(t)$ denotes the deflection of the microtubule end at a given time t after release ($t=0$), it is given by

$$\frac{y(t)}{y(0)} = e^{-\frac{\beta}{\gamma}t} \tag{13}$$

with $y(0)$ as the maximum initial deflection. The free end deflection decreases in time with the relation of the elastic constant β and the hydrodynamic friction constant γ.

The hydrodynamic forces acting on the relaxing microtubule are given by a velocity distribution along the microtubule length reflecting the shape of the microtubule at each instant of relaxation (as described by an equation similar to Eqs. (6), (7) or (8). The relative velocity is zero at the fixed micro-

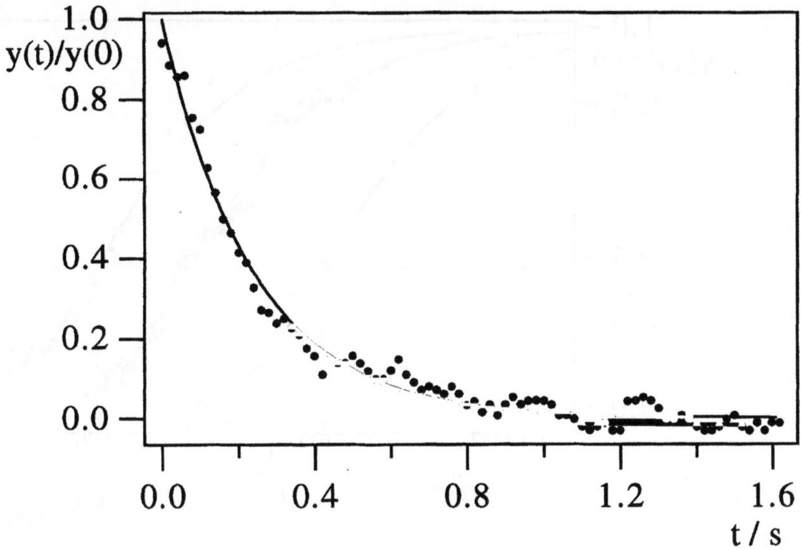

Fig. 5.4. Relaxation of a taxol-treated microtubule. The *circles* show the recorded end positions (normalized to the initial deflection) of the microtubule over time in a typical relaxation experiment. The *line* is an exponential function fitted to that movement

tubule end and increases to a maximum velocity at the free, relaxing end. We therefore approximate this velocity distribution as increasing linearly from the axoneme ($v=0$) to the free microtubule end ($v = v_{max}$). Then the shape of the microtubule may be approximated by Eq. (8) with a maximum deflection as in Eq. (9). The hydrodynamics of the relaxation in this simplification is described by Eqs. (11) and (12) under that assumption the decay Eq. (13) may be rewritten as

$$\frac{y(\tau)}{y(0)} = e^{-EI\tau} \tag{14}$$

with

$$\tau = \frac{60 \ln(L/2d)}{11\pi\eta L^4} t . \tag{15}$$

Figure 5.4 shows the relaxation movement of the end of a taxol-treated microtubule (symbols) and a fit to an expontial function (line). From the fit and Eqs. (14) and (15) the flexural rigidity of the microtubule can be determined.

A prerequisite for the use of linear elastic theory in equation 1 is a small deflection of the microtubule end. Expressed in another way, the cosine of the corresponding initial deflection angle should be close to 1. The microtubules measured varied in length between 6.1 and 14.0 µm. A taxol microtubule of 11 µm length was maximally deflected 5 µm at its free end. In that extreme example the cosine of the corresponding initial deflection angle is 0.90. In the case of pure tubulin and MAP microtubules, the microtubules

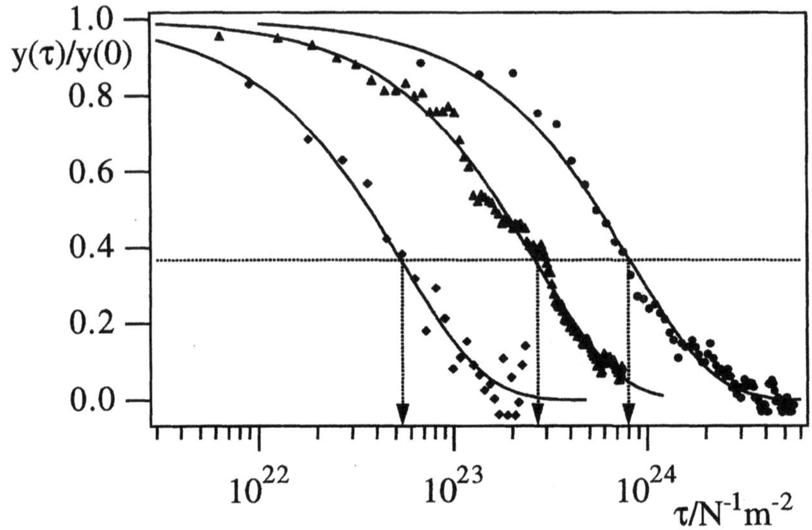

Fig. 5.5. Relaxation of three different microtubules. The *circles* show the end positions of the same microtubule from Fig. 5.4 [$L=7.0$ μm, $y\,(0)=3.1$ μm]. The transformation of the time axis to a quantity τ [see Eq. (15)] enables the direct comparison of the movement to that of an untreated microtubule [*triangles*, $L=13.2$ μm, $y\,(0)=4.6$ μm] and that of a MAP-stabilized microtubule [*squares*, $L=12.0$ μm, $y\,(0)=1.4$ μm]. The inverse of τ at the positions shown with *arrows* directly gives the flexural rigidity of the respective microtubules

were deflected less far (0.94 and 0.99 for the cosine of the angle, respectively).

In order to be able to compare the relaxation of different microtubules graphically, the different lengths of the microtubules used have to be taken into account. Therefore the time axis t was replaced by an axis τ as given in equation 15. Figure 5.5 shows the result for the same taxol-stabilized microtubule as in Fig. 5.4 and, in addition, for a microtubule untreated after polymerization and another one stabilized with MAPs. In this way the relaxation curves can be compared directly. The ordinate is the normalized deflection, so that 1 and 0 correspond to the deflected and the relaxed position, respectively. The flexural rigidities can be directly determined as the inverse of τ at the points where the relaxation curves have reached $y(\tau)/y(0)=1/e\approx 0.37$ (arrows). These three examples clearly show that taxol makes microtubules more flexible, and the binding of MAPs stiffens microtubules (see Felgner et al. 1996, for further details).

As demonstrated in this application, optical tweezers are useful tools for the manipulation of biological structures in the range of submicroscopic organelles and macromolecular complexes. Undoubtedly the future will bring further cell biological applications within the reach of these versatile instruments.

References

Ashkin A, Dziedzic JM (1987) Optical trapping and manipulation of viruses and bacteria. Science 235:1517–1520

Ashkin A, Dziedzic JM (1989) Internal cell manipulation using infrared laser traps. Proc Natl Acad Sci USA 86:7914–7918

Aufderheide KJ, Du Q, Fry ES (1992) Directed positioning of nuclei in living *Paramecium tetraurelia*: use of the laser optical force trap for developmental biology. Dev Genet 13:234–240

Berns MW, Wright WH, Tromberg BJ, Profeta GA, Andrews JJ, Walter RJ (1989) Use of a laser-induced optical force trap to study chromosome movement on the mitotic spindle. Proc Natl Acad Sci USA 86:4539–4543

Block SM (1990) Optical tweezers: a new tool for biophysics. In: Foskett JK, Grinstein S (eds) Noninvasive techniques in cell biology. John Wiley, New York, pp 375–402

Buican TN (1991) Automated cell separation techniques based on optical trapping. Am Chem Soc Symp Ser 464:59–72

Doi M, Edwards S (1986) The theory of polymer dynamics. Clarendon, Oxford

Dye RB, Fink SP, Williams RC (1993) Taxol-induced flexibility of microtubules and its reversal by MAP-2 and tau. J Biochem 268:6847–6850

Felgner H, Frank R, Schliwa M (1996) Flexural rigidity of microtubules measured with the use of optical tweezers. J Cell Sci 109:509–516

Feynman RP, Leighton RB, Sands M (1964) The Feynman lectures on physics II. Addison-Wesley, Reading

Finer JT, Simmons RM, Spudich JA (1994) Single myosin molecule mechanics: piconewton forces and nanometre steps. Nature 368:113–119

Gittes F, Mickey B, Nettleton J, Howard J (1993) Flexural rigidity of microtubules and actin filaments measured from thermal fluctuations in shape. J Cell Biol 120:923–934

Kuo SC, Sheetz MP (1992) Optical tweezers in cell biology. Trends Cell Biol 2:116–118

Kuo SC, Sheetz MP (1993) Force of single kinesin molecules measured with optical tweezers. Science 260:232–234

Kurachi M, Hoshi M, Tashiro H (1995) Buckling of a single microtubule by optical trapping forces: direct measurement of microtubule rigidity. Cell Motil Cytoskel 30:221–228

Kurz JC, Williams RC (1995) Microtubule-associated proteins and the flexibility of microtubules. Biochem 34:13374–13380

Landau LD, Lifschitz EM (1986) Theory of elasticity, 3rd edn. Pergamon, Oxford

Leitz G, Weber G, Seeger S, Greulich KO (1994) The laser microbeam trap as an optical tool for living cells. Physiol Chem Phys Med NMR 26:69–88

Mandelkow EM, Hermann M, Rühl U (1985) Tubulin domains probed by limited protolysis and subunit-specific antibodies. J Mol Biol 185:311–327

Mandelkow E, Mandelkow E-M (1994) Microtubule structure. Curr Opinion Struct Biol 4:171–179

Mandelkow E, Mandelkow E-M (1995) Microtubules and microtubule-associated proteins. Curr Opinion Cell Biol 7:72–81

Martenson C, Stone K, Reedy M, Sheetz MP (1993) Fast axonal transport is required for growth cone advance. Nature 366:66–69

Perkins TT, Smith DE, Larson RG, Chu S (1995) Stretching of a single tethered polymer in a uniform flow. Science 268:83–87

Schiff PB, Fant J, Horwitz SB (1979) Promotion of microtubule assembly in vitro by taxol. Nature 277:665–667

Seeger S, Manojembashi S, Hutter K-J, Futerman G, Wolfrum J, Greulich KO (1991) Application of laser optical tweezers in immunology and molecular genetics. Cytometry 12:497–504

Shelanski ML, Gaskin F, Cantor CR (1973) Assembly of microtubules in the absence of added nucleotides. Proc Natl Acad Sci USA 70:765–768

Svoboda K, Schmidt CF, Schnapp BJ, Block SM (1993) Direct observation of kinesin stepping by optical trapping interferometry. Nature 365:721–727

Venier P, Maggs AC, Carlier M-F, Pantaloni D (1994) Analysis of microtubule rigidity using hydrodynamic flow and thermal fluctuations. J Biochem 269:13353–13360

Weber G, Greulich KO (1992) Manipulation of cells, organelles, and genomes by laser microbeam and optical trap. Int Rev Cytol 133:1–41

Witman GB (1986) Isolation of *Chlamydomonas* flagella and flagellar axonemes. Methods Enzymol 134:280–290

CHAPTER 6

Dynamics of Single Protein Polymers Visualized by Fluorescence Microscopy

Josef Käs* · Jochen Guck · David Humphrey

Contents

Center for Nonlinear Dynamics and Program for Cellular and Molecular Biology, Physics Department, University of Texas at Austin, Austin, Texas 78712, USA

* Corresponding address: Center for Nonlinear Dynamics and Program for Cellular and Molecular Biology, Physics Department, University of Texas at Austin, Austin, Texas 78712, USA, Tel.: +1-512-475-7646, Fax: +1-512-471-1558, e-mail: kas@chaos.ph.utexas.edu

6.1 Introduction

Filamentous protein networks are one of the main structural elements in nature. This report describes novel approaches in video-enhanced fluorescence microscopy to visualize and analyze the molecular motions of single constituent filaments of these networks. Common examples of these biopolymer networks are the collagen matrix of connective tissue, cartilage and bones, fibrinogen networks with intercalated blood platelets, and the cytoskeleton of cells. The collagen matrix is a chemically cross-linked network of collagen fibers which consist of tropocollagen filaments. An *in vitro* example of a network of fibrinogen fibers with embedded platelets is displayed in Fig. 6.1. These fibrinogen gels, which form in wounds, are slowly contracted by the platelets facilitating wound closure during healing.

The cytoskeleton is a mesh of several different biopolymers, namely actin filaments (F-actin), microtubules, and intermediate filaments, spanning the cytoplasm. In simplified terms, the cytoskeleton and cell membrane can be characterized as a lipid bilayer coupled to a net of biopolymers (see Fig. 6.2). However, the membrane and the cytoskeleton form together a compound material with unusual elastic properties, that can not be explained by these two parts independently. While lipid membranes in the fluid state are highly flexible (about 1000× more flexible than a polyethylene shell of the same thick-

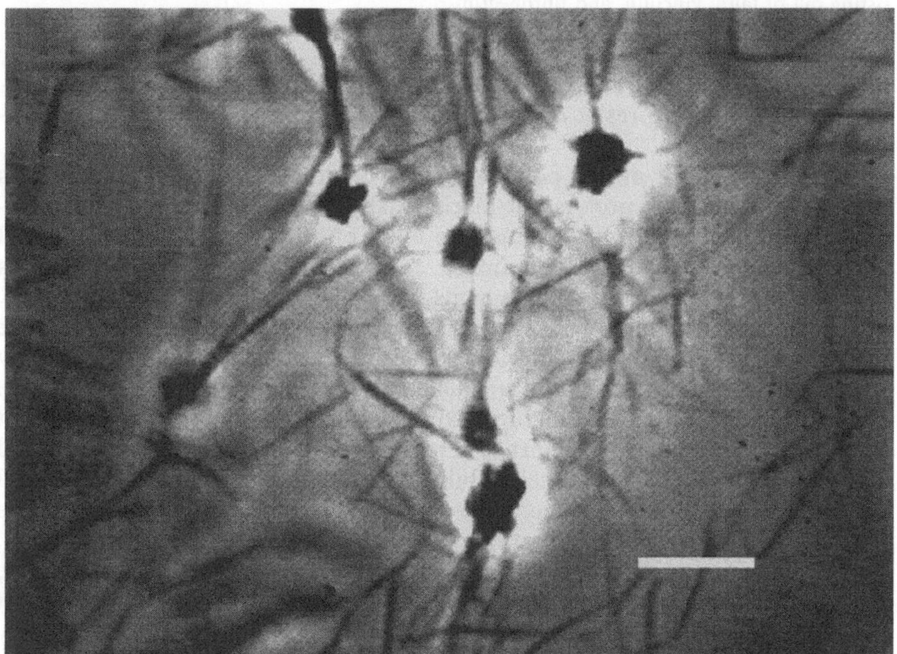

Fig. 6.1. Phase contrast picture of a network of fibrinogen fibers with embedded activated platelets

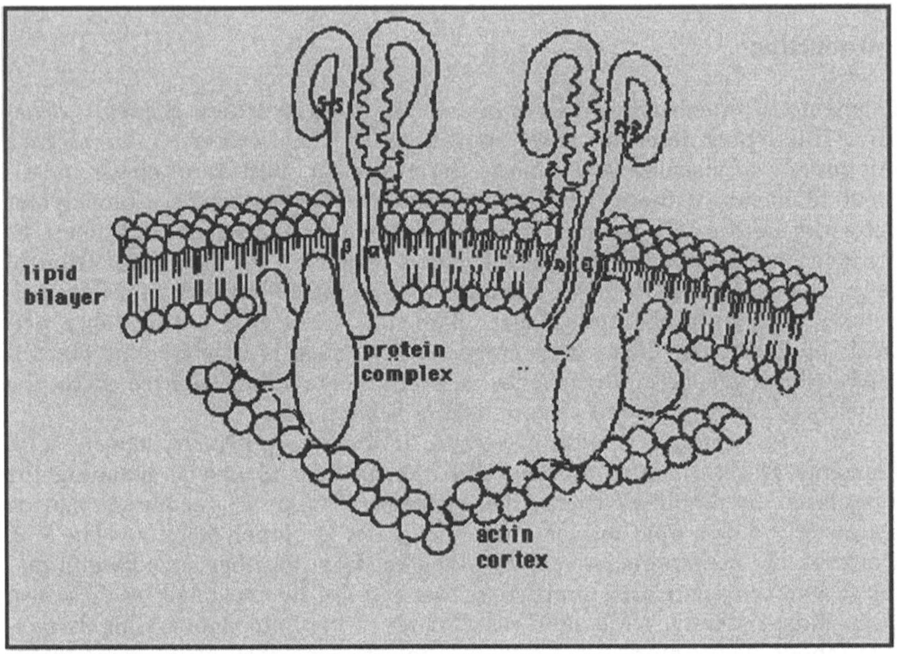

Fig. 6.2. Schematic drawing of the cell membrane and the underlying cytoskeletal rim (actin cortex). The actin filaments are coupled to the lipid membrane by a protein complex consisting out of talin, vinculin, and fibronectin

ness), when coupled to the cytoskeleton they have amazing mechanical stability - high enough to resist osmotic pressures of up to 1000 bar without rupture.

Why does nature utilize polymers as structural elements? The reason is that they form highly porous but mechanically strong structures. Thus diffusive transport of molecules through the mesh of biopolymers is not inhibited and mechanical support is guaranteed. A good example of this behavior is the way the actin cortex complements the cell membrane. The membrane, which is only ~5 nm thick, is a perfect barrier controlling the exchange between cell exterior and interior. It occupies little space and is highly impermeable, which prevents unregulated transport through the membrane. However, the membrane is highly flexible as mentioned above and does not provide any mechanical support. Therefore it is coupled to the cytoskeletal rim which provides mechanical stability to cells. Actin filaments of the cell cortex are semiflexible polymers (see Fig. 6.3), which allow nature to create particularly strong networks at low volume fractions (MacKintosh 1995). F-actin networks can be up to 1000× stronger than synthetic networks made out of flexible polystyrene chains of the same volume fraction. Since it provides mechanical support to the cell membrane and is able to quickly polymerize and depolymerize with the help of accessory proteins, the actin cortex plays a key role in cell shape (Elson 1988; Janmey et al. 1991) and motility (Stossel 1993).

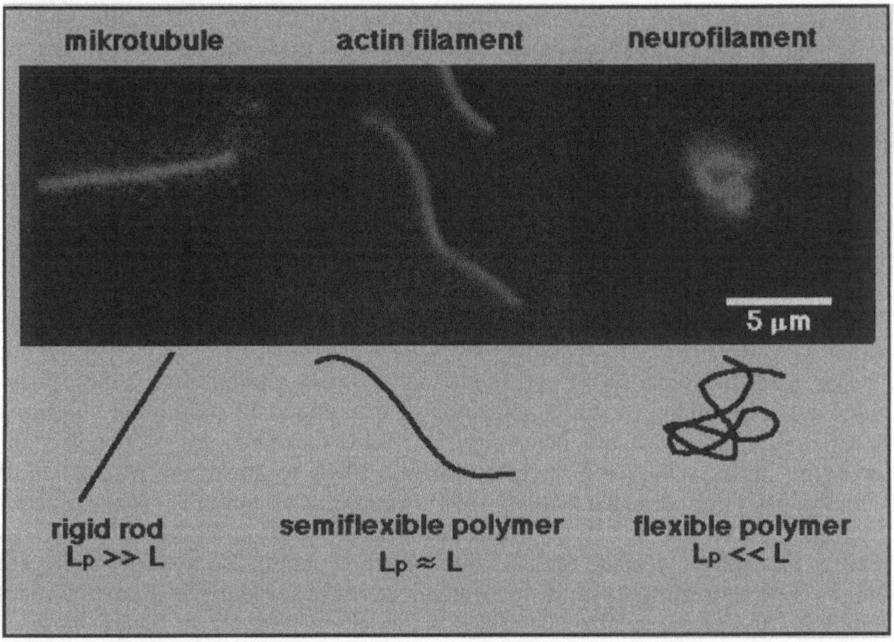

Fig. 6.3. Classification of polymer stiffness by the persistence length L_p. This concept is based on the idea that the correlation between the orientations of the local tangent, $t(s)$, decays with the distance s along the filament contour according to $\langle t(s)\, t(s')\langle = \exp(-|s-s'| / L_p)$. The rhodamine-labeled microtuble, which has a persistence length L_p that exceeds the filament length L, is a typical example of a stiff rod. The semiflexible character of the rhodamine-phalloidin labeled actin filament can be expressed by the fact that L_p and L are comparable. In case of flexible polymers, like the rhodamine-labeled neurofilament shown above, a *random coiled shape*, which is dominated by entropy, can be observed. The high flexibility means that the persistence length L_p is remarkably smaller than the filament length L

In addition to their biological importance, biopolymers such as actin and DNA have become useful systems to study fundamental aspects of polymer physics. Bulk properties of synthetic polymers have been extensively analyzed, but even the largest synthetic polymer is too small to be observed individually by methods such as light microscopy which could show directly the dynamics of single chains. Some types of biopolymers on the other hand exhibit *in vitro* a length of several microns and are easy to visualize by fluorescence labeling. This provides a novel approach to polymer physics (Ishijima et al. 1991; Smith et al. 1992; Volkmuth and Austin 1992, Käs et al. 1994a; Perkins et al. 1994) by observing the dynamics of single chains on a millisecond to hour time scale.

Nature provides model polymers for all three stiffness regimes usually distinguished in polymer science – stiff rods, semiflexible polymers and flexible polymers – as shown in Fig. 6.3. In polymer physics the stiffness of a polymer chain is usually described in terms of the characteristic length on which the polymer can be bent independently in two different directions. This length is called the persistence length. For rigid rodlike polymers the persistence length is much larger than the polymer length, for semiflexible chains both lengths are comparable, and for flexible polymers the persistence length is much smaller than the length of the polymer.

The semiflexible case seems to be the most suitable for video microscopy. Stiff rods like microtubules show translational and rotational diffusion, but no internal motions. Flexible polymers such as neurofilaments or DNA (Perkins et al. 1994) exhibit an overlapping random coil configuration which prevents detection of internal dynamics of the chain unless the chain is stretched by external forces. In contrast, the interplay between bending stiffness and entropy provides an elongated shape for semiflexible chains such as F-actin, and internal motions are easy to follow in the fluorescence microscope.

This text will focus on the visualization of single rhodamine-phalloidin labeled actin filaments in dilute solutions or embedded in semidilute solutions and nematic phases of unlabeled F-actin. Actin filaments are particularly interesting, because they are an informative system to study the properties of semiflexible polymers (Sackmann 1994). Semiflexible polymers are the most abundant polymers in nature. However, until recently, polymer physics was mainly focused on rodlike and flexible polymer chains. As a result, there are only a few experiments and theoretical models for semiflexible polymers found in the literature. The experiments described here try to fill this gap.

Actin monomers (G-actin, MW 42 000) polymerize in physiological salt solutions to form double stranded helical filaments. *In vitro* solutions of F-actin have a polydisperse length distribution which depends on the kinetics of polymerization and is altered by the presence of contaminating actin-binding proteins (Casella and Torres 1994). A typical solution of highly purified actin can have a range of filament lengths visible by fluorescence microscopy from about 1 μm to 70 μm with a mean length of ~22 μm as shown in Fig. 6.4 (Kaufmann et al. 1992). *In vivo* actin length is controlled by several actin-binding proteins and the filaments found in cells are not much longer than

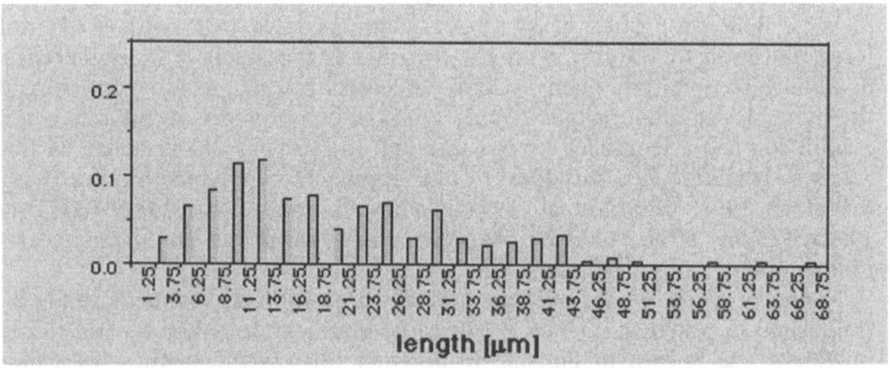

Fig. 6.4. *In vitro* length distribution of polymerized actin. The filaments showed a very broad length distribution of up 69 μm in length and an average length of 22 μm. The length distribution was determined with rhodamine-phalloidin labeled F-actin in the fluorescence microscope

one micron. The typical actin concentration within cells is about 10–20 mg/ml. Depending on the state of the cell, the fraction of monomeric actin is varied by so-called capping proteins, which prevent addition of monomers to one end of the actin filament. Actin filaments are in a steady state with monomeric actin, constantly polymerizing on one end while depolymerizing from the other (Sheterline et al. 1995). During this process, called treadmilling, ATP bound in a high-affinity binding pocket within the actin subunits is hydrolyzed to adenine nucleotide diphosphate (ADP). However, given that actin also polymerizes in the presence of ADP, the exact reason for ATP-hydrolysis is not entirely understood. Several theories have been proposed for ATP-hydrolysis that include promoting treadmilling (Wegner 1975), modulating F-actin structure (Janmey et al. 1990) or mediating interactions with actin-binding proteins (Carlier 1993). The studies described in this article focus on the properties of F-actin as a semiflexible chain and ignore the polymerization kinetics of actin filaments. Therefore the actin filaments were stabilized by phalloidin (as will be discussed later), which inhibits treadmilling and keeps nearly all of the actin in a polymeric state.

The thermally driven polymer dynamics of single actin filaments in dilute solutions or embedded in semidilute solutions and nematic phases of F-actin will be described. A normal mode analysis of single filaments in the dilute concentration regime allowed the determination of the bending stiffness of F-actin. Analysis of diffusion in semidilute solutions supported (Käs et al. 1994a) the tube model (Edwards 1967) and the concept of reptation (de Gennes 1971), which explain the diffusion of a polymer chain in an isotropic solution of entangled polymers. In more concentrated F-actin solutions the filaments were observed in a state of aligned and entangled domains due to a transition from an isotropic phase to a nematic-isotropic coexistence. Formation of aligned domains was predicted by classical theories of stiff polymers (Onsager 1949; de Gennes and Prost 1994) and has previously been demonstrated experimentally with F-actin (Suzuki et al. 1991; Coppin and Leavis 1992; Furukawa et al. 1993; Käs et al. 1996). By visualizing the motions of individual chains within the aligned domains it could be shown that they retain a very high degree of diffusional freedom along the direction of alignment and therefore were able to attain thermodynamic equilibrium between the isotropic and nematic phases (Käs et al. 1996). This first direct observation of the dynamics of semiflexible polymers in a nematic phase also indicated that the chain interactions accelerate the diffusion of the filaments and make this diffusion nearly independent of filament length.

6.2
Method

Limits of Optical Resolution. A general misconception is that the smallest objects which can be seen in a light microscope are at least as large as half the wavelength of the light source used. The range of visible light is about 300–800 nm, which means that visible structures have to be of the order of several hundreds of nanometers. However, a single microtubule which only has a diameter of ~120 nm can be seen with video enhanced DIC micro-

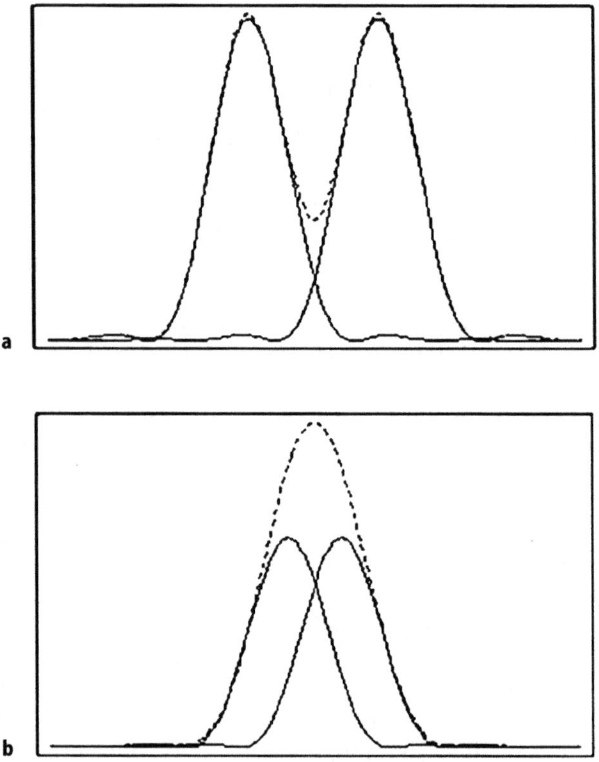

Fig. 6.5 a–b. Diffraction patterns of two point sources. **a** The point sources are sufficiently separated_ so that the two maxima can be resolved. **b** The point sources are so close together that the two maxima appear as one

scopy, and fluorescently labeled actin filaments and intermediate filaments with a diameter of ~10 nm can be seen in the fluorescence microscope. This raises the question why these objects can be seen in the light microscope.

The visualization of objects in the light microscope which are only several nanometers in diameter, can be explained by taking a closer look at the criteria generally referred to as optical resolution. This optical law has been derived by Helmholtz and Abbe to determine the closest distance at which two point sources can be resolved as separate objects. Each object in the microscope generates a diffraction pattern. If the distance between the objects is smaller than $0.61(\lambda/A)$ (λ: wavelength, A: numerical aperture) the main maxima of the diffraction patterns overlap to such a degree that they appear as one (see Fig. 6.5). This criterion does not make any predictions about the limits of visibility of single objects in the microscope, which is limited by the signal-to-noise ratio. The maximum of the diffraction pattern has to be distinctively above the background noise level. Even particles of the size of a few nanometers can be seen in the microscope if they provide a strong optical signal and are at least several hundred nanometers apart.

In case of actin filaments a good signal-to-noise ratio was achieved by the combination of labeling each monomer of the filament with rhodamine-phalloidin and using an appropriate fluorescence filter that guarantees a low background light level. The diffraction pattern that was emitted by each

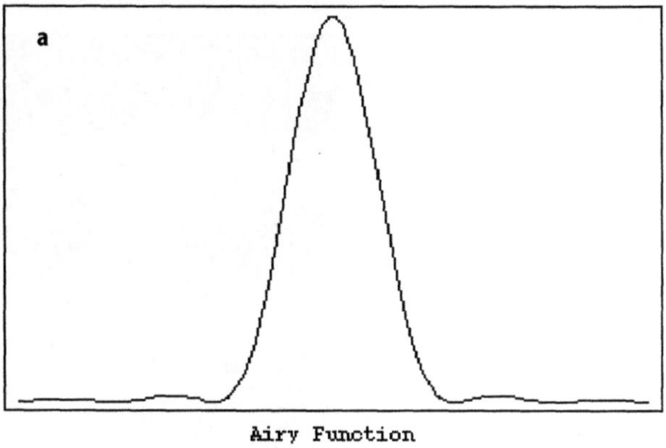

Fig. 6.6. a The theoretically predicted intensity profile of a fluorescence dye in a light microscope can be described by a so-called Airy function. **b** Fluorescence micrograph of a rhodamine-phalloidin labeled actin filament. The *white circle* denotes the area of which a two-dimensional intensity profile is displayed in c. **c** Intensity profile of a rhodamine-phalloidin labeled actin filament. The intensity is plotted in arbitrary units

fluorophore was equal to the intensity profile, which is produced by a light source with the wavelength of the fluorophore ($\lambda = 580$ nm) behind a pin hole of 7 nm diameter. This profile can be described by a so-called Airy function (see Fig. 6.6a). As can be seen in Fig. 6.6b, c the filaments were visible and the main maximum of the Airy function was clearly above the noise level.

By fitting the main maximum of an Airy-function (which can be roughly described by a parabola) to the intensity profile, the position of an actin segment, i.e. the maximum of the diffraction pattern, could be typically determined with an accuracy of $\sim\pm10$ nm. In a similar way it is possible to determine the position of highly reflective gold particles with diameters of ~5 nm in a DIC microscope. By attaching these particles to biological macromolecules the motion of these molecules can be followed and analyzed. This technique is usually referred to as nano-particle tracking.

In general, it is possible to follow molecular motions with a spatial resolution of ≈10 nm in the light microscope if the optical signal is strong enough and its source can be clearly identified. This does not mean that the molecular details of a protein can be seen in the light microscope, because the macromolecule generates only a diffraction pattern without internal structure. But its position can be determined and followed with this accuracy. Therefore this novel technique allowed us to follow the dynamics of a single protein and complements established techniques like electron microscopy, which provide static pictures of the molecular shape.

Sample Preparation. Monomeric actin (G-actin) was purified from rabbit skeletal muscle by the method of Spudich and Watt. Further purification by gel filtration chromatography using a Sephacryl S200 column did not show any differences in the performed experiments and was consequently dropped. G-actin was rapidly frozen in liquid N_2 and stored at −80 °C. On the day of use aliquots of actin were quickly thawed at 37 °C. Actin was polymerized for 2 h at room temperature or for 12 h at 5 °C in F-buffer (2 mM Tris pH 7.5, 0.5 mM ATP or 1.0 mM ATP in the case of the 12-h polymerization time, 0.2 mM $CaCl_2$, 150 mM KCl, 2 mM $MgCl_2$).

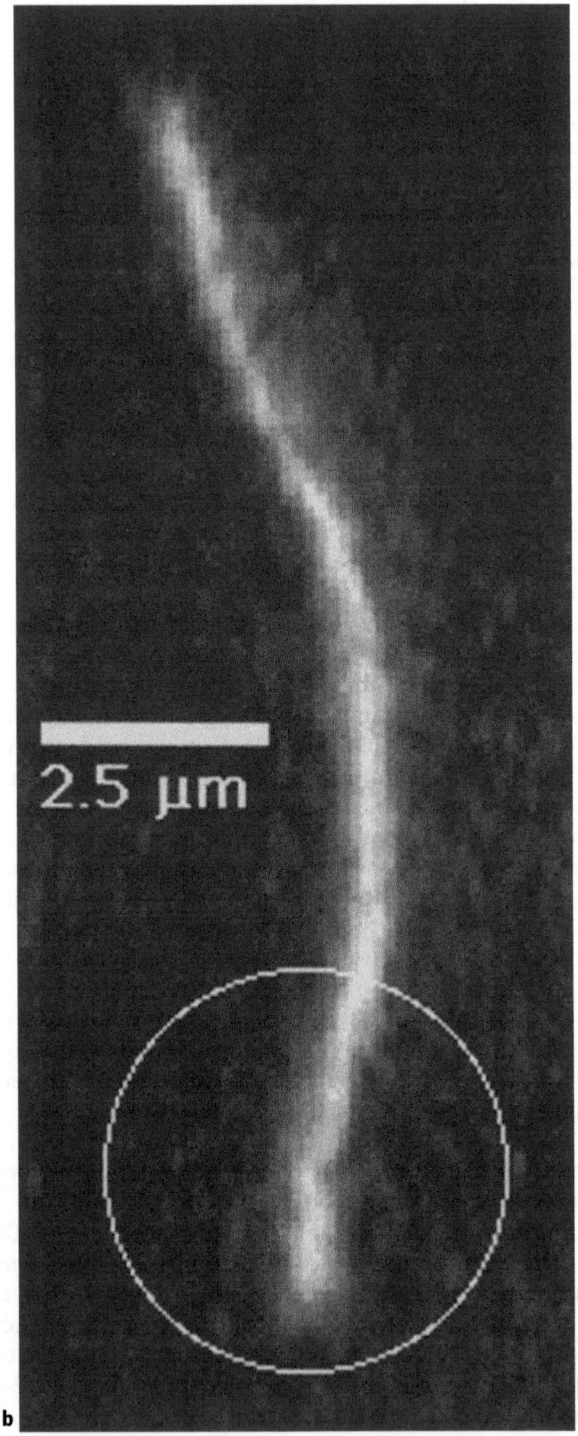

b

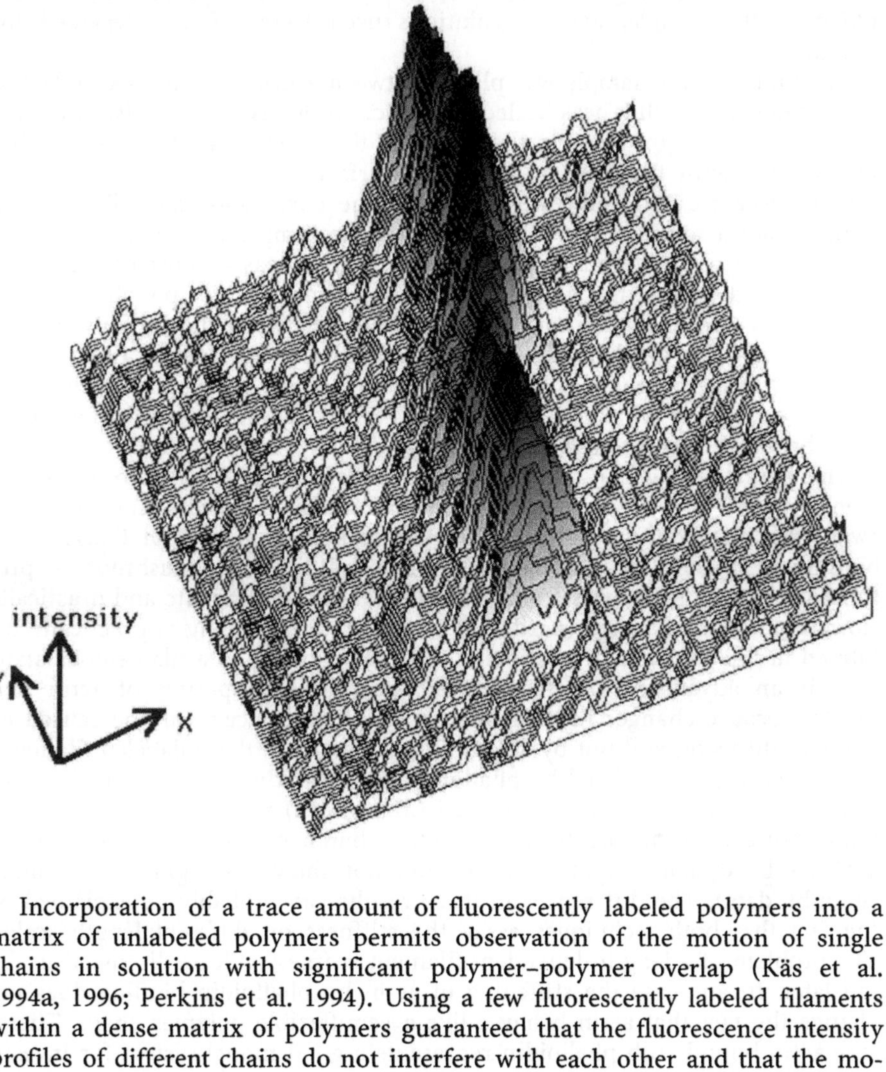

c

Incorporation of a trace amount of fluorescently labeled polymers into a matrix of unlabeled polymers permits observation of the motion of single chains in solution with significant polymer–polymer overlap (Käs et al. 1994a, 1996; Perkins et al. 1994). Using a few fluorescently labeled filaments within a dense matrix of polymers guaranteed that the fluorescence intensity profiles of different chains do not interfere with each other and that the motion of single chains could be resolved. For this purpose two actin solutions were polymerized: a 5 μM F-actin solution labeled with rhodamine-phalloidin at a molar ratio 1:1, and an unlabeled solution of the desired concentration for the polymer matrix. To adjust the final concentration of the labeled filaments in the sample to 2.5 nM/10 nM (fluorescence microscope/confocal microscope) a small fraction of the labeled solution was added to the unlabeled solution and mixed by repeated pipetting. The pipette tip (1 ml Eppendorf) was cut to a broader diameter ($\sim$7 mm) and pipetting was done slowly to prevent filament breakage (Janmey et al. 1994). At higher actin concentrations – due to the higher polymerization rates – the unlabeled matrix was also polymerized around the labeled filaments instead of mixing the two solutions. To reduce bleaching of the fluorescent dye, 4 μg/ml catalase, 0.1 mg/

ml glucose, 20 μg/ml glucose oxidase, and 0.05 vol% mercaptoethanol were added to the sample, and all solutions used were initially degassed for 30 min.

After mixing, the sample was placed between a glass microscope slide and a cover glass. The slide was sealed with vacuum grease, which also provided a spacing of 5–80 μm between the slide and the cover glass. To prevent adhesion of the actin filaments at the glass surfaces, the surfaces were coated with monomeric actin (by pre-incubating the cover glass and slide with a 25 μM G-actin solution for ~15 min). After the sample was sealed, the polymer solution was allowed to relax for 2 h. This released most of the initial strain caused by mixing. A longer relaxation time was not possible because of the slow exchange of rhodamine-phalloidin between the initially labeled filaments and the unlabeled filaments. The maximal observation time, due to bleaching under permanent radiation, was ≈20 min before the labeled filaments became too faint to be analyzed by the image processing system. Filament breakage was rarely seen.

The high binding constant of rhodamine-labeled phalloidin to actin filaments makes it easy to fluorescently label actin. However, one has to be aware that the bound phalloidin changes certain properties of F-actin. The binding of phalloidin – the famous poison of the death mushroom – prevents nucleotide hydrolysis from the high-affinity binding site and drastically slows monomer exchange (Barden et al. 1987). Treadmilling is practically inhibited in the presence of phalloidin and most actin is in a filamentous state. This is an advantage for studying the polymeric properties of actin filaments, because changes in the position of a filament can only be caused by filament diffusion and not by treadmilling. Therefore the unlabeled filaments were usually also stabilized by phalloidin. Some studies of the Brownian motion of single actin filaments reported (Maggs etc.) that the binding of phalloidin stiffens the filaments. On the other hand, measurements of filament stiffness by dynamic light scattering did not show any significant changes (unpubl. data). The differences in these results can probably be explained by the fact that both techniques assay the stiffness on different length scales. The problem of a length-dependent filament stiffness will be discussed in detail later. In any case the stiffening of actin by phalloidin binding does not change the fact that actin behaves like a semiflexible polymer chain. Therefore actin bound with phalloidin represents a good model system for investigating the polymer aspects of the actin cortex.

Microscopy and Image Processing. Fluorescence microscopy was performed with an inverted Zeiss microscope (Axiovert) equipped with a filter set for rhodamine fluorescence, a Zeiss Plan Neofluar 63×Ph3 objective ($nA = 1.4$) and Zeiss HBO100 light source. For documentation the microscope was connected to a SIT-camera by a 4× coupler. The images were recorded on videotape by a SVHS recorder.

A Biorad MRC 600 Confocal Imaging System (Biorad, Hemel Hempstead, England) attached to a Zeiss Axiovert with a Zeiss Plan Neofluar 100× objective (nA = 1.4) was used for confocal microscopy. Images were obtained in the slow scanning (1 frame/s), photon-counting mode. The interslice distance for a z-series was 0.2 μm. Confocal microscopy of very dilute F-actin was not possible because the image was blurry due to the fast thermal undulations of the filaments and the slow scanning rates of the microscope. However, when the labeled filaments were embedded in a network with a mesh size smaller than the resolution of the microscope ($\approx$300 nm), the undulations could not be detected, and a sharp picture was obtained.

Fluorescence images were digitized and analyzed with a modified version of the image processing software *Image* (Wayne Rasband, National Institutes of Health, Bethesda, Maryland). The spatial coordinates of an actin filament were determined using only filaments that were entirely in the focal plane (±0.1 μm). The contours of the filaments ($x(s)$, $y(s)$) (s: arc length) were measured by the following method. First, the two ends of the filament were marked interactively to estimate where the tracing algorithm terminated. Next, a point in the middle of the filament was chosen. The intensity profile around this point was used to fit a quadratic test function. The fit was done for a range of different tangent angles and the angle was chosen for which the value for the steepness of the intensity profile was maximal. The maximum of the fitted profile was stored as a coordinate point. By fitting a test function the positions (x(s), y(s)) could be located with an accuracy of ±11 nm (which was limited only by the signal-to-noise ratio). The determined tangent angle was used to direct the search matrix for the neighboring intensity maximum. The minimal distance for the next maximum between two neighboring pixels is 1.42 times the dimension of a square pixel if the tangent angle is 45°. This fact was used to jump at least one pixel. For this maximum the fitting procedure described above was repeated and the next data point was stored. This procedure was repeated until an intensity drop indicated that the filament had ended.

Three-dimensional reconstruction of confocal images was performed using Voxblast software (Unix version, Vaytek) on a Silicon Graphics Indigo 2 computer (Silicon Graphics Inc., Mountain View, California).

6.3 Polymer Dynamics of Actin Filaments

Polymer physics distinguishes between three typical concentration ranges for semiflexible polymers. The term dilute regime applies to concentrations which are low enough that the polymers in a solution do not overlap. The concentration range where the polymer chains start to sterically interact and form entangled networks is called semidilute. At even higher concentrations the filaments start to align due to their stiffness and form liquid crystalline phases.

Mode Analysis of the Thermally Excited Bending Motions in Dilute Solutions. A very dilute solution of rhodamine-phalloidin labeled F-actin allows visualization of the Brownian motions of a single filament, which is not interacting with neighboring filaments. Figure 6.7 shows a typical time sequence of the unrestricted thermally driven motions of an actin filament in a dilute solution of labeled F-actin with a monomer concentration $c = 2.5$ nM.

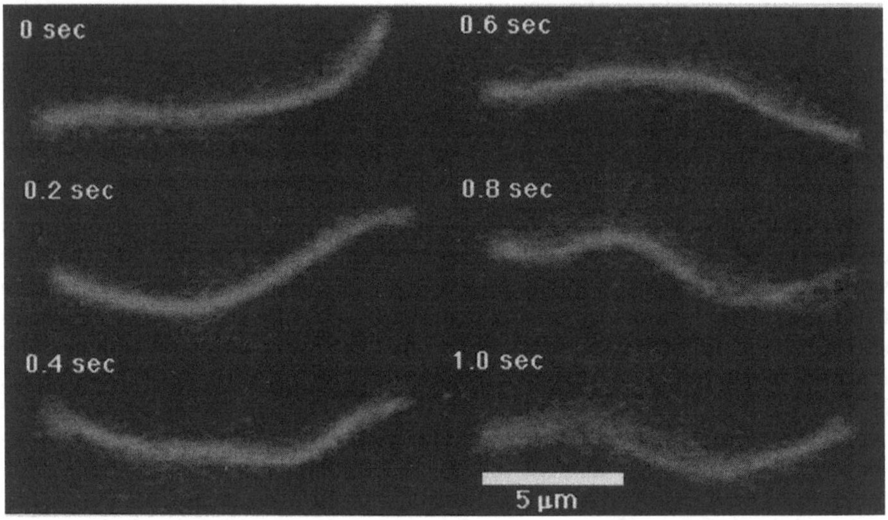

Fig. 6.7. Time sequence of the thermal undulations of a single, free rhodamine-phalloidin labeled actin filament. The bending undulations of the filament are clearly visible. Torsional modes could not be detected because of the broadness of the intensity profile. In the last picture of the sequence the filament started to diffuse out of the focal plane

The pictures were taken at time intervals of 0.2 s. On this time scale of a few tenths of seconds the filament exhibited bending undulations around a straight mean shape. With the exception of small bending undulations perpendicular to the focal plane (the bending motions occurred in all three dimensions), the filament remained in focus for a period of 1–2 s. At longer intervals the filament left the focal plane by rotational and translational diffusion. This type of bending motion around a straight mean shape confirmed the semiflexible character of F-actin. This also means that the filaments exhibit no spontaneous curvatures caused by defects under these conditions. Kinks such as those recently reported (Käs et al. 1993) did not appear in the presence of compounds used to reduce photobleaching of the sample.

In the case of small undulations around a straight shape the total bending energy H_{bend} of the filament could be expressed by (Landau and Lifshitz 1980)

$$H_{\text{bend}} = \frac{k_c}{2} \int_0^L \left(\frac{\partial \vartheta}{\partial s} \right)^2 ds\,, \tag{1}$$

where the integration extends over the filament length L, k_c denotes the bending modulus and ϑ is the tangential angle along the contour (see Fig. 6.8).

For a normal-mode analysis of the thermal bending excitations, a Fourier decomposition of the tangential angle $\vartheta(s)$ was performed:

$$\vartheta(s) = \sum_q a_q \cos(qs)\,, \tag{2}$$

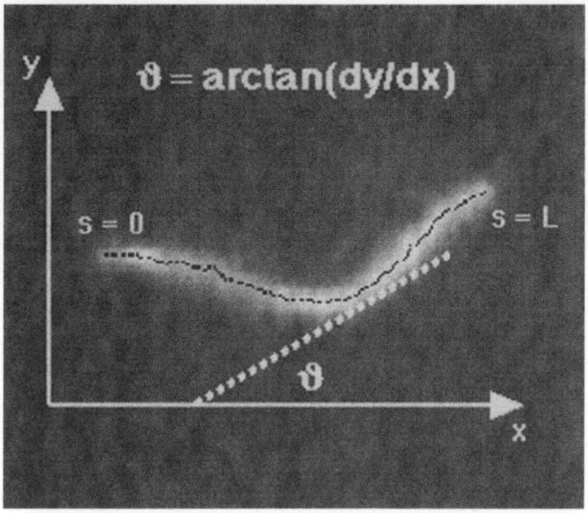

Fig. 6.8. Fluorescence micrograph of a rhodamine-phalloidin labeled actin filament. The *black line* shows the spatial coordinates of the filament determined by image processing. The *white lines* illustrate the concept of the tangent angle

where $q = \pi n/L$ ($n = 1, 2, 3...$) was the wave vector, which is related to a wavelength of $\lambda = \pi/q$. It should be noted that a long wavelength corresponds to a short wave vector. For free filament ends, the boundary conditions $(\partial\vartheta/\partial s)_{s=0;L} = 0$ reduced the Fourier expansion for $\vartheta(s)$ to a cosine expansion. By studying the normal modes of a filament its motions on different length scales could be analyzed separately. The length scale was given by the wavelength (i.e. wave vector) of the normal mode.

In these experiments bending of filaments was only due to their Brownian motion and not to other forces. Therefore the average bending energy stored in a filament was equal to the thermal energy. Using the equipartition theorem, the mean bending energy of each mode could be related to the thermal energy:

$$\langle H_{\text{bend}}(q)\rangle = \frac{k_c}{2}\left\langle \int_0^L \left(\frac{\partial}{\partial s} a_q \cos(qs)\right)^2 \mathrm{d}s \right\rangle = \frac{k_c}{4} L q^2 \langle a_q^2 \rangle = \frac{k_B T}{2} \,. \tag{3}$$

By solving this equation the following result for the mean square amplitudes $\langle a_q^2 \rangle$ could be derived

$$\langle a_q^2 \rangle = \frac{2 k_B T}{k_c L q^2} \,. \tag{4}$$

This result implies that the mean square amplitudes $\langle a_q^2 \rangle$ scale with the wave vector q as $\langle a_q^2 \rangle \sim q^{-2}$. Equation (4) has been derived for unrestricted three-dimensional bending undulations of a semiflexible filament. Unfortunately, the analyzed filaments were confined in a narrow space (~5 μm) between two glass slides. Otherwise the filament diffused out of the focal plane before enough configurations could be sampled for the calculation of the mean square amplitudes a_q. However, each bending mode can independently pro-

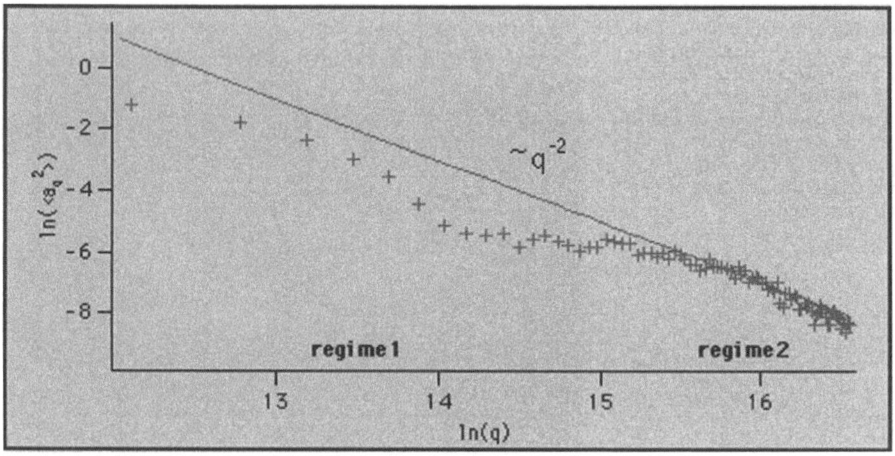

Fig. 6.9. Double logarithmic plot of the mean square of the Fourier amplitudes $\langle a_q^2 \rangle$ vs the wave vector q. In the lower wavelength regime (*regime 2*) the amplitudes scaled with a coefficient of -2 as expected for bending motions. In *regime 1* $\langle a_q^2 \rangle$ scaled with a lower power of q indicating that the thermal motions were dominated by excitations other than pure bending modes

pagate in a random plane. Therefore confining and sampling the motions of an actin filament in only one plane should not alter the results. Figure 6.9 shows a typical double logarithmic plot of $\langle a_q^2 \rangle$ versus q for a confined rhodamine-labeled filament. For the short wavelength regime the plot shows the expected scaling behavior with a scaling coefficient of -2.03 ± 0.07. But for long wavelengths – larger than ~ 1 µm–$\langle a_q^2 \rangle$ scales with a lower power of q.

Another characteristic of independent random bending motions is that their amplitudes have a Gaussian distribution function. Coupling to other excited modes results in a non-Gaussian behavior. This is another criterion to investigate the unexpected behavior of the Brownian motions of F-actin in the long wavelength regime. Gaussian distribution of the Fourier amplitudes a_q was tested by checking if the condition $\langle a_q^4 \rangle / \langle a_q^2 \rangle^2 = 3$ is satisfied in the studied wave vector regime. In the low wave vector regime, $q < 5 \times 10^6\ \mathrm{m}^{-1}$, the value of $\langle a_q^4 \rangle / \langle a_q^2 \rangle^2$ was systematically below 3 and improving the statistics by increasing the number of traced filaments did not change this result. For higher q-values the result $\langle a_q^4 \rangle / \langle a_q^2 \rangle^2 = 3 \pm 0.5$ proved the expected Gaussian character for bending undulations.

These results show that the excited modes were not only free bending motions. The mode spectra could be changed by tensions restricting the undulations or the influence of other excited motions. For short wavelengths the mode spectrum of the filament was dominated by bending modes as expected for wormlike chains (Doi and Edwards 1986). In the long wavelength regime they exhibited a different scaling behavior than for pure bending motions. This could be due to the confinement between two glass plates (Hendricks et al. 1995), which could restrict the long modes. Coupling of the bending modes to other modes could also cause deviations from the expected scaling behavior. A third possibility is that Eq. (4) which has been derived for small undulations does not apply to the long wavelength regime. The approximation that bending undulations have only a component perpendicular to the filament axis may no longer be valid.

The persistence length L_p, which can be expressed in terms of the bending modulus k_c by the formula $L_p = k_B T / k_c$, could be calculated using Eq. (4). The average persistence length in the high wave vector regime (=short wavelength) – $5 \times 10^6\ m^{-1} < q < 12 \times 10^6\ m^{-1}$ – was $L_p = 1.8 \pm 0.3\ \mu m$ (Käs, 1996). This result is in agreement with values determined by end-to-end distance measurements in semidilute solutions (Käs et al. 1994). This value of the persistence length applies to F-actin in a 0.5 mM ATP solution, irrespective of whether the filaments are assembled from ATP- or ADP-containing subunits. However, F-actin in a 0.5 mM ADP solution had nearly twice the stiffness.

In the low wave vector regime the persistence length showed a large variance as a function of the wave vector and exhibited a local maximum around $q = 2 \times 10^6\ m^{-1}$ (Käs et al. 1996). The values for the persistence length were consistently higher in this regime, which resulted in a mean value of $L_p \sim 14 \pm 5\ \mu m$. The fact that the persistence length was no longer constant for $q < 5 \times 10^6\ m^{-1}$ expresses again the observation that in the long wavelength regime other modes besides bending motions were excited. Further, the decrease in bending stiffness in the small wavelength regime ($1.0\ \mu m > \lambda > 500\ nm$) was not an artifact of the limited time resolution of the video camera. Beyond the time resolution the maximal amplitudes of the undulations are underestimated, which results in an apparent stiffening. In fact, this happened for $\lambda < 390\ nm$, a wavelength distinctly below the smallest wavelength ($\lambda = 500\ nm$) analyzed in these experiments. The onset wavelength for this apparent stiffening shifted to smaller wavelengths if the viscosity of the solution is increased.

The general character of F-actin as a semiflexible polymer was observed at all concentration ranges tested, but there were significant differences between the apparent filament stiffness in the dilute and semidilute regimes. Free actin filaments in dilute solution showed only small bends and the persistence length determined by end-to-end distance measurements was comparable to the length of the filament ($\sim 22\ \mu m$). This is in disagreement with the finding of a persistence length of 1.8 μm measured by an analysis of the thermal undulations and the value which was previously measured in semidilute solutions (Käs et al. 1994a). It is also inconsistent with the observation that a diffusing filament in a semidilute solution showed minimal radii of curvature of ~2 μm implying a persistence length of a few microns. This leads to the conclusion that the free filaments showed an apparent increased stiffness on long length scales, possibly due to coupling of the bending modes to other modes such as torsion (Prochniewicz et al. 1995), a restricted sliding motion of the two helical strands of one filament against each other (Bremer et al. 1991), or a partial untwisting of the two strands. Unfortunately, motions like these could not be detected in the fluorescence microscope, because the width of the fluorescence intensity profile of an actin filament was $\approx 0.5\ \mu m$, which is ≈ 100 times the real diameter of the filament. The Debye screening length was $\approx 1\ nm$ under the typical buffer conditions for an F-actin solution. The short screening length ruled out stiffening and long range electrostatic interactions for actin filaments. This unexpected behavior of a semi-flexible chain is potentially interesting and can be further investigated for example by scattering techniques like dynamic light scattering and neutron scattering. Considering the complex double helical structure of F-actin it is not surprising that the concepts of a simple linear bending elasticity do not apply perfectly. A good estimate for the stiffness of F-actin is also important for answering the question whether a polymerizing actin filament could deform a cell membrane by a thermal ratchet-like mechanism (Mogilner and Oster 1996).

Semidilute Solutions and Filament–Filament Contact. *In vitro* solutions of actin filaments of the length distribution shown in Fig. 6.4 were dilute at concentrations below 40 nM. Above this concentration the solution crossed over from dilute to semidilute conditions in which significant filament–filament overlap began to occur. This transition could be detected by determining the concentration at which rhodamine-labeled filaments started to overlap as shown in Fig. 6.10 (Käs et al. 1996).

By modeling the polymer solution as a solution of stiff rods (Doi 1975) the onset concentration of the semidilute regime c^* is

$$c^* = B(M_p/N_A)(1/L^3) \qquad (5)$$

where M_p is the molecular weight of the polymer, N_A is Avogadro's number, L is the chain length and B is a numerical constant for which Doi predicted $B = 1$. Using the average chain length of 22 μm in Eq. (5) resulted in an onset

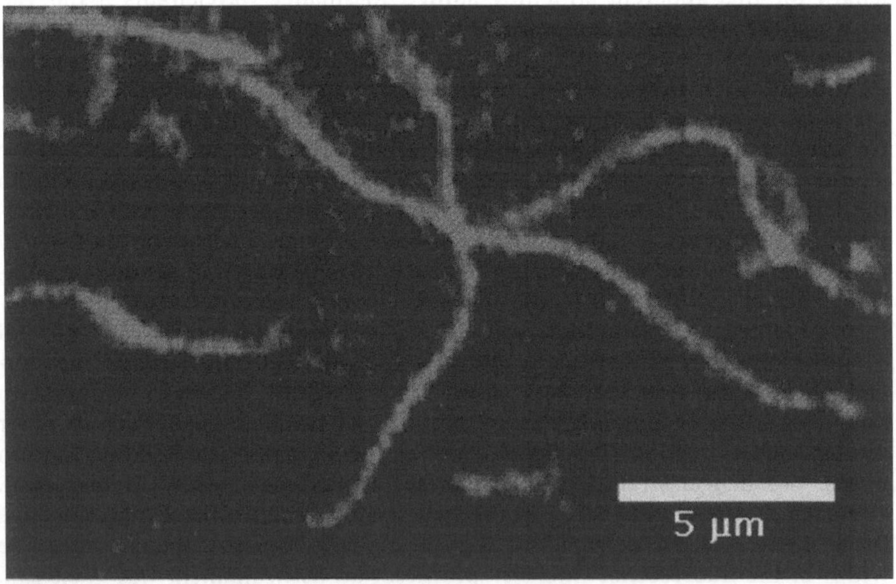

Fig. 6.10. Onset concentration of the semidilute regime. In the 40 nM solution of rhodamine-phalloidin labeled F-actin the single filaments started to overlap and sterically restrict each other

concentration of 1.3 nM. More recent computer simulations showed that $B \sim 30-70$ (Magda et al. 1986). The experimental value of 40 nM was in good agreement with their predictions and other data obtained by rheological experiments (Janmey et al. 1986).

Entanglements of F-Actin in Semidilute Solutions. Semidilute F-actin solutions have length-dependent viscoelastic properties. Solutions of very long filaments can have shear moduli of several hundred Pa at low volume fractions over a wide frequency range of exciting shear oscillations (Müller et al. 1991; Janmey et al. 1994). This shear resistance is caused by steric interactions between long actin filaments. In order to study the structure of these entanglements, labeled filaments were embedded in a matrix of unlabeled filaments at a ratio 1:3500 and the entanglements of the labeled filaments were examined by confocal microscopy.

Figure 6.11 displays a three dimensional reconstruction of a partially labeled entangled network (rotated 60° with respect to the z-axis). The actin solution ($c = 1.4$ mg/ml) contained $\sim 2.32 \times 10^{12}$ filaments/ml, assuming an average filament length of 22 µm, and the ratio of labeled to unlabeled filaments was 1:3300. The fluorescently labeled filaments formed U-turns and single loops around neighboring filaments. The smallest radii of curvature which were found for these loops and U-turns were about 2±0.2 µm. None of the actin filaments observed were twisted into knots or around each other.

The physics, which causes the viscoelastic behavior of F-actin solutions, is quite different from the so-called rubber elasticity of conventional flexible polymer networks. The microscopic mechanism causing elastic behavior of flexible polymer solutions can be pictured as follows: under fast shear motions, entanglements between polymer chains act like knots and the random coiled polymer segments between two knots are stretched out. This reduces the configurations which a polymer segment can assume and results in an entropy loss. Thus a tension builds up resisting the shear force. The three-dimensional reconstruction of the filament matrix (Fig. 6.11) showed that gel-like elasticity of F-actin can exist in the absence of specific cross-links or even tightly bent entanglements characteristic of flexible polymer networks. Actin filaments exhibited an almost straight shape between contact points within the mesh. The absence of knot-like entanglements and random coiled polymer segments in F-actin solutions ruled out a rubber-like mechanism to generate elastic behavior.

Recently, several models have been suggested to explain the viscoelastic behavior of semiflexible polymers (MacKintosh, Maggs, Frey). All models pointed out that in contrast to flexible polymers the shear elasticity depended critically on single filament properties of the polymers forming the solution. It was namely the filament stiffness which modulated the elastic strength of actin networks. In particular, the model by F.C. MacKintosh et al. (1995) provided predictions of the shear modulus of actin networks, which agreed well with rheological data. In this model the restoring elastic force is due to the entropic tension, which is caused by pulling out the thermal undulations of the filament. The interplay between entropy and bending causes that this tension and therefore the elastic strength depends strongly on the

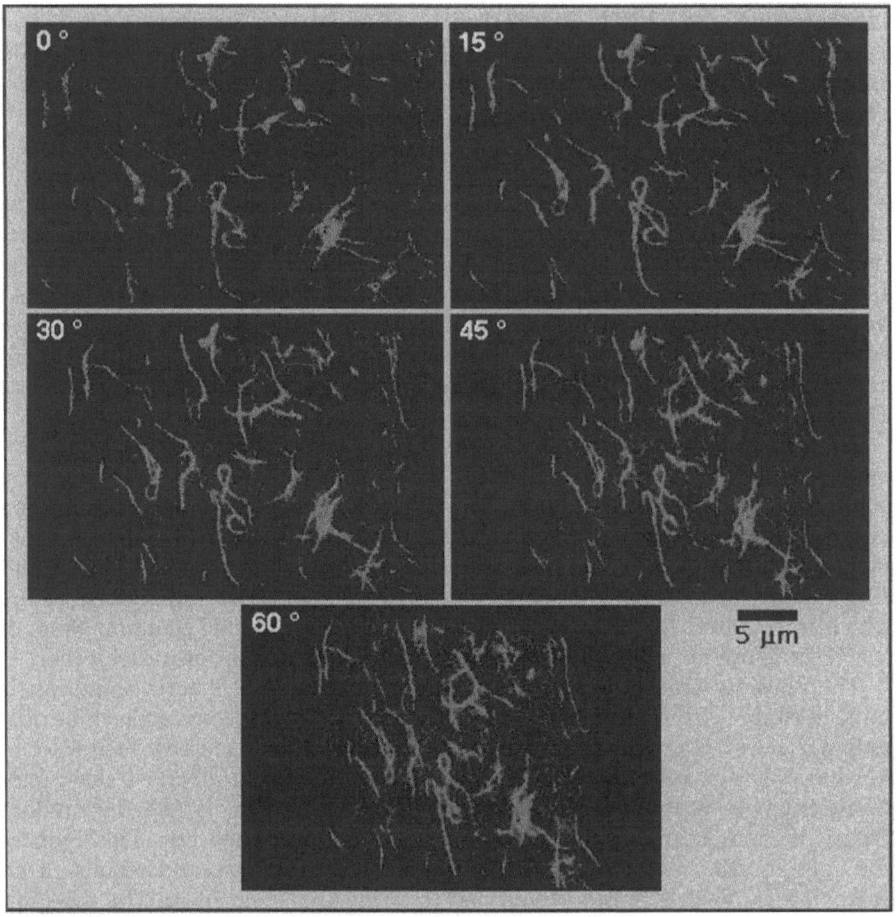

Fig. 6.11. Entangled mesh of actin filaments. To image the network, fluorescently labeled filaments were embedded in a matrix of unlabeled filaments (actin monomer concentration $c_A = 1.4$ mg/ml, ratio of unlabeled to labeled filaments 3500:1). The three-dimensional reconstruction was obtained from a z-series of confocal images and the sequence shows a rotation of 60°

bending stiffness of the filaments. In addition, this mechanism for resisting shear allows F-actin to generate networks which are a 1000× stronger than conventional networks of flexible polymers at the same volume fraction. These results, together with the finding that adenine nucleotides changed filament stiffness, suggest that cells are able to modulate the elasticity of their outer cytoskeltal rim by local changes in filament stiffness.

The precise character of the entanglements in an F-actin solution, which hold back an actin filament under shear, is still unknown and appears only as a phenomenological parameter in MacKintosh's model. It is expected that the steric interactions, which allow an F-actin solution to resist shear, are of a collective nature and require multifilament interactions. However, the question about the nature of entanglements does not arise in the actin cortex of cells, because in this case, the F-actin is cross-linked by actin-binding proteins and the model can be applied without ambiguity. The cross-linking by accessory proteins is necessary, because the combination of a persistence length of $\approx 2\ \mu m$ and a maximal filament length of $1\ \mu m$, in vitro, makes bends which are necessary to form entanglements impossible.

Restricted Bending Undulations in Semidilute F-Actin Solutions. In a semidilute F-actin solution the available space for a single filament was highly constrained by its neighbors. Rhodamine-labeled filaments embedded in semidilute F-actin solutions exhibited a large drop in the amplitudes of bending oscillations due to filament–filament contacts (see Fig. 6.12). For flexible chains these steric interactions with the surrounding chains were described as a tube around the filament formed by the neighboring polymers (Edwards 1967). This idea also applied to semiflexible filaments with the modification that the finite bending stiffness of the filaments created a characteristic minimal length between two points where the filament touches the tube (Odijk 1983). This distance is called the deflection length.

The thermal undulations of a filament define the tube diameter (Käs et al. 1994a, 1996). By superimposing a sufficient number of transient traces of a filament (~64 traces were taken at time intervals of 0.1 s) the tube was imaged and its diameter was measured as the maximum deflection along the

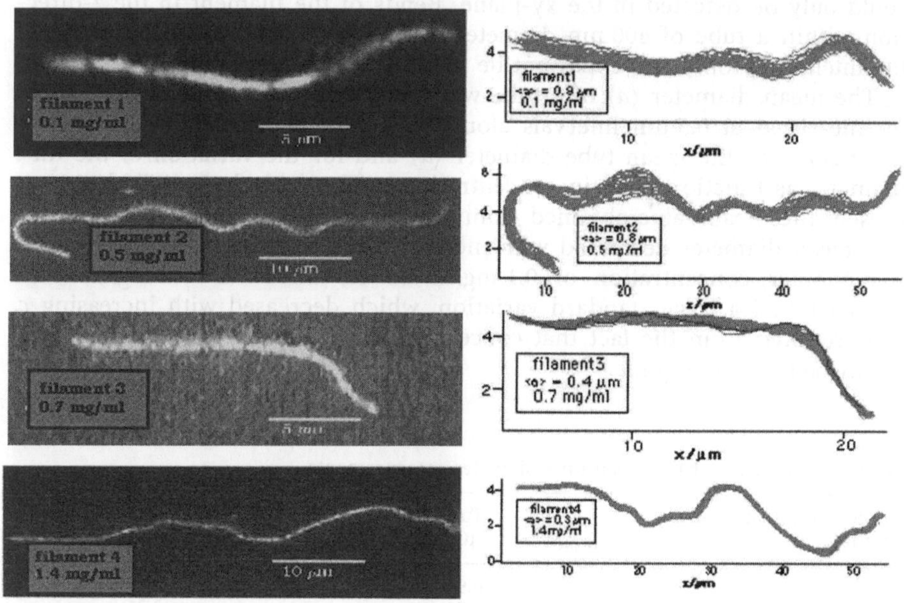

Fig. 6.12. Direct visualization of the tubes around actin filaments in F-actin solutions of different concentrations. The pictures were obtained by superimposition of 64 traces of the contour taken at time intervals of 0.1 s. The left sides of the graphs display a snapshot of the filament which was confined in the tube below. Below the tube the mean tube diameter $\langle a \rangle$ and the actin concentration of the surrounding F-actin matrix is denoted. The tube diameter decreased with increasing actin concentration and showed a large variation which is typical for the semidilute concentration regime

contour. Figure 6.12 confirmed that the restrictions to the thermal undulations of a polymer chain in a polymer solution can be described by a surrounding tube. This description was valid within a concentration range of 0.1–4.0 mg/ml (including the coexistence regime between nematic and isotropic domains). At lower concentrations this concept broke down due to the increasing role of constraint release as will be shown below (see Fig. 6.16). The gray line in Fig. 6.12c shows a snapshot of the undulating chain within the tube. For the filament shown in Fig. 6.12c the deflection length λ_d was estimated to $\lambda_d \sim 3.8 \pm 0.6$ μm by measuring the average distance between two contact points of transient contours of the filament with the tube.

These results prove Odijk's wormlike tube model (Odijk 1983) and show that other models – like the fuzzy cylinder model (Sato et al. 1991) – do not apply. Theoretical considerations predicted the following dependence of the deflection length λ_d on the mean tube diameter $\langle a \rangle \sim 0.4$ μm and the persistence length $L_p \approx 1.8$ μm (Odijk 1983):

$$\lambda_d = \langle a \rangle^{2/3} L_p^{1/3} \,. \tag{6}$$

By using Eq. (6) a deflection length of 0.7 μm was calculated. This is considerably smaller than the measured value. This means that the measurements overestimated the deflection length or $L_p \gg 1.8$ μm. The first possibility seems to be more probable because contact points of the chain with the tube could only be detected in the xy-plane. Bends of the filament in the z-direction within a tube of 400 nm diameter were only visible as a broadening of the intensity profile and could not be identified as contact points.

The mean diameter $\langle a \rangle$ of a tube was determined by averaging its diameter measured at 0.5 μm intervals along the tube axis. Table 6.1 summarizes the results for the mean tube diameter $\langle a \rangle$ and for the variation of the tube diameter as function of actin concentration c. The $\langle a \rangle$-values of Table 6.1 represent the mean value obtained from four filaments per each concentration. The mean diameter decreased with increasing concentration from 0.9 μm for a monomer concentration of 0.1 mg/ml to 0.3 μm for $c = 1.4$ mg/ml. The diameter had a large standard variation, which decreased with increasing c. This resulted from the fact that concentration fluctuations are characteristic for the semidilute regime.

Table 6.1. Summary of the mean tube diameter measurements

Actin concentration c_A (mg/ml)	Tube diameter $\langle a \rangle$ (μm)	$\Delta a/\langle a \rangle$ (%)
0.1	0.94	28
0.4	0.71	26
0.5	0.79	22
0.7	0.43	33
1.4	0.30	13
4.0 Nematic domain	0.10	6
4.0 Entangled domain	0.18	9

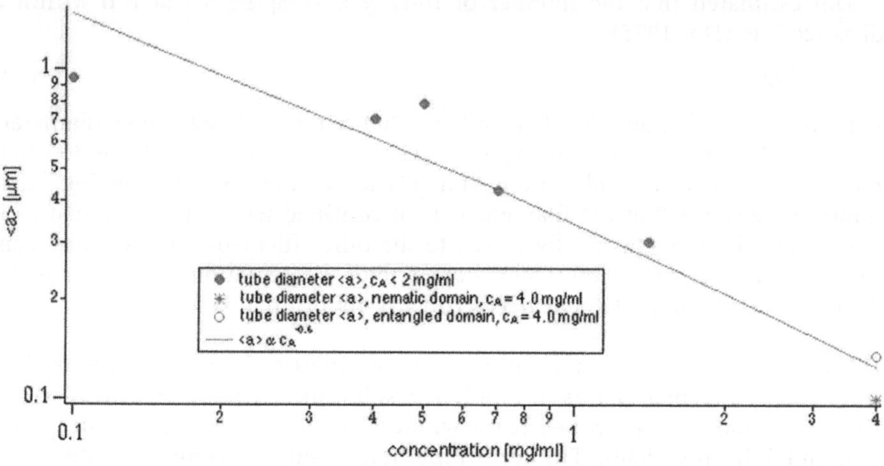

Fig. 6.13. Mean tube diameter $\langle a \rangle$ as a function of the actin monomer concentration c. Below 2 mg/ml the mean tube diameter scaled approximately as $\langle a \rangle \sim c^{-3/5}$. For 4.0 mg/ml the average tube diameter was 130 nm in the entangled domains and 100 nm in the nematic domains

Semenov pointed out that the tube, which is probed by an undulating semiflexible chain, is smaller than the mesh size ξ of the network because the chains can only bend by a certain degree (Graessley 1980; Semenov 1986). This can be expressed by the following formula:

$$\langle a \rangle \sim \xi(\xi/L_p)^{1/5}, \tag{7}$$

where L_p is the persistence length. Assuming that the concentration dependency of the mesh size can be approximately described by a scaling law for stiff rods, $\xi \sim c^{-1/2}$ (de Gennes et al. 1976), the following equation was obtained:

$$\langle a \rangle \sim c^{-3/5}. \tag{8}$$

The five values displayed in Table 6.1 were not sufficient to establish the scaling law which was predicted in Eq. (8). But the plot in Fig. 6.13 at least shows that the data are in good agreement with Eq. (8). Previous mesh size measurements of F-actin were also consistent with these data (Schmidt et al. 1989).

The scaling argument of Eq. (8) also agreed with predictions by Odijk, which can be expressed by the following formula (Odijk 1983):

$$\rho \langle a \rangle^{5/3} L_p^{1/3} L \gg 1, \tag{9}$$

where ρ is the filament density and L the filament length. [For example, for an average filament length of 22 µm and for an actin monomer concentration of 1.4 mg/ml (=filament density of 2.33×10^{12} filaments/ml), Eq. 9 predicts an average tube diameter larger than 14 nm.]

Doi estimated that the number of rods N enveloping a test rod within a distance b is (Doi 1975)

$$N \approx bL^2\rho\,, \tag{10}$$

For an actin solution of 1.4 mg/ml, $b=300$ nm ($\equiv$ average tube diameter) and $\rho=2.33\times10^{12}$/ml, a filament of $L=22$ µm was surrounded by ~300 filaments which form the tube. Even though each contact may be fleeting, their multitude ensures that the filament is kept confined within the tube just as it would be if it were chemically linked to the other filaments. The confinement in a tube applies on a time scale shorter than the reptation time. This is the characteristic time a filament needs to leave its original tube.

Diffusion of Single Actin Filaments in Semidilute F-Actin Solutions. The diffusion of polymer chains in a polymer solution is usually described by the reptation model of the Nobel prize winner P.G. de Gennes (de Gennes 1971; Doi and Edwards 1986). The term reptation originates from the Latin word *reptare* (=to creep) and describes a snake-like motion of a polymer through the entangled mesh. The model has been developed for flexible chains (Edwards 1967; de Gennes 1971) and later modified for semiflexible, so called wormlike chains (Odijk 1983). Despite the success of the model to explain the behavior of polymer solutions, direct proof was missing until the techniques described in this chapter allowed us to follow the motion of a single polymer (Käs et al. 1994a). The following Figs. 6.14 and 6.15 confirm that the diffusion of actin filaments in F-actin solutions can be described by the reptation model.

Figure 6.12 shows that actin filaments remained in their original tube for a period of a few seconds, which is a prerequisite for reptation. At longer times the filament was driven out of the tube by its thermal undulations. The reptation motion of two labeled filaments of different length in a matrix of unlabeled filaments (at 0.1 mg/ml or 2.0 mg/ml, respectively) is shown in Fig. 6.14a. The speed of this motion depended heavily on the length of the filament. In order to guide the eye, thin white lines are drawn around the filaments to symbolize a tube with approximately three to four times the diameter of the tubes surrounding the filaments. In a time span of 48 s the filament of ≈40 µm length explored new tube segments by its fingering motion out of the tube. At first the filament slid ≈3 µm out of the tube at the right side. Then it retracted ≈2 µm back into the tube. Aside from the correlated sliding motion of the filament ends the filament remained within the

→

Fig. 6.14. a Time sequences ($\Delta t=8$ s) of the reptation motion of actin filaments of 40.0 µm (*right side*) and 6.9 µm length (*left side*) in a matrix of F-actin of 0.1 mg/ml or 2.0 mg/ml respectively. The *thin white lines*, drawn to guide the eye, are three to four times larger than the mean diameter of the tube. The shorter filament diffused visibly faster than the long one. The reptation time for the short filament was 2.2 and 61.4 min for the long filament. **b** Time sequence of the reptating filament shown on the *right side* of **a** recorded at time intervals of 1 s. The *thin white lines* again symbolize the surrounding tube and the *black dashed lines* along the filament show the traced contour. The shorter time intervals between the snapshots demonstrate that the filament slid back and forward in order to leave its original tube

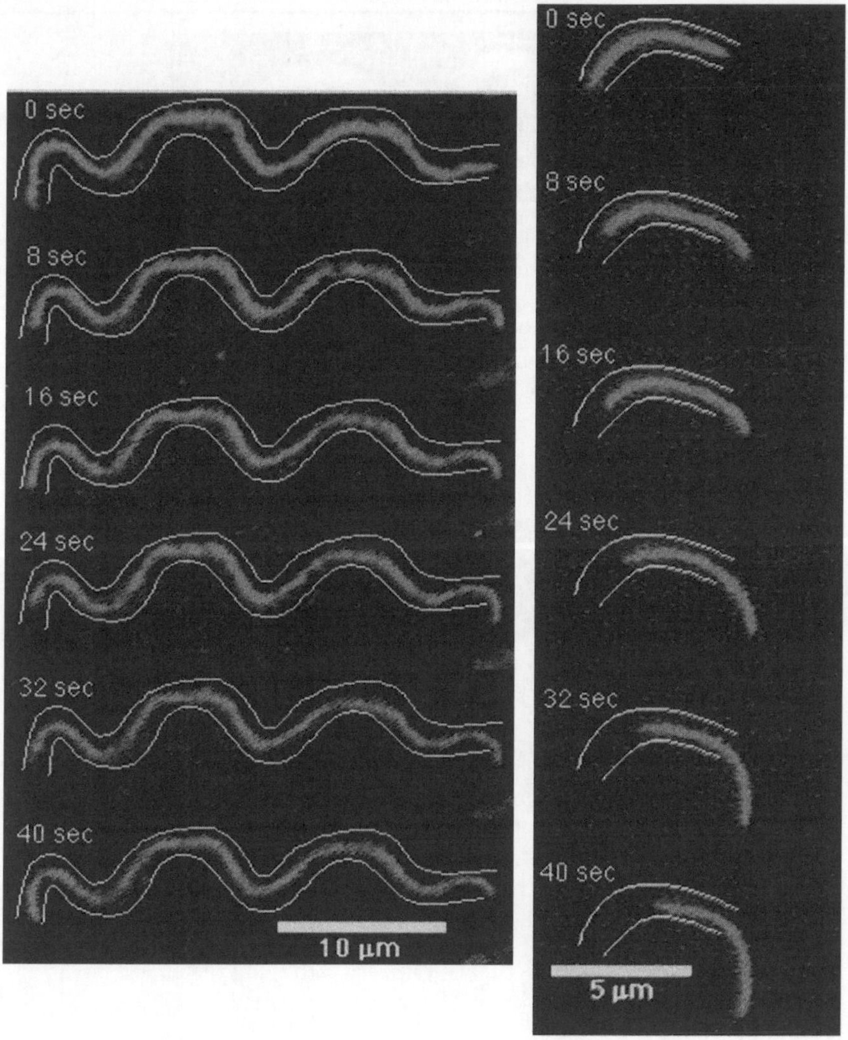
0 sec
8 sec
16 sec
24 sec
32 sec
40 sec
10 µm
0 sec
8 sec
16 sec
24 sec
32 sec
40 sec
5 µm

a

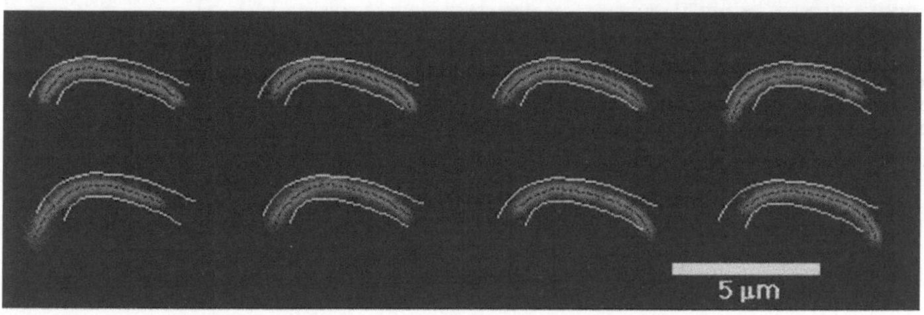
5 µm

b

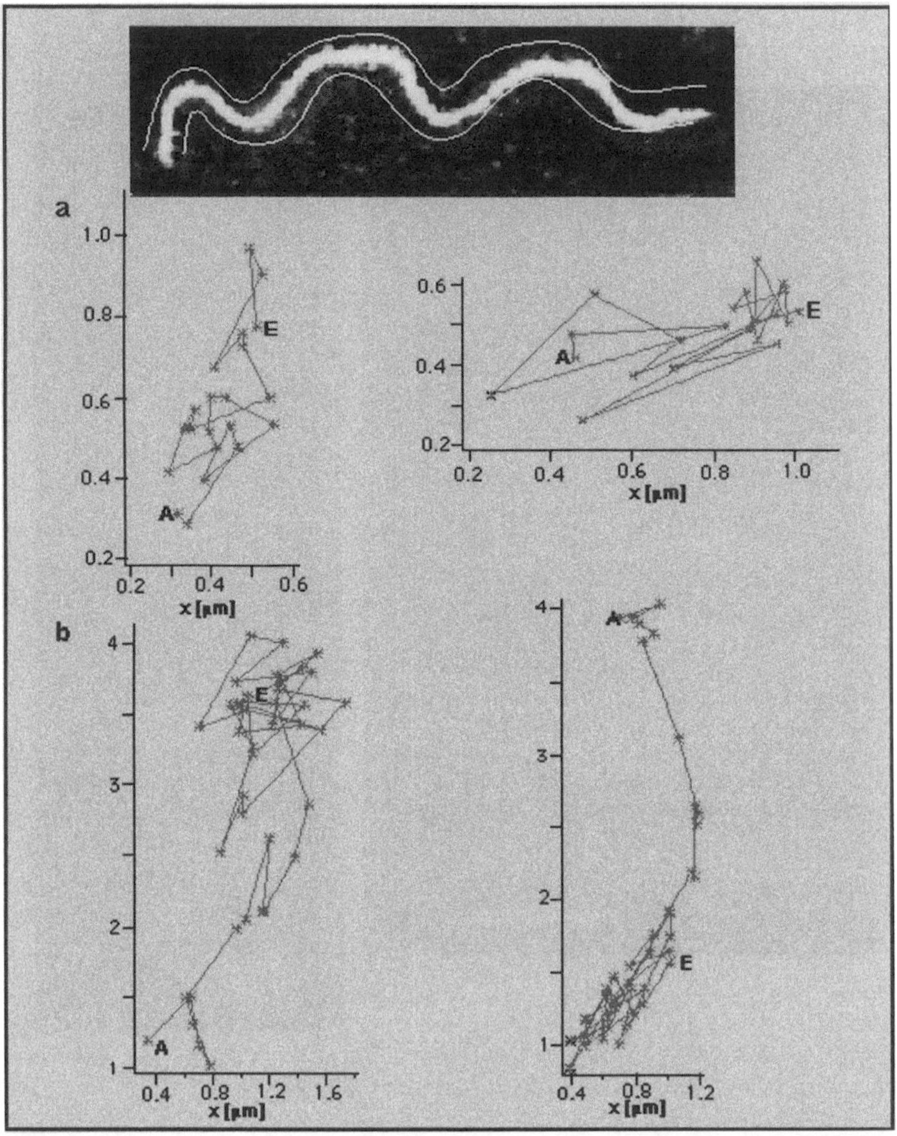

Fig. 6.15 a–b. Track of the movement of the filament ends of the filament displayed on the left side of Fig. 6.14a in steps of 0.12 s (**a**) and 0.84 s (**b**) respectively. The y-axes were oriented parallel to the local axes of the tube at the two ends. The time sequences started at points A and ended at points E. The two ends of a filament performed an uncorrelated random motion (**a**), which proceeded to a correlated sliding motion at longer time scales (**b**)

original tube. In the same time frame the ~7-μm-long filament reptated almost completely out of its original tube. This filament left its tube by repeatedly sliding back and forth by the same type of fingering motion as seen in the time sequence for the 40-μm-long filament. But the fingering occurred on a shorter time scale, as shown in Fig. 6.14b. The amplitude of the fingering motion of the filament ends out of the tube was typically ≈1–2 μm. The minimal radius of curvature, which a reptating filament end of this length can take, was ≈2 μm. Therefore the accessible part of the mesh for the fingering motion of the filament ends can be described as a cone around the end of the tube, which is restricted by how far a filament bends over a length where the filament fingers out of the original tube.

On short time scales the ends of a filament performed an uncorrelated random motion which proceeded to a correlated sliding motion on longer time scales, allowing the filament to leave its original tube. This is illustrated in Fig. 6.15 by recording the motion of the filament ends of the 40.0 μm long filament displayed in Fig. 6.14 in time steps of 0.12 s (Fig. 6.15a) and 0.84 s (Fig. 6.15b). In a time interval of 2.5 s the ends moved uncorrelated and in different directions (see Fig. 6.15a). Theories of polymer physics often refer to this random motion as tube fluctuations (Doi and Edwards 1986). On a longer time span of 34.4 s a transition to a correlated sliding of the filament out of the right side of the tube had been observed (see Fig. 6.15b). This correlated sliding does not mean that both ends moved exactly the same distance. The elastic undulating filament must not necessarily transmit the whole motion of one end to the other end. On average the motion of the ends was coupled to the motion of the center of mass of the filament due to the semiflexible character of the chain and its low longitudinal compliance. Further, it should be noted that the crossover from an uncorrelated to a correlated motion of the filament ends occurred later for longer or stiffer filaments (data not shown).

The diffusion coefficient of a chain along the tube $D_{||}$ was determined by measuring the diffusive motions of the filament ends on time scales where the motions were correlated. The chain end positions (x_i, y_i) were recorded at time intervals $\Delta t=8$ s. The diffusion coefficient of the ends parallel to the tube was calculated as $D_i=1/(N\text{-}1)\sum_{j=2}^{N}\{(y(j)-y(j-1))/(2\times\Delta t)\}$ where N was the number of steps and in most measurements $N=30$. $D_{||}$ was the arithmetic mean of the diffusion coefficients of the filament ends D_i of four independent time sequences (30×8 s). The results of these measurements are summarized in the graph of Fig. 6.16.

Ignoring minor corrections which arise from the length dependence of the friction coefficient ζ, $D_{||}$ is predicted to increase linearly with inverse filament length L in isotropic suspensions (Doi and Edwards 1986):

$$D_{||} = (k_B T)/\zeta L\,. \tag{11}$$

The data for filaments in the entangled phase shown in Fig. 6.16 confirmed the prediction of Eq. (11). $D_{||}$ decreased linearly with increasing chain length L. It has been expected that hydrodynamic effects cause deviations from the predicted length dependence of $D_{||}$ and that hydrodynamic screening slows the diffusion down with increasing actin concentration (Muthkumar and

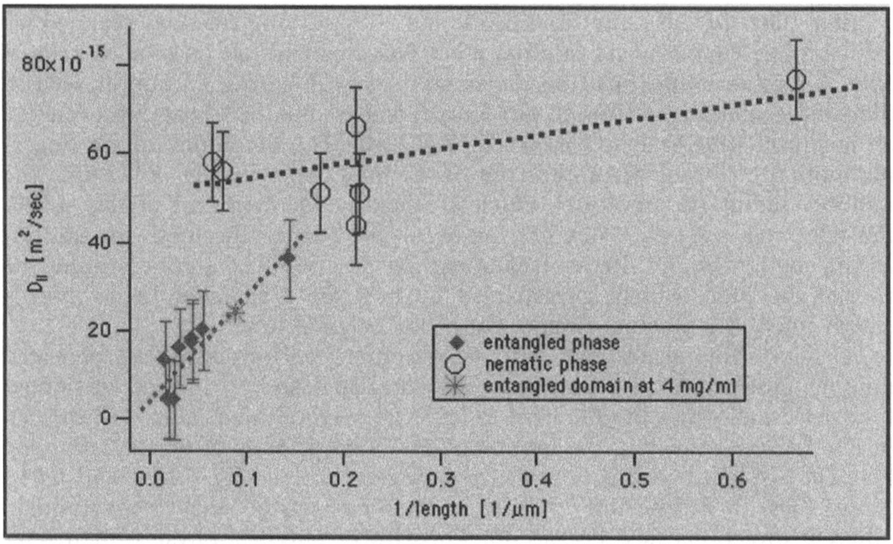

Fig. 6.16. Length dependency of the diffusion coefficient $D_{||}$. In the pure entangled phase the diffusion coefficient along the tube increased linearly with the inverse filament length. The measured diffusion coefficient of the filament in the entangled domain at 4 mg/ml indicated that this scaling behavior also applies for the entangled phase in the coexistence regime. The diffusion coefficient $D_{||}$ parallel to the axis of alignment in the nematic domains at 4 mg/ml was nearly length independent and showed a large variance. The error bars were calculated from the statistical error between the four independent time sequences which were used to obtain the average

Edwards 1983; Odijk, 1986). Both effects were weak and were not detectable within the experimental error of $\pm 10\times 10^{-15}$ m^2/s and within the small accessible concentration and length regime. It should be further noted that the typical time, which a free actin filament of 1 μm length in a cell needed to diffuse over a distance of 1 μm, can be extrapolated from Table 6.2 as 1.5 min. This is within the time frame which is relevant to cellular processes. However, it seems to be improbable that filamentous diffusion plays a significant role in cells because most of the F-actin is cross-linked.

A detailed theory for the diffusion coefficient $D_{||}$ of a reptating polymer only exists for stiff rods and flexible polymers (Doi and Edwards 1986). The semiflexible actin filaments are closer to the case of stiff rods. Therefore it seems reasonable to compare the data with the formula predicted for stiff rods: $D_{||}=(k_B T \ln(L/b))/(2\pi\eta L)$, where $b=7$ nm is the diameter of the filament and η ($\approx\eta_{\text{water}}$) is the viscosity of the buffer. The experimental values were smaller by a factor of ≈ 20 than the values calculated for stiff rods indicating that semiflexible chains diffuse more slowly in semidilute solutions. This finding contradicts a theoretical estimate (Doi 1975) which concludes that semiflexible chains reptate faster than stiff rods in an isotropic phase.

The reptation time τ_D already mentioned is defined as the time a chain needs to leave its original tube entirely. The reptation time can be calculated from $D_{||}$ by $\tau_D = L^2/D_{||}\pi^2$ (Doi and Edwards 1986). This equation applies for flexible chains. The reptation time, which has been predicted for wormlike

Table 6.2. Summary of the measured diffusion coefficients $D_{||}$ and the resulting reptation times τ_D, which is the time a filament needs to leave its original tube. The first line in the table separates the data below 2.5 mg/ml from the data at 4.0 mg/ml in a nematic domain. The last value in the table is for a filament in an entangled domain at 4.0 mg/ml

Actin concentration c_A (mg/ml)	Filament length, L (μm)	Diffusion coefficient $D_{\|\|}$ (m²/s)	Reptation time $\tau_D = L^2/\pi^2 D_{\|\|}$
0.1	22.7	1.7×10^{-14}	51.1 min
	40.0	4.4×10^{-14}	61.4 min
0.4	22.7	1.8×10^{-14}	48.3 min
0.5	59.5	1.3×10^{-14}	7 h 42 min
	50.5	4.6×10^{-15}	15 h 36 min
0.7	18.2	2.0×10^{-15}	28.0 min
	32.2	1.6×10^{-14}	109.4 min
1.4	52.1	4.3×10^{-15}	17 h 48 min
2.0	6. 9	3.6×10^{-14}	2.2 min
4.0	1.5	7.7×10^{-14}	3.0 s
	4.6	5.1×10^{-14}	42.0 s
	4.7	4.4×10^{-14}	50.6 s
	4.7	6.6×10^{-14}	33.9 s
	5.7	5.1×10^{-14}	64.5 s
	13.3	5.6×10^{-14}	5.3 min
	15.7	5.8×10^{-14}	7.2 min
4.0	11.2	2.4×10^{-14}	8.8 min

chains, is practically the same (Keep and Pecora 1985). The results of this calculation are listed in Table 6.2. The calculated reptation time increased from 2.2 min for the filament of 6.9 μm length displayed in Fig. 6.14 to 15 h 36 min for a filament of 50.5 μm length. The calculated value of τ_D for the 6.9 μm long filament agreed with the value of 2.0 min which was obtained by direct observation for τ_D (Fig. 6.14).

Due to the fact that all filaments were able to diffuse in the solution a filament could leave its original tube not only by its own reptation motion, but also by changes in the structure of the tube if chains of the surrounding polymer matrix reptated away. This process is called constraint release (Lodge et al. 1993). In the concentration range of 0.1–2.0 mg/ml constraint release was a very infrequent event due to the tightly packed meshwork. In a solution of semiflexible chains, the tube changed its shape due to constraint release only if the mesh size locally exceeded the persistence length. At lower concentrations of ~0.01 mg/ml constraint release began to dominate the diffusion in the loose mesh of filaments. Fig. 6.17 illustrates that at higher actin concentrations diffusion of filaments was determined by reptation and that the tube model broke down at low concentrations. At 0.4 mg/ml the F-actin solution formed a tight mesh of ~0.7 μm mesh size and the tube around the filament changed its contour mostly at its ends, because the filament was only able to leave its original position by reptation (Fig. 6.17). In more dilute solutions ($c = 0.05$ mg/ml) the mesh was so loose that rotational diffusion started to be an important contribution and the constraining filaments diffused away within several minutes.

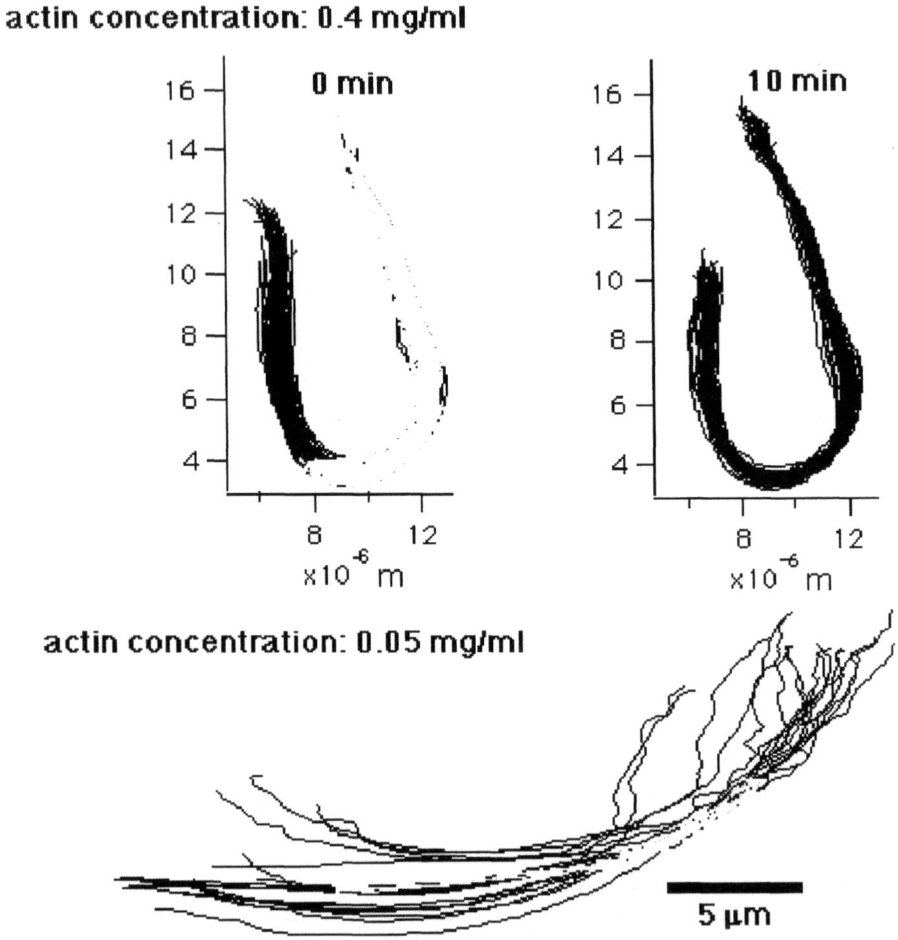

Fig. 6.17. Validity of the tube model. *Upper panel*, development of the tube around a filament at a concentration of 0.4 mg/ml. The contour of the tube was obtained in the same way as described above. After 10 min the filament left the original tube at both ends by repeatedly sliding backward and forward which can be seen by the changes in the contour of the tube at both ends. The central part of the tube remained unchanged because the mesh is too dense for constraint release to occur. *Lower panel* 30 transient contours taken at time intervals of 2 s at a lower concentration ($c = 0.05$ mg/ml). The steric interactions could no longer be described as a tube. Constraint release now played a major role

In rheological experiments the complete dissipation of shear stress defines the terminal relaxation time of an F-actin solution under shear (Ferry 1980; Ruddies et al. 1993). Assuming that this relaxation process is dominated by reptation the longest relaxation time is equal to the reptation time of the polymer chains for a polymer solution with monodisperse length distribution. According to Ruddies et al the terminal relaxation time of an F-actin solution was ≈ 160 min. Assuming a similar length distribution of the F-actin solutions as shown in Fig. 6.4 this result indicates that the relaxation of shear in a solution of actin filaments was dominated by filaments which are longer than average. The reptation time, which was calculated from the measurements of the diffusion coefficients shown in Table 6.2, was $\approx$50 min for a filament of average length.

Alignment of Filaments at Higher Actin Concentrations. It has been reported that F-actin solutions undergo a transition to a nematic phase above actin concentrations of 2.0 mg/ml (Suzuki et al. 1991; Coppin and Leavis 1992; Furukawa et al. 1993). The alignment of filaments, which has been detected by birefringence and anisotropic light scattering, is characteristic for this liquid crystalline phase. In a true nematic phase – distinct from a bundle of actin filaments – aligned filaments should be able to diffuse along the axis of alignment as in a fluid.

Alignment of F-actin by itself does not necessarily result from a thermal equilibrium state representing a nematic phase (Käs et al. 1996). A frozen glass-like state of shear-aligned filaments would be also birefringent over long time periods. In experiments, anisotropic parallel orientation began to occur at 1.4–2.0 mg/ml shortly after the actin solution was pipetted onto the cover glass. This non-equilibrium state was caused by the high sensitivity of actin filaments to shear alignment. After 2–3 h the filaments relaxed into an isotropic entangled network (Fig. 6.18). At concentrations higher than 2.5 mg/ml the aligned areas in the F-actin solutions remained stable and some of the initially isotropic regions of entangled filaments aligned within 1.5–2 h (see Fig. 6.18). The actin solution used in these experiments was characterized by a maximal filament length of 44 µm and an average length of 14 µm. These time sequences showed that metastable aligned or entangled domains relaxed on a time scale of several hours. This result is in agreement with the reptation times, which were calculated in Table 6.2.

Solutions of F-actin $\geq$2.5 mg/ml showed clear birefringence between crossed polarizers and remain birefringent for more than 2 weeks. The solutions exhibited a very broad coexistence between the isotropic and the nematic phase. At concentrations from 2.5 to 6.0 mg/ml a mixture of aligned and entangled domains of a size of about 60 to 140 µm was always found. A total demixing into purely aligned and isotropic phases did not occur even after 1 week. Only after spinning the samples in a table-top centrifuge for several hours did total phase separation take place.

Figure 6.19 shows the phase boundary between an aligned domain and an isotropic domain. Despite the parallel orientation of the filaments in the nematic domain, this image also demonstrates imperfections in the alignment indicating a low order parameter. The filaments exhibited bends away from the axis of alignment and even U-turns – so-called hairpins (de Gennes 1982; Williams and Warner 1990) – as deviations from their main orientation.

Longer filaments were found mainly within the nematic domains, whereas short filaments accumulated in the isotropic phase. To illustrate this length-dependent phase separation, length distributions of 40 filaments in a nematic or an isotropic domain are shown in Fig. 6.20. The enrichment of long filaments in the nematic phase was predicted by Vroege and Lekkerkerker

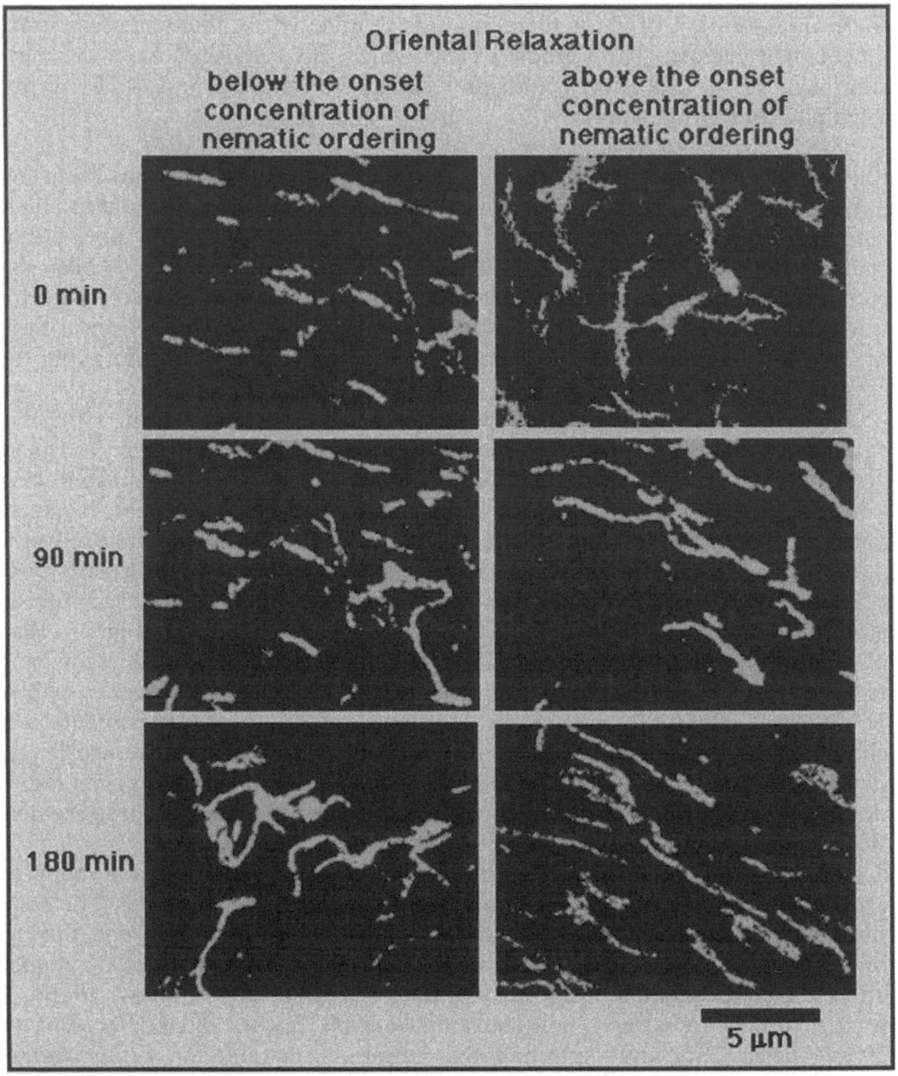

Fig. 6.18. Relaxation of actin filaments from non-equilibrium states. *Left side* Relaxation of a shear-aligned F-actin solution (1.4 mg/ml) into an entangled phase. *Right side* Relaxation of an F-actin solution (4.0 mg/ml) from an entangled, non-equilibrium state into a nematic domain. Confocal pictures were taken at 0 min, 1.5 and 3 h. The snapshots clearly prove that on a time scale of several hours F-actin solutions were able to relax from non-equilibrium states

(1992) and can be explained by the earlier nematic onset concentration for longer filaments.

The technique of fluorescence imaging of single filaments allowed for the first time measurements of the dynamics of polymer chains in a liquid crystalline phase. As expected for polymers in a nematic phase the actin filaments stayed aligned during diffusion. No rotational, only translational diffu-

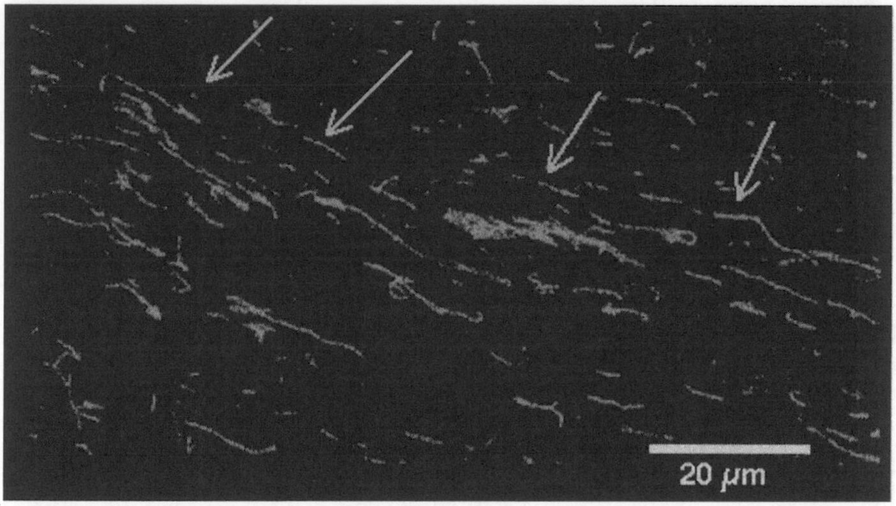

Fig. 6.19. Micrograph of an aligned and an entangled domain of actin filaments at 4.0 mg/ml. The *gray arrows* mark the phase boundary between the nematic and the isotropic domain. The fluorescently labeled filaments showed obvious alignment. The picture also indicates that the longer filaments were preferentially in the nematic domains

sion occurred. The time sequence of Fig. 6.21 illustrates diffusion of three filaments of different lengths embedded in a nematic domain of a 4.0 mg/ml F-actin solution. The filaments slid quickly back and forth along the axis of alignment, and the short filament suddenly disappeared out of the focal plane without a change in orientation. As long as the filaments were oriented parallel to their axis of alignment they diffused as if in a viscous fluid. In the nematic domains $D_{||}$ was measured by the method described above, and the results are summarized in Fig. 6.16. $D_{||}$ was nearly length independent and, for the longest filaments, higher than the diffusion coefficient along the tube in the entangled phase. The diffusion coefficient along the tube, measured for a filament in a coexisting isotropic domain, was in agreement with the values measured at lower concentrations in which only the isotropic phase was formed.

Surprisingly, $D_{||}$ was nearly length independent. One might expect that the diffusion coefficient scales as it does in the isotropic phase, which is in first order inversely proportional to filament length. Recently, Radzihovsky and Frey (1993) presented a theory for the hydrodynamics of flux lines in superconductors, which also applies to aligned polymer melts. Two cases are distinguished in this theory. Whereas for aligned non-interacting chains the diffusion coefficient is predicted to scale with $1/L$, the relaxations in the interacting aligned phase are independent of the filament length L. In the interacting theory the 1/L-dependence of the diffusion coefficient gets suppressed by the interaction between the chains with increasing filament density. The chain interactions even speed up the dynamics of the chains. If these findings also apply to longitudinal diffusion they could explain the weak length dependence and the increased diffusion coefficients compared to the isotro-

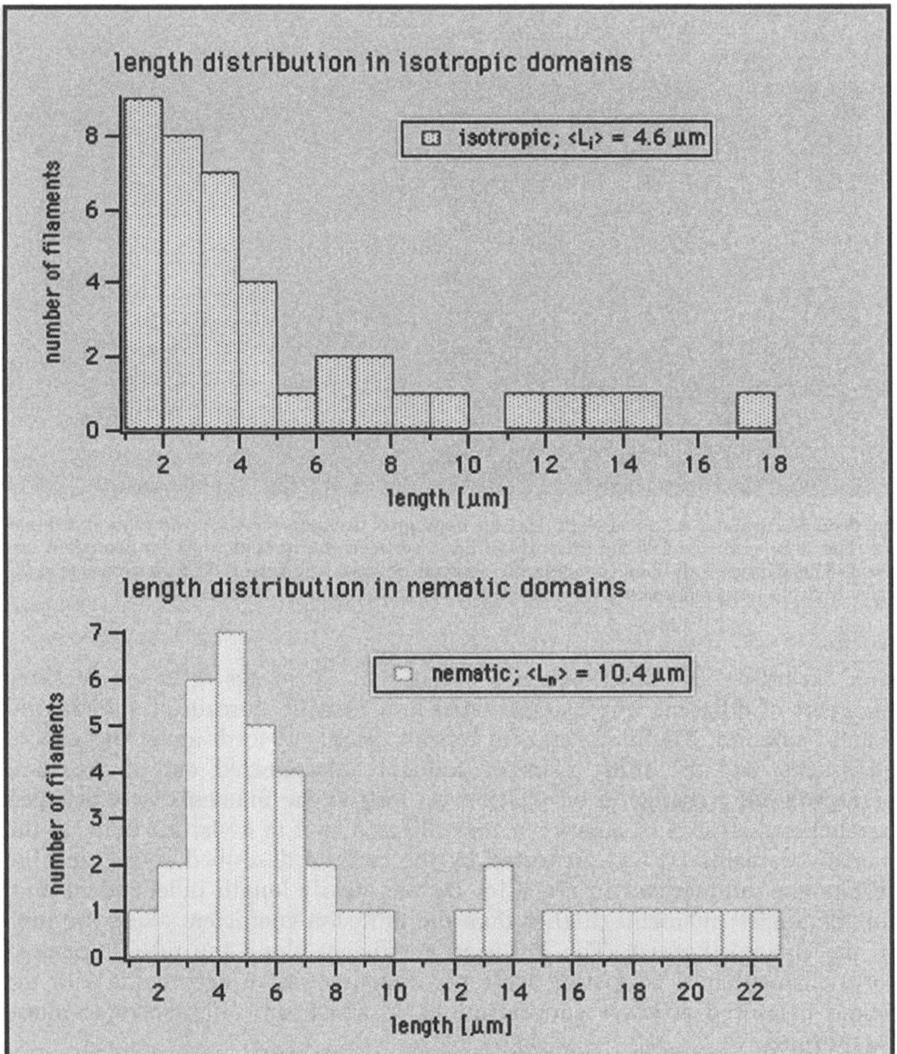

Fig. 6.20. Length distribution of rhodamine-phalloidin labeled actin filaments in the nematic and isotropic domains at an actin monomer concentration of 4.0 mg/ml

pic phase. To prove that this theory really applies to these measurements further experiments at different filament densities are required. If the enhanced diffusion parallel to the axis of alignment also applies for macromolecules embedded in parallel F-actin bundles, stress fibers might serve as transport pathways within cells in addition to microtubules.

The free space around a filament in a nematic domain can also be pictured as a tube as shown in Fig. 6.22. At an actin concentration of 4.0 mg/ml the tube was very narrow and showed very little variation from a mean diameter of 104 nm. The average diameter $\langle a_n \rangle$ for four filaments was

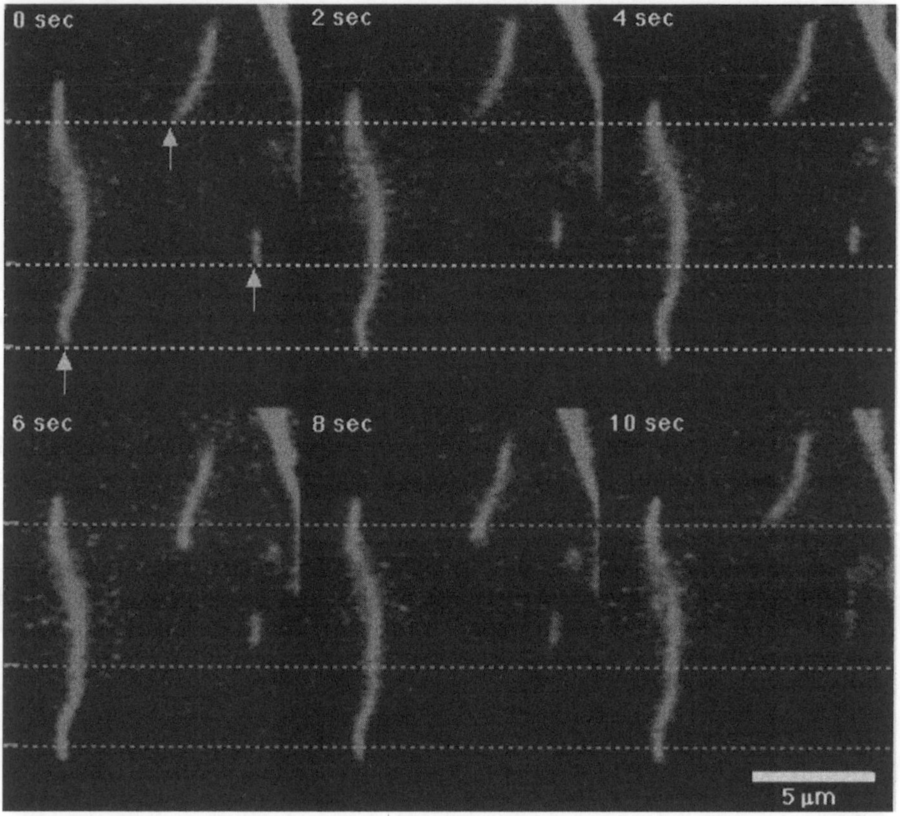

Fig. 6.21. Diffusion of actin filaments in a nematic domain (actin monomer concentration $c_A = 4.0$ mg/ml). The *gray arrows* and *dashed lines* (y-coordinate) mark the original position of the diffusing filaments. The filaments moved along the axis of alignment. The small filament on the right side diffused out of the focal plane after 10 s without changing its orientation

(100±10) nm, whereas in the entangled domains $\langle a_i \rangle$ was (130±10) nm. The relative density difference of the isotropic and nematic domains W can be defined by the following equation (Chen 1993)

$$W \equiv c_n / c_i - 1 \,, \tag{12}$$

where c_n is the number of filaments per volume in the nematic phase and c_i in the isotropic phase. Assuming that filament end effects can be neglected the relative density difference was estimated by

$$W \equiv c_n / c_i - 1 \approx \langle a_i \rangle^2 / \langle a_n \rangle^2 - 1 \,, \tag{13}$$

where $\langle a_i \rangle$, $\langle a_n \rangle$ were the mean tube diameters in the isotropic and nematic phase. The difference in the diameters implied a filament density difference of 60% between the nematic and the isotropic phase. Equation 13 is just an estimate based on the simplified assumption that the volume per filament can be described by a cylinder with diameter $\langle a \rangle$.

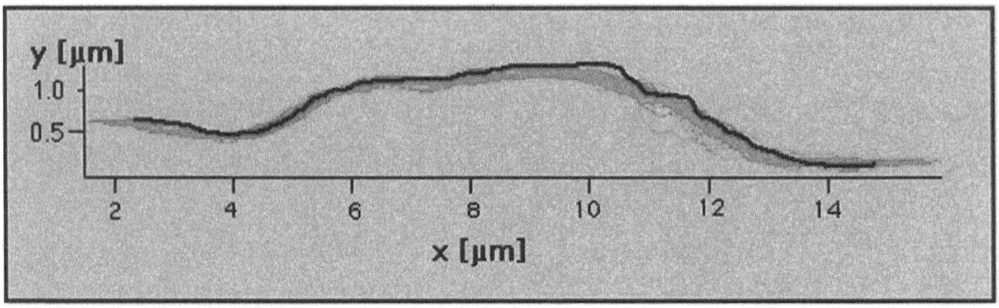

Fig. 6.22. Tube-like free space around the long filament of Fig. 6.21 (left side) in a nematic domain. The tube diameter was very narrow and had a mean of 104 nm. The fluctuations in the diameter were very smal

Theoretical calculations for semiflexible polymers predict a density difference of ~7% at the onset of the coexistence regime (Chen 1993; Vroege and Odijk 1988). Interestingly, the relative concentration difference nearly gets length independent for length-to-persistence length ratios larger than 1 (Chen 1993). The experimental value clearly exceeded the theoretical value. But taking into consideration that in polydisperse systems the relative density difference is expected to be greater (Vroege and Lekkerkerker 1992) and that the actin concentration (4 mg/ml) was nearly twice the onset concentration (2 mg/ml) (Furukawa et al. 1993) for nematic ordering, a larger density difference seems reasonable.

6.4 Conclusions

Fluorescently labeled actin filaments represent a versatile system to study semiflexible polymers in a wide concentration range from dilute solutions to concentrated, nematic solutions. The data obtained provide a better understanding of F-actin solutions *in vitro* and allow therefore better predictions for the function of actin filaments in cells. The main conclusions that may relate to actin function *in vivo* are that purified actin filaments do not display specific chemical interactions with each other, but can form viscoelastic networks because of steric interactions. Due to filament stiffness, the structure of actin networks is quite different from networks comprised of conventional flexible polymers.

The transition from isotropic to nematic structures that occurs over a small concentration range may also be relevant to the formation of such structures as filipodia and stress fibers. Although it seems unlikely that phases attain thermodynamic equilibrium in cells, the thermodynamic drive to form aligned domains and the surprising motility of filaments within such domains may act in concert with specific actin-binding proteins to produce the variety of actin bundles observed *in vivo.*

References

Barden JA, Miki M, Hambly BD, Dos Remedios CG (1987) Eur J Biochem 162:583–588

Bremer A, Millonig RC, Sutterlin R, Engel A, Pollard TD, Aebi U (1991) The structural basis for the intrinsic disorder of the actin filament: the "lateral slipping" model. J Cell Biol 115:689–703

Carlier M-F (1993) Dynamic actin. Curr Biol 3(5):321–323

Casella JF, Torres MA (1994) Interaction of Cap Z with actin. The NH2-terminal domains of the alpha 1 and beta subunits are not required for actin capping, and alpha 1 beta and alpha 2 beta heterodimers bind differentially to actin. J Biol Chem 269:6992–6998

Chen ZY (1993) Nematic ordering in semiflexible polymer chains. Macromolecules 26:3419–3423

Coppin CM, Leavis PC (1992) Quantitation of liquid-crystalline ordering in F-actin solutions. Biophys J 63:794–807

Doi M (1975) Rotational relaxation time of rigid rod-like macromolecule in concentrated solution. J Phys (Paris) 36:607–617

Doi M (1985) Effect of chain flexibility on the dynamics of rodlike polymers in the entangled state. J Polym Sci Polym Symp 73:93–98

Doi M, Edwards SF (1986) The theory of polymer dynamics. Clarendon, Oxford

Edwards SF (1967) Proc Phys Soc 92:9–13

Elson E L (1988) Cellular mechanics as an indicator of cytoskeletal structure and function. Annu Rev Biophys Chem 17:397–430

Ferry JD (1980) Viscoelastic properties of polymers. John Wiley, New York

Furukawa R, Kundra R, Fechheimer M (1993) Formation of liquid crystals from actin filaments. Biochemistry 32:12346–12352

de Gennes PG (1971) Reptation of a polymer chain in the presence of fixed obstacles. J Chem Phys 55:572–579

de Gennes PG (1982) Kinetics of diffusion-controlled processes in dense polymer systems. II. Effects of entanglements. In: Ciferri A, Krigbaum WR, Meyer RB (eds) Polymer liquid crystals, chap 5. Academic Press, New York

de Gennes PG, Prost J (1994) The physics of liquid crystals. Clarendon, Oxford

de Gennes PG, Pincus P, Velasco RM, Brochard F (1976) Remarks on polyelectrolyte conformation. J Phys (Paris) 37:1461–1473

Graessley WW (1980) Some phenomenological consequences of the Doi-Edwards theory of viscoelasticity. J Polym Sci 13:27–34

Hendricks J, Kawakatsu T, Kawasaki K, Zimmermann W (1995) On confined semiflexible polymer chains. Phys Rev E51:2658–2661

Ishijima A, Doi T, Sakurada K, Yanagida T (1991) Sub-piconewton force fluctuations of actomyosin in vitro. Nature 352:301–306

Janmey PA, Peetermans J, Zaner KS, Stossel TP, Tanaka T (1986) Structure and mobility of actin filaments as measured by quasielastic light scattering, viscometry, and electron microscopy. J Biol Chem 261:8357–8362

Janmey PA, Hvidt S, Oster GF, Lamb J, Stossel, TP, Hartwig JH (1990) Effect of ATP on actin filament stiffness. Nature 347:95–99

Janmey PA, Euteneuer U, Traub P, Schliwa M (1991) Viscoelastic properties of vimentin compared with other filamentous biopolymer networks. J Cell Biol 113:155–160

Janmey PA, Hvidt S, Käs J, Lerche D, Maggs A, Sackmann E, Schliwa M, Stossel TP (1994) The mechanical properties of actin gels. J Biol Chem 269:32503–32513

Käs J, Strey H, Bärmann M, Sackmann E (1993) Direct measurement of the wave-vector-dependent bending stiffness of freely flickering actin filaments. Europhys Lett 21:865–870

Käs J, Strey H, Sackmann E (1994a) Direct visualization of reptation for semiflexible actin filaments. Nature 368:226–229

Käs J, Laham LE, Finger DK, Janmey PA (1994b) Solution ATP affects the bending elasticity of actin filaments implying a low affinity ATP-binding site on F-actin. Mol Biol Cel. 5:157a

Käs J, Strey H, Tang JX, Finger D, Ezzell R, Sackmann E, Janmey PA (1996) F-actin, a model polymer for semiflexible chains in dilute, semidilute and liquid crystalline solutions. Biophys J 70:609–625

Kaufmann S, Käs J, Goldmann WH, Sackmann E, Isenberg G (1992) Talin anchors and nucleates actin filaments at lipid membranes – a direct demonstration. FEBS Lett 314: 203–205

Keep GT, Pecora R (1985) Reevaluation of the dynamic model for rotational diffusion of thin, rigid rods in semidilute solution. Macromolecules 18:1167–1173

Khokhlov AR, Semenov AN (1982) Susceptibility of liquid-crystalline solutions of semiflexible macromolecules in an external orientational field. J Phys A 15:1361–1367
Landau LD, Lifshitz EM (1980) Statistical physics, part 1, 3rd edn. Pergamon, Oxford
Lodge TP, Rotstein NA, Prager S (1993) Dynamics of entangled polymer liquids: do linear chains restate? Adv Chem Phys 79:1–132
MacKintosh FC, Käs J, Janmey PA (1995) Elasticity of semiflexible biopolymer networks. Phys Rev Lett 75(24):4425–4428
Magda JJ, Davis HT, Tirrell M (1986) The transport properties of rod-like particles via molecular dynamics. I. Bulk fluid. J Chem Phys 85:6674–6685
Mogilner A, Oster G (1996) Cell motility driven by actin polymerization. Biophys J 71:3030–3045
Müller O, Gaub HE, Bärmann M, Sackmann E (1991). Viscoelastic moduli of sterically and chemically cross-linked actin networks in the dilute to semidilute regime - measurements by an oscillating disk rheometer. Macromolecules 24:3111–3120
Muthukumar M, Edwards SF (1983) Screeining of hydrodynamic interaction in a solution of rodlike macromolecules. Macromolecules 16:1475–1478
Odijk T (1983) On the statistics and dynamics of confined or entangled stiff polymers. Macromolecules 16:1340–1344
Odijk T (1986) Translational friction coefficient of hydrodynamically screened rodlike macromolecules. Macromolecules 19:2073–2074
Onsager L (1949) The effects of shape on the interaction of colloidal particles. Ann NY Acad Sci 51:627–659
Perkins TT, Smith DE, Chu S (1994) Direct observation of tube-like motion of a single polymer chain. Science 264:819–822
Prochniewicz E, Zhang Q, Janmey PA, Thomas DD (1995) Cooperativity in F-actin: binding of gelsolin at the barbed end affects the structure of the whole filament. Biophys J 68:A248
Radzihovsky L, Frey E (1993) Kinetic theory of flux-line hydrodynamics: liquid phase with disorder. Phys Rev B 48:10357–10381
Ruddies R., Goldmann WH, Isenberg G, Sackmann E (1993) The viscoelasticity of entangled actin networks: the influence of defects and modulation by talin and vinculin. Eur Biophys J 22:309–322
Sackmann E (1994) Intracellular and extracellular macromolecular networks - physics and biologicla function. Macromol Chem Phys 194:7–28
Sato T, Takada Y, Teramoto A (1991) Dynamics of stiff-chain polymers in isotropic solution. 3. Flexibility effect. Macromolecules 24:6220–6226
Schmidt CF, Bärmann M, Isenberg G, Sackmann E (1989) Chain dynamics, mesh size, and diffusive transport in networks of polymerized actin. A quasielastic light scattering and microfluorescence study. Macromolecules 22:3638–3649
Semenov AN (1986) Dynamics of concentrated solutions of rigid-chain polymers, part 1. Brownian motion of persistent macromolecules in isotropic solution. J Chem Soc, Faraday Trans 2(82):317–329
Sheterline P, Clayton J, Sparrow J (1995) Actin. Protein Profile 2(1):1–103
Smith BS, Finzi L, Bustamente C (1992) Direct mechanical measurements of the elasticity of single DNA molecules by using magnetic beads. Science 258:1122–1126
Stossel TP (1993) On the crawling of animal cells. Science 260:1086–1094
Suzuki A, Maeda T, Ito T (1991) Formation of liquid crystalline phase of actin filament solutions and its dependence on filament length as studied by optical birefringence. Biophys J 59:25–30
Volkmuth WD, Austin RH (1992) DNA electrophoresis in microlithographic arrays. Nature 358:600–602
Vroege GJ, Odijk T (1988) Induced chain rigidity, splay modulus, and other properties of nematic polymer liquid crystals. Macromolecules 21:2848–2858
Vroege GJ, Lekkerkerker HNW (1992) Phase transitions in lyotropic colloidal and polymer liquid crystal. Rep Prog Phys 55:1241–1315
Wegner A J (1975) Head to tail polymerization of actin. Mol Biol 108:139–150
Williams DRM, Warner M (1990) Statics and dynamics of hairpins in worm-like main chain nematic polymer liquid crystals. J Phys Fr 51:317–339

CHAPTER 7

The Interaction of Proteins with Membrane Surfaces at Molecular Resolution: The Neutron Reflection Method

Thomas M. Bayerl* · Andreas P. Maierhofer

Contents

Department of Biophysics, Institute for Experimental Physics, University of Würzburg, 97074 Würzburg, Germany
*Corresponding address: Department of Biophysics, Institute for Experimental Physics, University of Würzburg, 97074 Würzburg, Germany, Tel.: +49 0931 8885853, fax: +49 0931 706297, e-mail: tbayer1@physik.tu-muenchen.de

7.1 Introduction – The Neutron Advantage

Neutron scattering represents a unique tool for studying the structure and the phase behavior of lipid membranes. Among the latest applications of neutron scattering in lipid membranes is the neutron reflection technique. This method proves as very powerful for the study of the interfacial structure of lipid monolayers at the air–water interface and its changes due to the presence of water-soluble proteins. Moreover, its sample quantity requirements are extremely low compared to other neutron techniques, and is generally in the microgram to milligram region. This makes it suitable for the neutron study of rare and expensive biological macromolecules. A particularly interesting feature of this method from the cell biology point of view is that it allows the acquisition of information about the penetration depth of extrinsic proteins into the monolayer and the associated changes of the lipid head group hydration. After giving an introduction into neutron reflection theory and the experimental setup, we will concentrate on this biological aspect by discussing results obtained from the study of monolayer interaction with different extrinsic proteins.

Cold neutrons with wavelengths λ in the range 0.4–3 nm, which covers the size range of molecules like membrane phospholipids, are used for the reflection experiments. Neutrons interact with atomic nuclei, unlike x-rays which interact primarily with atomic electrons, and so neutrons give a different, complementary picture to that given by x-rays. This interaction, which results in the scattering or reflection of a part of the incident neutron beam from its original direction in space, is remarkably different for hydrogen and deuterium, and so provides a unique tool which is not available for x-rays. This tool, the contrast variation method, provides the basis for all neutron reflection experiments in membrane research. It allows the modification of the total reflectivity of the lipid monolayer by partial deuteration of the lipids as well as by variation of the D_2O content of the subphase. In the case of protein-adsorption studies the reflectivity contribution of the latter may be suppressed as well.

A limiting factor in the use of neutron scattering is the availability of high flux neutron beams. Typically, cold neutrons come from a large nuclear reactor or from a spallation source (which is driven by a powerful particle accelerator) which is a national resource shared by many users, so "beam time" is scarce. Since the neutron flux is low compared to synchrotron radiation sources used for x-ray experiments, experiments can be long and tedious. However, a distinct advantage of the neutron method is that sample damage as a result of the exposure of the monolayer sample to the cold neutron beam is usually negligible, which is not true for synchrotron radiation.

7.2
The Neutron Reflection Experiment

7.2.1
Experimental Setup

The very first application of neutron reflection to phospholipid systems was devoted to the study of the structure of a DMPC/DMPG monolayer (DMPC: 1,2-dimyristoyl-sn-glycero-3-phosphatidylcholin; DMPG: 1,2-dimyristoyl-sn-glycero-3-phosphatidylglycerol) at the air/water interface as a function of the lateral pressure (Bayerl et al. 1990). Similarly, the DPPC (1,2-dipalmitoyl-sn-glycero-3-phosphatidylcholin) monolayer structure was studied by this technique and the data were discussed in the light of previously obtained synchrotron radiation reflection experiments (Vaknin et al. 1991a) and its hydration (Naumann et al. 1995) and headgroup conformation (Brumm et al. 1994) were determined as a function of monolayer lateral pressure. Further steps toward the application of neutron reflection in membrane biophysics were studies of the interaction of nonionic surfactants as models for extrinsic proteins with DPPC monolayers (Naumann et al. 1995) and of real proteins like spectrin and polylysine with electrically charged phospholipid monolayers (Johnson et al. 1991). Furthermore, recognition processes at a functionalized lipid monolayer were observed using this technique (Vaknin et al. 1991b).

The general experimental setup used for these studies is shown for a time-of-flight type reflectometer like CRISP (Rutherford Appleton Laboratory, Chilton, UK) in Fig. 7.1 (Penfold and Thomas 1990; Russel 1990). It comprises essentially the neutron beam impinging at the sample (lipid monolayer in a Langmuir trough) at a low glancing angle θ_1 after passing previously a number of neutron optical elements (collimators, apertures, neutron guides, mirrors etc.) and, beyond the sample, the neutron detector which can be either one or two dimensional to detect the neutrons specularly reflected from the sample. For two-dimensional detectors, the diffuse reflection of the surface can be monitored as well, which can provide additional information under certain conditions. In order to pick up the specular reflected part only, the angle under which the detector is arranged with respect to the monolayer surface must be identical to the angle of incidence θ_1 of the neutron beam. The neutrons can be either polychromatic, i.e. consisting of a distribution of neutron wavelengths, or monochromatic. In the former case, a time-of-flight analysis of the neutrons will be required to obtain the desired spatial resolution (time-of-flight neutron reflectometer) while for monochromatic neutrons this is achieved by variation of the angle of incidence (fixed wavelength neutron reflectometer). Both approaches can provide neutron reflectivities up to 10^{-7}, allowing an accessible Q-range (which ultimately determines the resolution) of up to 0.2 $Å^{-1}$.

Since the proteins under study are usually available in very low quantities only, a normal Langmuir trough is often not suitable because of its large subphase volume, which would dilute the dissolved proteins beyond dectectability and require exceedingly long equilibration times after the protein in-

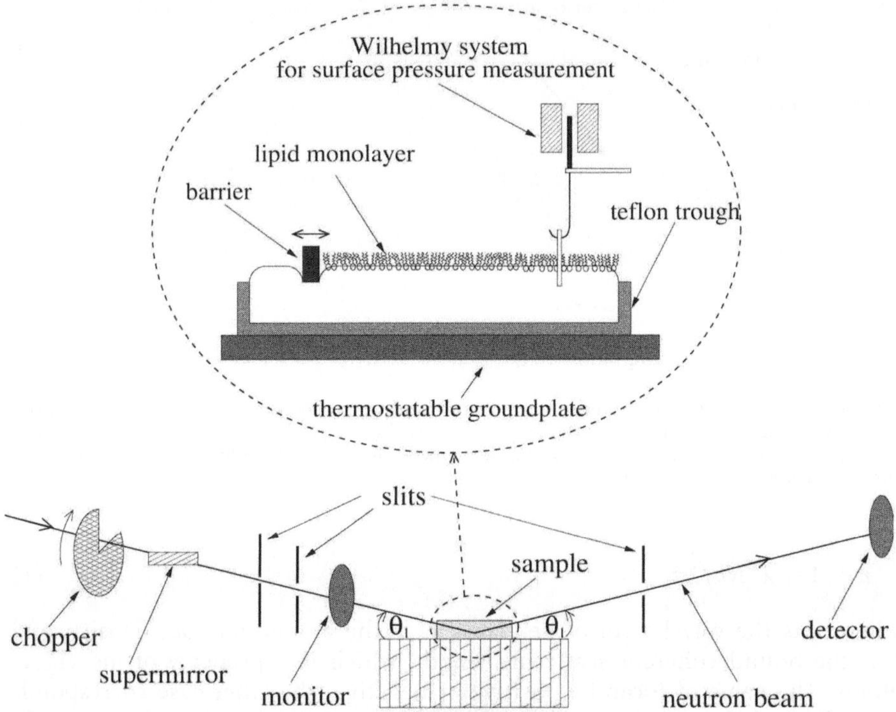

Fig. 7.1. Schematic representation of the experimental setup for a neutron reflection experiment on a time-of-flight reflectometer. The *insert* shows the sample site consisting of a Langmuir trough with the phospholipid monolayer on top of the subphase. For protein – monolayer interaction studies, the water soluble proteins are dissolved by injection into the subphase below the lipid monolayer (cf. Fig. 7.2).

jection into the subphase. Therefore a small Teflon trough (surface area: 64.55 cm^2 and subphase volume: 40 ml) is inserted into the large trough (dimension $15\times45\times0.5$ cm^3) of the film balance (Fig. 7.2) in such a way that the monolayers are spread using a special technique which implies the application of a lipid solution on the water surface in either compartment. After spreading the lipid monolayer and adjusting the lateral pressure, the channel link between the two compartments is closed by a removable barrier and the proteins are injected into the subphase of the insert by syringes. The lateral pressure is measured in the small compartment by a Wilhelmy system (accuracy: 0.1 mN/m).

7.2.2
Theory – Basic Concepts of Neutron Reflection at Interfaces

Because of the analogy between reflection of neutrons and of light, one can apply the fundamentals of optics to the phenomenon of neutron reflection. For non-adsorbing samples, the neutron refractive index n is slightly less than 1 and is given, to within a good approximation, by

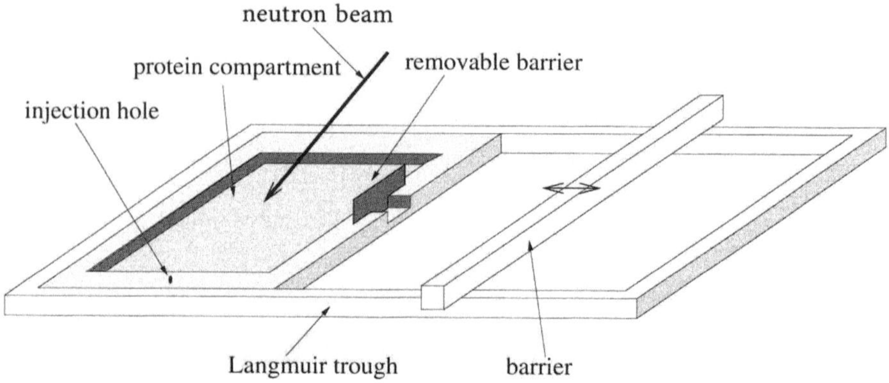

Fig. 7.2. The dedicated Langmuir trough for neutron reflection measurements of lipid monolayers on a subphase of proteins dissolved in buffer medium consisting of inner and outer compartments connected by a channel link with a removable barrier for lateral pressure adjustment.

$$n = 1 - \lambda^2 Nb/2\pi \tag{1}$$

where λ is the wavelength of neutrons, N is the atomic number density, and b is the bound coherent scattering length, which is a property of individual atoms. The value of b can be positive or negative (the latter case corresponds to a phase shift by π due to the scattering). In particular, there is a remarkable difference of b for protons (negative b) and deuterons (positive b). Hence, by isotopic substitution of protons by deuterons, n can be modified and large scattering contrasts can be obtained or adjusted according to the experimental requirements (contrast variation technique).

The simplest case of specular reflection occurs at an interface separating air or vacuum (refractive index $n_1 = 1$) from a solid or liquid (refractive index n_2) where the refraction angle θ_2 is given by the grazing angle of incidence θ_1, according to Snell's law

$$\cos\,\theta_2 = (1/n_2)\,\cos\,\theta_1 \tag{2}$$

Specular reflection is then observed at incident angles $\theta_1 \geq \theta_c$, the critical angle, where $\cos\,\theta_c = n_2$. The refractive index profile (or scattering length density profile) normal to an interface can be determined by the measurement of the specular reflection. Hence, reflectivity is measured as a function of momentum transfer $Q_{z,0}$ normal to the surface,

$$Q_{z,0} = (4\pi/\lambda)\,\sin\,\theta\,, \tag{3}$$

where $\theta = \theta_1$ is the grazing angle of incidence. In a medium i with a neutron scattering length density ρ_i with

$$\rho_i = \sum_a N_{i,a}\, b_a \tag{4}$$

where N_{ia} is the atom number density and b_a the bound coherent scattering length of atoms of type a, the normal component of the momentum transfer is modified such that

$$Q_{z,0} = (Q_{z,0}^2 - Q_{c,i}^2)^{1/2} \tag{5}$$

where $Q_{c,i}$ is the critical value below which total reflection occurs. Thus, specular reflection measurements can be performed either by varying θ_1 at fixed neutron wavelength λ or at various λ, keeping θ_1 fixed.

For a more complicated system consisting of several distinct layers j of different scattering length density ρ_j and thickness d_j, reflection occurs at each interface and the resulting reflectivity profile is modulated periodically by interference. For real reflectivities, this corresponds to a cosine modulation of the reflectivity and the distance between two successive minima is $\Delta Q_{z,0}$. From the latter, the average layer thickness $\langle d \rangle$ can be estimated as $\langle d \rangle = \pi / \Delta Q_{z,0}$. The value of $\langle d \rangle$ can serve to test the reliability of the models applied in the analysis of the reflectivity data. This analysis is often done using the optical matrix method (Born and Wolf 1970), where the interfacial structure is divided into a series of flat layers with sharp boundaries. A characteristic matrix can be defined for each layer, and for the jth layer, we have

$$M_j = \begin{vmatrix} \cos \beta_j & -(\mathrm{i}/p_j) \sin \beta_j \\ -\mathrm{i}p_j \sin \beta_j & \cos \beta_j \end{vmatrix} \tag{6}$$

where $p_j = n_j \sin\theta_j$ and $\beta_j = (2\pi/\lambda) n_j d_j \sin \theta_j$ where d_j is the thickness of the layer. For l layers, the characteristic matrices are multiplied: $\mathbf{M}_\mathrm{R} = [\mathbf{M}_1][\mathbf{M}_2]\cdots[\mathbf{M}_l]$. The reflectivity R is calculated from

$$R = \left| \frac{(m_{11} + m_{12}\, p_s)\, p_a - (m_{21} + m_{22})\, p_s}{(m_{11} + m_{12}\, p_s)\, p_a + (m_{21} + m_{22})\, p_s} \right|^2 \tag{7}$$

where m_{ij} designates the elements of the matrix $\mathbf{M}_\mathrm{R}$, and subscripts s and a refer to substrate and air, respectively. In applying the optical matrix method, one begins by specifying a model for the interfacial structure, and then proceeds to solve Eq. (7) for reflectivity as a function of momentum transfer Q, adjusting the model as necessary to achieve good fits to the data.

In general, there are two important limitations of the method. One is that a real interface is not infinitely sharp, leading to an interfacial roughness or a continuous variation of ρ_i near the interface. Second, since the square of the reflectivity amplitude is measured in a reflection experiment, phase information is lost and thus it is not possible to obtain a unique density profile which fits a single set of reflectivity data.

In order to cope with the above limitations, it is important to conduct neutron reflection experiments at more than one contrast in scattering length density (contrast variation). For each layer in the model, there are two parameters to fit: thickness of the layer and its scattering length density. Thus for a two-layer model, such as we use to describe the lipid monolayer/subphase system, there are four unknown parameters provided that interfacial roughness and surface roughness are already known. There are essentially two ways in obtaining different scattering contrasts in reflection studies of lipids. One is to change the contrast to the monolayer subphase by using either pure H_2O or D_2O (subphase contrast variation). By mixing both waters at the appropriate ratio, one can obtain a match of the scattering length density of the subphase water to that of air. Then only the lipids would contribute to the reflectivity. The second method for the variation of the scattering contrast is the application of selectively or totally deuterated phospholipids or of isotopic mixtures of them (inverse contrast variation). This enables the screening of parts of the monolayer such that the number of interfaces contributing to reflectivity is reduced and the model applied can be checked for consistency.

7.2.3 Analysis of Lipid Monolayer Neutron Reflection Data

A meaningful data analysis in terms of the mass density distribution in the normal direction of the layer system requires a systematic application of contrast variation procedures by partial deuteration of the lipid and/or of the subphase, combined with simulations of the reflectivity curves for the various contrast settings. The simplest models for the interpretation of the data are a two-layer model for the case of the lipid/protein layer on D_2O-buffer and a two- or three-layer model for the lipid/protein layer on a subphase of null reflecting water (i.e. contrast matched to air, CMA). The major problem is that the two-layer model is determined by a minimum of four parameters and the three-layer model by at least six parameters. Without additional information about the structural parameter of the lipid and other constraints, the data analysis of the $R(Q_{z,0})$-vs-$Q_{z,0}$ curves would not yield a unique set of parameters. A possible strategy to tackle these problems is as follows:

1. Each experiment of lipid/protein coupling is started by recording reflectivity-vs-$Q_{z,0}$ curves for the pure lipid monolayer.

The reflectivity curves can be interpreted in terms of a one-layer model in the case of a deuterated lipid monolayer on D_2O since the contrast between subphase and head group region is very small. The fitting of the experimental $R(Q_{z,0})$-vs.-$Q_{z,0}$ curve by two free parameters, namely the scattering length density and the thickness, leads to an unique result.

In the case of the deuterated lipid monolayer on CMA-buffer, the monolayer has to be described by a two-layer model which implies four parameters (two values of scattering length density and the thickness of each layer). Moreover, the molecular area obtained from the fitted parameters A_L(NR) must agree with the molecular area A_L(FB) measured independently by a film balance experiment. By taking into account this constraint, one has only two free parameters left (thicknesses of each layer), which again leads to an unique solution. The validity of this procedure was checked in several previous studies (Naumann et al. 1994, 1995).

2. In the case of the lipid/protein systems, where the protein may exhibit a partial penetration into the lipid monolayer, we can make two further assumptions.

The first is that addition of protein does not cause a change of the thickness of the lipid layer.

For proteins that may interact both electrostatically and hydrophobically with the membrane like hisactophilin (see example below), the second assumption is that only the hydrophobic anchor of the protein is able to penetrate into the hydrophobic interior of the lipid monolayer.

For the case of the protein hisactophilin discussed below, we assume that when a lipid monolayer with an area per molecule of about 60 $Å^2$ is saturated with this particular protein which exhibits an area of about $A_P = 1000$ $Å^2$ (Habazettl et al. 1992), about one anchor (a myristic acid chain) should be present for each 17 phospholipid molecules. The incorporation of a protonated protein chain into the fully deuterated lipid chain interior should therefore cause a maximum decrease in scattering length density of the hydrophobic membrane region of $\Delta\rho = 0.17 \times 10^{-6}$ $Å^{-2}$. We therefore conclude that addition of DIC-HIS decreases the scattering length density of the hydrophobic interior of the lipid by not more than $\Delta\rho = 0.2 \times 10^{-6}$ $Å^{-2}$.

For the lipid/protein layers on both D2O-buffer and CMA-buffer, we can take this value of $\Delta\rho$ as a restricted parameter exhibiting a range of variations of $\Delta\rho \leq 0.2 \times 10^{-6}$ $Å^{-2}$. In order to take into account incorporation of the hydrophobic anchor of hisactophilin, we varied this parameter by incremental steps of $\Delta\rho = 0.1 \times 10^{-6}$ $Å^{-2}$.

Based on the above procedures the following fixed and restricted parameters are used.

For lipid/protein layers on D_2O-buffer we have to consider one restricted parameter (scattering length density of the hydrophobic region of the lipid) and one fixed parameter (thickness of the hydrophobic region of the lipid). The reflectivity data of lipid/protein layers on CMA-buffer can be accounted for by one restricted parameter (scattering length density of the hydrophobic region of the lipid) and two fixed parameters (thickness of the lipid chain region and thickness of the hydrophilic moiety of the lipid).

Finally we are left with two free floating parameters for the case of the lipid/protein layer on D_2O-subphase and three free parameters for the lipid/protein layer on CMA-buffer. The statistical errors are obtained by a systematic χ^2-analysis of each free floating parameter taking into account that all free floating parameters depend on each other.

In some cases (e.g. in the presence of two proteins) the $R(Q_{z,0})$-vs.-$Q_{z,0}$ curves could not be fitted by the above procedure in a satisfactory way. The fitting could, however, be considerably improved by considering contributions by the interfacial roughness σ_i. The following procedure was applied:

1. Each experiment was firstly analyzed by neglecting interfacial roughness as described above ($\sigma_i = 0$).
2. The original fitting parameters were kept constant while the interfacial roughnesses σ_i were varied until best fit was achieved.
3. Subsequently, the original parameters were varied again while the σ_i were kept constant.

4. We repeatedly applied procedures 2 and 3 in a successive manner until the minimum χ^2-value was achieved.

A systematic error (χ^2) analysis showed that the interfacial roughness can be detected to an accuracy of $\Delta\sigma_i = \pm 3$ Å.

7.3 Applications

7.3.1 Interaction of Cytochrome c with Anionic Phospholipid Monolayers: The Role Of The Monolayer Phase State

The aim of this study was to establish the limits of the detectability of the electrostatic coupling of a well-known, most likely purely extrinsic protein, to a phospholipid monolayer at the air–water interface. Since cytochrome c has a pI of 9.6, it exhibits positive excess charges at neutral pH and thus can be expected to interact strongly with anionic lipid surfaces via Coulomb interaction. Furthermore, the amount of protein adsorbed may depend on the phase state of the monolayer since domains in a two-component lipid mixture where one component is anionic and the other neutral will inevitably be enriched or depleted of one component and thus the charge distribution over the surface will be heterogeneous (Loidl-Stahlhofen et al. 1996). To test this assumption for monolayers, neutron reflection measurements were performed at different lateral pressures π of the mixed DPPC/DMPG monolayer corresponding to the solid-like s-phase, the fluid-like LE-phase and to the coexistence region. In order to modulate the amount of cytochrome c bound to the monolayer, the amount of the anionic lipid component in the monolayer was varied from 10–50%.

As an example, Fig. 7.3 A shows reflectivity curves ($R(Q_{z,0})$-vs.-$Q_{z,0}$) for the case of an equimolar DPPC/DMPG monolayer in the s-phase with and without the presence of cytochrome c in the subphase while the same is shown for the fluid-like LE phase in Fig. 7.3B. The subphase was in both cases D_2O with 20 mM Hepes buffer at pH = 7.0. The amount of protein added to the subphase was up to 20 n_o, where n_o corresponds to the amount of cytochrome c that is required to form a single protein monolayer, calculated from the known cytochrome c diameter (15.5 Å) assuming spherical geometry and packing.

In general, the curves show a rapid decrease of $R(Q_{z,0})$ with $Q_{z,0}$, which falls off more rapidly than the theoretically expected $1/Q_{z,0}^4$-dependence for a perfectly smooth interface, and levels off in the incoherent background at momentum transfers $Q_{z,0} > 0.2$ Å^{-1}. Furthermore, interference fringes between the layers are not much pronounced because the low monolayer thickness of less than 30 Å would require a much wider Q-range covered to see them more clearly. Nevertheless, the modulations of the curve are still sufficient for a data analysis along the strategy outlined above and the differences of the curves with and without cytochrome c are remarkably strong. The full lines in Fig. 7.3 represent two-layer model fits of the data. The corresponding

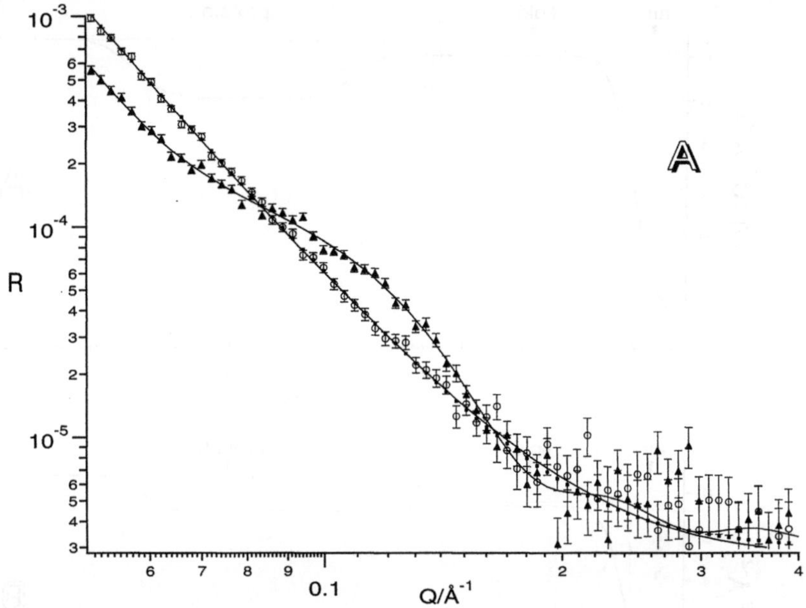

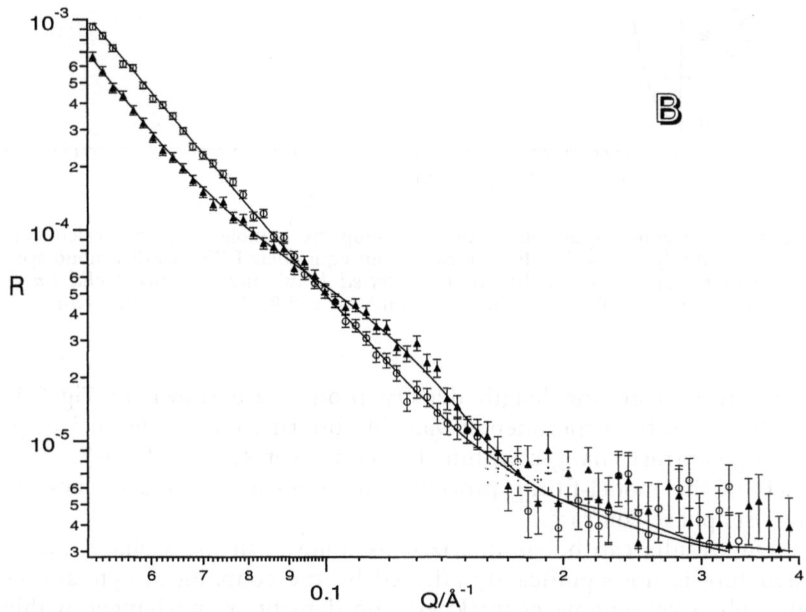

Fig. 7.3. Neutron reflectivity curves for an equimolar DPPC/DMPG monolayer at a subphase of buffered D_2O in the absence (*circles*) and presence (*triangles*) of cytochrome c in the subphase at two monolayer phase states. **A** s-phase at 30 mN/m and **B** LE-phase at 14 mN/m. The *full lines* are obtained from two-layer model fits of the data.

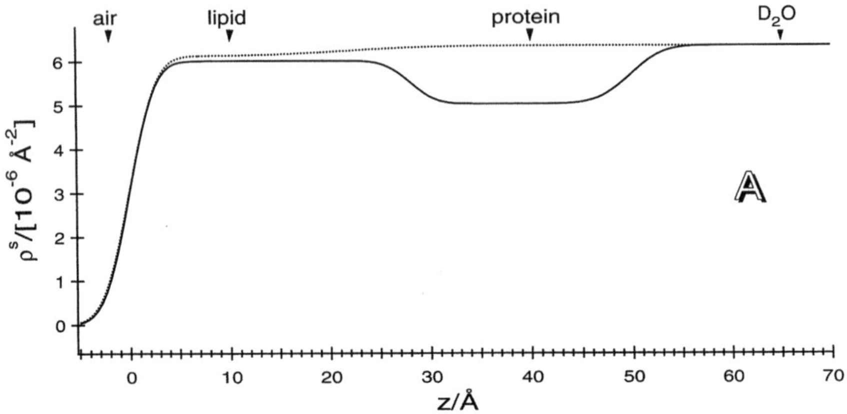

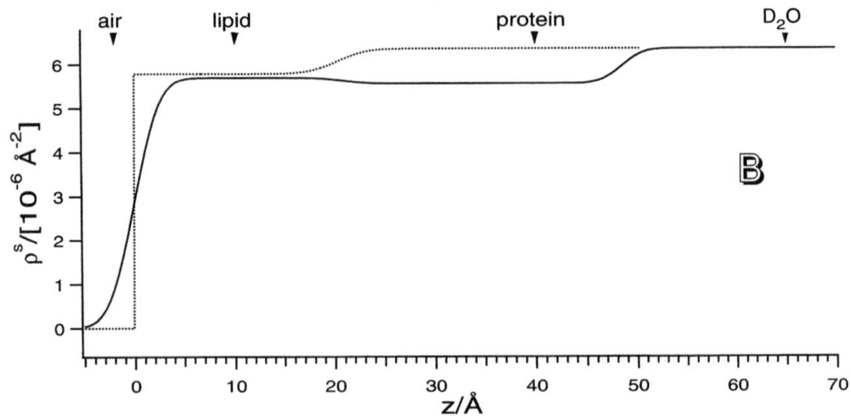

Fig. 7.4. Scattering length density profiles along the monolayer normal calculated from the fitting results (cf. Table 7.1) for the case of an equimolar DPPC/DMPG monolayer at a subphase of buffered D_2O in the absence (*dotted lines*) and presence (*solid lines*) of cytochrome c in the subphase. **A** s-phase at 30 mN/m and **B** LE-phase at 14 mN/m

theoretical scattering length density profiles are shown in Fig. 7.4. Such a profile gives the experimenter valuable information on the loci of the dominating contrast changes within the monolayer system. Table 7.1 lists the results of the model-fitting procedure for one- and two-layer models and for two subphase contrasts.

Our results can be summarized as follows. In the s-phase the monolayer structure is not significantly affected by the coupling of cytochrome c from the subphase; monolayer thickness and density are unchanged within the experimental error. However, in the presence of cytochrome c the formation of a protein monolayer adjacent to the lipid headgroups can be clearly observed. The thickness of this layer (29±2 Å) agrees well with the diameter of cytochrome c known from crystal structure analysis. The degree of protein coverage of this monolayer in the s-phase depends on the proportion of anionic DMPG in the monolayer. For 50% DMPG we obtained a coverage of

Table 7.1. Results of the data fitting procedure according to one- and two-layer models for equimolar and 7:3 DPPC/DMPG monolayers without and with cytochrome c in a D_2O or H_2O subphase. The index 1 denotes the lipid monolayer, the index 2 the region of the protein monolayer

System	Phase	Buffer	d_1 (Å)	ρ_1 (10^{-6} Å^{-2})	d_2 (Å)	ρ_2 (10^{-6} Å^{-2})
1:1 DPPC/DMPG	s	H_2O	24.0±0.2	4.70±0.03	–	–
+6 n_0 CytC			20.5±0.5	4.65±0.07	28.7±0.8	0.23±0.03
1:1 DPPC/DMPG	LE	D_2O	20.1±1.3	5.77±0.07	–	–
+6 n_0 CytC			21.0±5.8	5.69±0.05	27.6±6.0	5.56±0.03
1:1 DPPC/DMPG	s		27.5±4.3	6.24±0.02	–	–
+6 n_0 CytC			28.1±2.5	6.01±0.05	21.8±3.4	5.03±0.11
7:3 DPPC/DMPG	s	H_2O	24.8±0.2	5.64±0.03	–	–
+6 n_0 CytC			24.7±0.6	5.03±0.08	28.9±1.6	0.07±0.07
7:3 DPPC/DMPG +6 n_0 CytC		D_2O	21.2±0.9	6.84±0.06	33.3±1.1	5.53±0.03

39%, whereas at 30% DMPG only 26% coverage was obtained. For 10% DMPG the cytochrome c coverage was less than 10%. Under the last conditions, the limit of detectability of a protein is reached. The above values for coverage were considered in terms of a hexagonal packing model of spherical cytochrome c molecules within the plane of the protein monolayer (Fig. 7.5). Since the gaps between the proteins are filled with buffer within this model, we can calculate a protein coverage of 60% for the case of ideal dense packing. Thus, even at 50% DMPG, the protein monolayer is still not uniform but may exhibit some holes and defects.

The dependence of protein coverage on the anionic DMPG content (coverage increases by a factor of 1.5 in the s-phase as the DMPG content increases from 30 to 50%) is a clear indication for the dominance of electrostatic interactions, since the above factor of 1.5 correlates well with the increase of monolayer surface charge density (factor of 1.45 assuming homogeneous charge distribution) accomplished by the increase of DMPG content. Thus, the number of anionic charges in the monolayer provided by DMPG per coupled cytochrome c molecule is rather constant in the s-phase up to the point where sterical effects and electrostatic screening ultimately prevent the coupling of further protein from the subphase. The data suggest that this limit is not reached at 50% DMPG in the monolayer.

For the LE phase of the equimolar mixture, we find indications for a correlation between the degree of protein coverage and the existence of domains. Here we observe significant deviations from the above mentioned correlation between monolayer surface charge density and protein coverage. At lateral pressures π where no domains can be observed by fluorescence film balance techniques (Fig. 7.6), the degree of cytochrome c coverage is signifi-

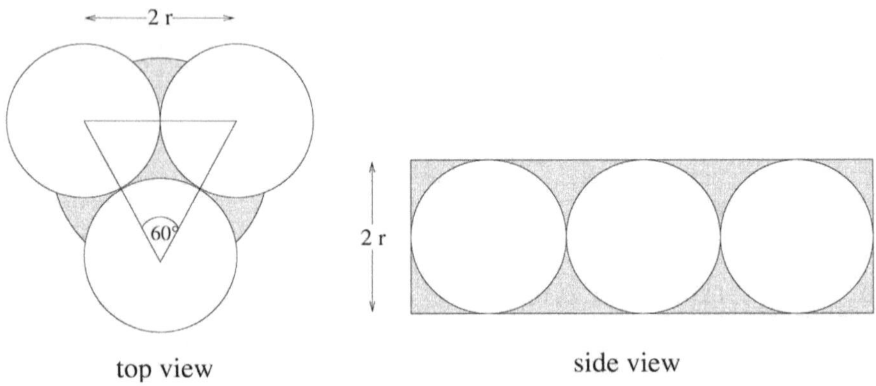

Fig. 7.5. Model of the hexagonal packing of cytochrome c within a monolayer of its own. **A** Top view. **B** Side view

cantly below the level expected for this amount of anionic DMPG present in the monolayer. In contrast at a only 4 mN/m higher lateral pressure where domains can be clearly observed by fluorescence, the coverage of cytochrome c was drastically increased. This is in agreement with recent supported bilayer studies, where the dispersion of domains at the phase transition from the gel to the fluid phase was shown to be an efficient way to decouple cytochrome c from the membrane surface (Loidl-Stahlhofen et al. 1996).

An often discussed and still controversial question is about a possible partial penetration of cytochrome c into the hydrophobic interior of the monolayer. As will be shown below, the neutron reflection method is very sensitive to the penetration of hydrophobic anchors of proteins into the monolayer as this process causes changes of the scattering length density of the monolayer. The most sensitive contrast to detect such changes is a fully deuterated lipid monolayer and non-deuterated proteins in a subphase of contrast matched water (CMA). Although our film balance measurements show a slight increase of the lateral pressure (up to 5 mN/m) after the injection of cytochrome c into the subphase in both s and LE phase, the analysis of the reflectivity data does not indicate any significant penetration of the protein into the monolayer. We have to conclude that for cytochrome c a partial penetration into the s-phase monolayer is rather unlikely. In the LE phase, there might be a certain low probability of fluctuations of parts of the protein into the monolayer interior, but it is certainly not an important mechanism in terms of protein coupling .

The cytochrome c measurements gave us significant clues about the detectability condition of proteins with the neutron reflection method and about the importance of the monolayer phase state for the degree of electrostatic protein coupling. These findings proved to be very useful for the interpretation of neutron reflection data obtained for a much more complex protein which includes a hydrophobic interaction component and is reported in the next section.

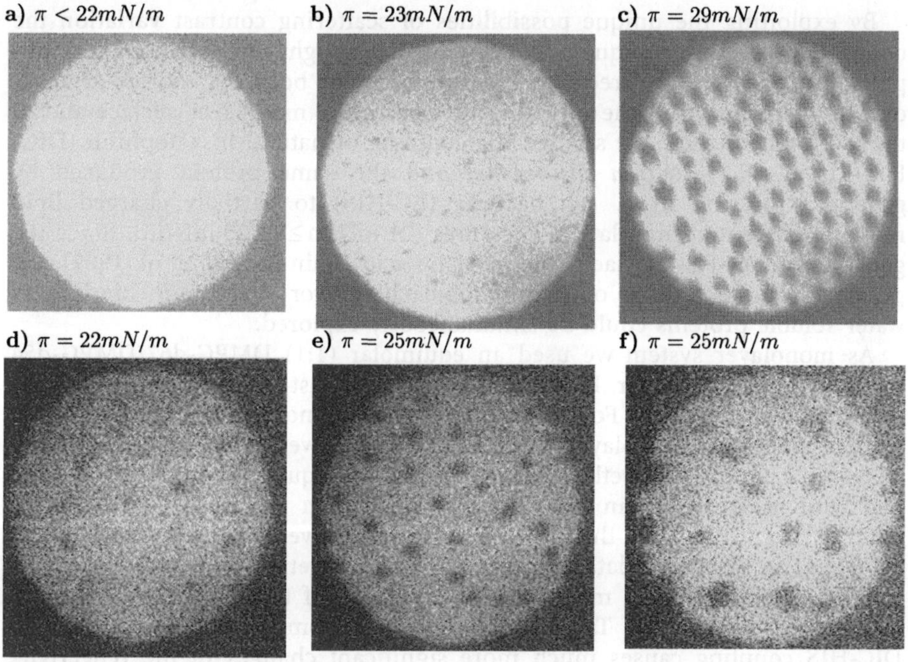

Fig. 7.6. Fluorescence film balance frames (frame diameter ≈100 μm) of an equimolar DPPC/DMPG mixture at different lateral pressures π in the absence (**a–c**) and in the presence (**d–f**) of cytochrome *c* in the subphase. Frame **f** was obtained by compressing the layer up to 25 mN/m and equilibration at this pressure for 30 min

7.3.2 The Function of a Hydrophobic Anchor for Protein Coupling: The Actin-Coupling Protein Hisactophilin

For this study, we selected the protein hisactophilin which is involved in mediating the binding between the cell membrane and its cytoskeleton (Kaufmann et al. 1992; Dietrich et al. 1993; Behrisch et al. 1995). This binding mechanism of the actin-based cytoskeleton to membranes is still widely unknown. Hisactophilin is a small, roughly cylindrical polypeptide of 13.5 kDa exhibiting at one face three histidine-rich loops while the opposite face exhibits both C- and N-terminal ends. A myristic acid chain is coupled to the N-terminus on the top of a beta-barrel structure (Hanakam et al. 1995). The histidine-rich face provides the binding site for actin while the fatty acid is supposed to facilitate the membrane anchoring. It has been shown previously by film balance and microfluorescence techniques (Behrisch et al. 1995) that the membrane binding of this protein is dominated by electrostatic interaction but that the protein shows additionally a partial penetration into the semipolar monolayer surface.

Moreover, it was demonstrated that the fatty acid chain is required for the functional orientation of the actin monomer binding protein.

By exploiting the unique possibilities of scattering contrast variation the neutron reflection technique may allow further insight into this complex coupling process by high precision measurements of both the molecular mass density distribution in the direction normal to the monolayer surface and of the layer thicknesses. We studied the coupling of natural hisactophilin (DIC-HIS) from *Dictyostelium discoideum* and the same protein produced by genetic expression in *E. coli* bacteria (EC-HIS) to partially charged lipid monolayers kept at high lateral pressures (21 mN/m $\geq \pi \geq$ 25 mN/m). Since the genetic product EC-HIS lacks the myristic acid chain (Scheel et al. 1989), the possible functional role of hydrophobic chains for membrane binding of water soluble proteins could be simultaneously explored.

As monolayer system we used an equimolar (1:1) DMPC-d67/DMPG-d54 on a subphase of either D_2O or on water contrast-matched to air (CMA) (Naumann et al. 1996). For testing the effect of monolayer charge on coupling, zwitterionic monolayers of pure DMPC-d67 were also studied.

As an example, the reflectivity curves of the equimolar mixture without and with hisactophilin in the subphase are shown in Fig. 7.7 for both the natural (Fig. 7.7 A) and the genetically modified version (Fig. 7.7 B) of the protein. The results of data analysis of Fig. 7.7 together with those obtained for the pure DMPC-d67 monolayer on the basis of two- and three-layer fitting models are listed in Table 7.2. It is obvious from the data in Fig. 7.7 that DIC-HIS coupling causes much more significant changes in the reflectivity curve of the equimolar mixture than the genetic product EC-HIS. According to Table 7.2, the scattering length density of the lipid head group for CMA-buffer is decreased by about 30% (from $\rho_2 = 3.5 \pm 0.2 \times 10^{-6}$ Å^{-2} to $\rho_2 = 2.5 \pm 0.2 \times 10^{-6}$ Å^{-2}). This strongly suggests that the natural protein DIC-HIS penetrates with amino acid side chains into the semipolar region of the lipid monolayer. A remarkable result is that this interaction mechanism is similar for charged and non-charged (pure DMPC-d67) lipid layers. Additionally, our experimental results indicate a partial penetration of the myristic chain of DIC-HIS into the hydrocarbon interior of the lipid layer although the degree of penetration is astonishingly small. The protein coverage of the monolayer with DIC-HIS was determined to be 18% with an area per DIC-HIS molecule of 5956 Å^2.

In a further series of neutron reflection experiments on CMA and D_2O subphases, we tested the ability of the natural DIC-HIS to mediate the coupling of actin to the charged monolayer. For that purpose monomeric actin (42 kDa) was added to the subphase after equilibration of the hisactophilin binding to the monolayer. Experiments were performed at concentrations of DIC-HIS of 1.5 mg/ml (0.11 μM) and actin of 3.75 μg/ml (0.09 μM). In order to avoid polymerization of actin, Ca^{2+} in the subphase was sequestered by EGTA. The addition of actin causes a significant reduction of the reflectivity of the curve shown in Fig. 7.7A for momentum transfers at $Q_{z,0} \leq 0.07$ Å^{-1} (data not shown) and a data analysis shows that the thickness of the protein layer (d_2 in Table 7.2) increases from 25±2 to 63±2 Å in the presence of actin while the scattering length density of this layer remains constant. This increase of thickness by about 40 Å agrees well with the dimensions of an actin monomer, leading to the conclusion that actin forms a single, monomole-

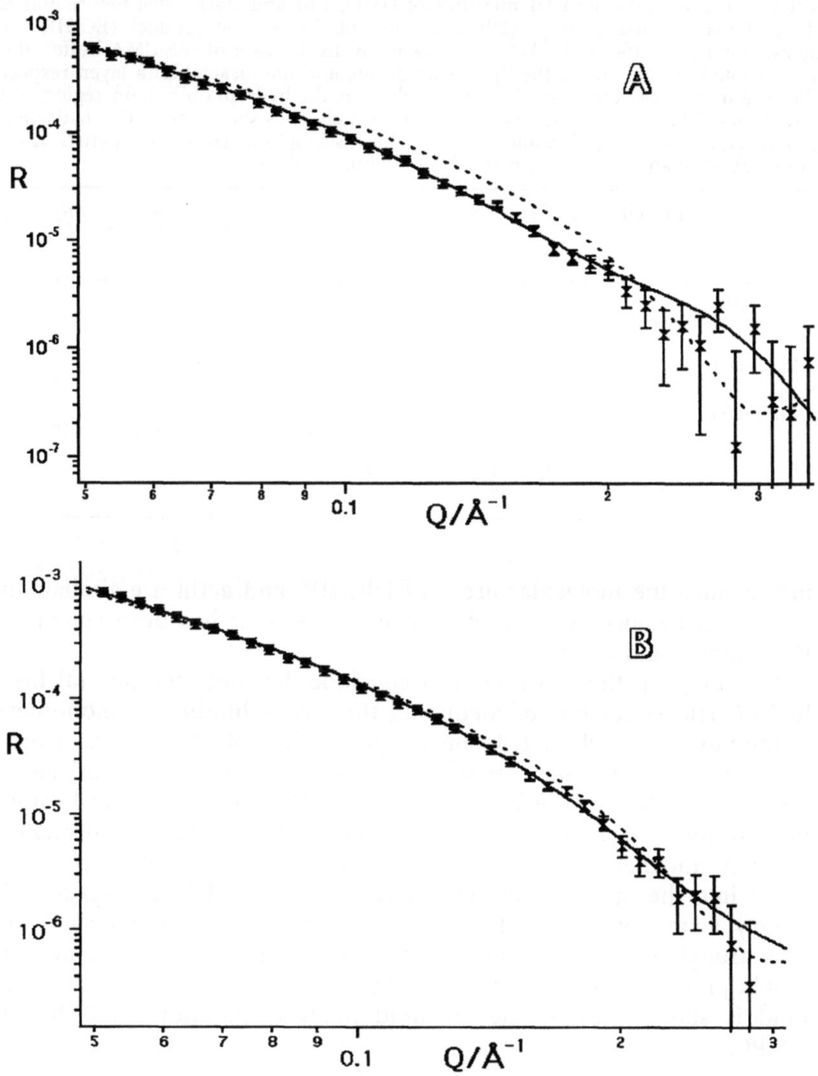

Fig. 7.7. Plot of reflectivity $R(Q)$ versus momentum transfer Q (*marker*) and fitting curve (*solid line*) for 1:1 mixture of DMPC-d67/DMPG-d54 in presence of 0.11 mM of **A** natural hisactophilin (DIC-HIS) and **B** fatty acid deficient hisactophilin (EC-HIC) on CMA-buffer at pH 6. For comparison the fitting curve obtained for the pure lipid monolayer (*dashed line*) is shown. The layer thicknesses d_j and scattering length densities r_j corresponding to the fitting curves are summarized in Table 7.2

cular layer. This indicates that indeed a specific binding of actin and not a mere accumulation of this protein in the vicinity of the monolayer takes place. The coverage of the monolayer with actin was determined to be 18% and thus very similar to that found for DIC-HIS. This suggests that about 65% of the membrane bound DIC-HIS is coupled to actin in a 1:1 stoichio-

Table 7.2. Comparison of fitted parameters of layer thicknesses d_i and scattering length densities ρ_i of DMPC-d67 and 1:1 mixtures of DMPC-d67 and DMPG-d54 before and after binding of natural hisactophilin (DIC-HIS) and of the genetic product (EC-HIS). Data are shown for D_2O-buffer and CMA-buffer at pH 6. In the case of the CMA buffer, the indices 1–3 denote the lipid chain, the lipid head group, and the hisactophilin layer, respectively. In the case of the D_2O contrast, the index 1 denotes the hydrophobic chain region of the lipid monolayer. The index 2 denotes the layer which consists of both the hydrophilic head group region of the lipid monolayer and the hisactophilin layer. The statistical errors obtained by χ^2 analysis are $< \pm 2$ Å for d_j and $< \pm 0.1 \times 10^{-6}$ Å^{-2}

System	Protein	Buffer	Number of layers	d_1 (Å)	ρ_1 ($\cdot 10^{-6}$ Å^{-2})	d_2 (Å)	ρ_2 ($\cdot 10^{-6}$ Å^{-2})	d_3 (Å)	ρ_3 ($\cdot 10^{-6}$ Å^{-2})
DMPC-d67	–	CMA	2	15	6.4	10.9	3.2	–	–
	DIC-HIS		3	15	6.3	10.9	2.9	12	0.15
DMPC-d67	–		2	14	5.5	8.8	3.5	–	–
+	DIC-HIS		3	14	5.4	8.8	2.5	12	1.1
DMPG-d54	–		2	14	6.0	9.9	3.3	–	–
	EC-HIS		3	14	6.0	9.9	3.3	10	0.8
	–	D_2O	1	14	5.5	–	–	–	–
	DIC-HIS		2	14	5.4	25	5.9	–	–

metry since the molecular areas of DIC-HIS and actin are similar. In a previous film balance study (Behrisch et al. 1995) it has been shown that EC-HIS cannot bind actin.

Summing up the result we can conclude that only the natural hisactophilin DIC-HIS is capable of mediating the strong binding of monomeric actin to the monolayer where it forms a layer of about 40 Å thickness, corresponding to the average diameter of actin monomers. Both DIC-HIS and EC-HIS bind tightly to negatively charged 1:1 DMPC-d67/DMPG-d54 monolayers, thereby forming a thin and most probably monomolecular protein layer of 12–15 Å thickness, but only the natural protein (DIC-HIS) partially penetrates into the lipid monolayer whereas the chain-deficient species (EC-HIS) merely forms an adsorbed layer. The coverage of the monolayer with DIC-HIS strongly depends on the presence of anionic DMPG in the monolayer. At a bulk protein concentration of 1.5 µg/ml the molar ratio of bound protein to lipid is about 1:45 for the 1:1 lipid mixture but only 1:420 for the pure DMPC.

7.3.3 Conclusions for the Application of Neutron Reflection in Biology

The above examples demonstrate that neutron reflection provides a powerful tool for detailed studies of protein–membrane interaction and protein–protein recognition at membranes. In particular, it can be applied to the large class of non-integral but membrane-associated proteins, which couple to lipid layers by combination of electrostatic or/and hydrophobic forces (in presence of fatty acid chains). These include phospholipases, kinases, G-proteins or cell contact site proteins of *Dictyostelium* cells. Since many of these proteins are only available in small quantities the high sensitivity of the present technique is a great advantage. Future applications of the technique may include additionally in-plane neutron scattering which allows the study of the distribution of proteins within the plane of the membrane.

References

Bayerl TM, Thomas RK, Penfold J, Rennie AR, Sackmann E (1990) Specular reflection of neutrons at phospholipid monolayers. Changes of monolayer structure and head group hydration at the transition from the expanded to the condensed phase state. Biophys J 57:1095–1098

Behrisch A, Dietrich Ch, Noegel AA, Schleicher M, Sackmann E (1995) Actin binding protein hisactophilin binds to partially charged membranes and mediates actin coupling to membranes. Biochemistry 34:15182–15190

Born M, Wolf E (1970) Principles of optics. Pergamon Press, Oxford

Brumm T, Naumann C, Sackmann E, Rennie AR, Thomas RK, Kanellas D, Penfold J, Bayerl TM (1994) Conformational changes of the lecithin head group in monolayers at the air/water interface. A neutron reflection study. Eur Biophys J 23:289–296

Dietrich C, Goldmann WH, Sackmann E, Isenberg G (1993) Interaction of NBD – talin with lipid monolayers, a film balance study. FEBS Lett 324/1:37–40

Habazettl J, Gondol D, Wiltscheck R, Otlewski J, Schleicher M, T. Holak A (1992) Structure of hisactophilin is similar to interleukin-1b and fibroblast growth factor. Nature 359:855–858

Hanakam F, Eckerskorn C, Lottspeich F, Müller-Taubenberger A, Schäfer W, Gerisch G (1995) The pH-sensitive actin-binding protein hisactophilin of *Dictyostelium* exists in two isoforms which both are myristylated and distributed between plasma membrane and cytoplasm. J Biol Chem 270:596–602

Johnson SJ, Bayerl TM, Weihan W, Noack H, Penfold J, Thomas RK, Kanellas D, Rennie AR, Sackmann E (1991) Coupling of spectrin and polylysine to phospholipid monolayers studied by specular reflection of neutrons. Biophys J 60:1017–1025

Kaufmann S, Käs J, Goldmann WH, Sackmann E,. Isenberg G (1992) Talin anchors and nucleates actin filaments at lipid membranes: a direct demonstration. FEBS Lett 314:203–205

Loidl-Stahlhofen A, Kaufmann S, Brumm T, Ulrich AS, Bayerl TM (1996) The thermodynamic control of protein binding to lipid bilayers for protein chromatography. Nature Biotechnol 14:999–1002

Naumann C, Dietrich C, Lu JR, Thomas RK, Rennie AR, Penfold J, Bayerl TM (1994) Structure of mixed monolayers of DPPC and polyethylene glycol monododecyl ether at the air/water interface determined by neutron reflection and film balance techniques. Langmuir 10:1919–1925

Naumann C, Brumm T, Rennie AR, Penfold J, Bayerl TM (1995) Hydration of DPPC monolayers at the air/water interface and its modulation by the nonionic surfactant $C_{12}E_4$: a neutron reflection study. Langmuir 11:3948–3952

Naumann C, Dietrich C, Berisch A, Bayerl T, Schleicher M, Bucknall D, Sackmann E (1996) Hisactophilin-mediated binding of actin to lipid lammelae: a neutron reflectivity study of protein membrane coupling. Biophys J 71:811–823

Penfold J, Thomas RK (1990) The apllication of the specular reflection of neutrons to the study of surfaces and interfaces. J Phys: Condensed Matter 2:1369

Russel TP (1990) X-ray and neutron reflectivity for the investigation of polymers. Material Sci Rep 5:171–271

Scheel J, Ziegelbauer K, Kupke Th, Humbel BM, Noegel AA, Gerisch G, Schleicher M (1989) Hisactophilin, a distidine-rich actin-binding protein from *Dictyostelium discoideum.* J Biol Chem 264(5):2832–2839

Vaknin D, Kjaer K, Als-Nielsen J, Lösche M (1991a) Structural properties of phosphatidylcholine in a monolayer at the air/water interface. Biophys J 59:1325–1332

Vaknin D, Als-Nielsen J, Piepenstock M, Lösche M (1991b) Recognition processes at a functionalized lipid surface observed with molecular resolution. Biophys J 60:1545–1552

Vist MR, Davis JH (1990) Phase equilibria of cholesterol/DPPC mixtures: 2H-NMR and DSC. Biochemistry 29:451–464

CHAPTER 8

The Study of Fast Reactions by the Stopped Flow Method

Wolfgang H. Goldmann[1] · Zeno Guttenberg[1] · Robert M. Ezzell[1]
Gerhard Isenberg[2*]

Contents

[1] Surgery Research Laboratories, Massachusetts General Hospital, Dept. of Surgery, Harvard Medical School, Charlestown, Massachusetts 02129, USA
[2] Technical University of Munich, Department of Biophysics E22, 85747 Garching, Germany
[*] Corresponding address: Technical University of Munich, Department of Biophysics E22, 85747 Garching, Germany, Fax: +49 089 289 12469

8.1 Summary

The mechanisms involved in protein reactions may be elucidated by investigating the reaction kinetics. There are two investigative approaches in ascertaining the combination of elementary steps which constitute these mechanisms: (1) steady state kinetics permits analysis of the overall reaction in which protein substrates are converted into products without examining the protein molecule itself; and (2) transient kinetics allows the direct measurement of each component in the overall reaction. In this latter case, attention is focused on changes occurring in the molecule upon binding to another protein whereas the study of the equilibrium (steady state) reaction does not examine the protein molecule directly. Equilibrium studies have wider applicability since they usually require only a small amount of protein and do not involve the use of special equipment. Necessarily, the information obtained is indirect and often ambiguous. Although transient kinetics requires special techniques for measuring the rates of fast reactions in solutions, it provides information which is far more direct and useful for elucidating complicated mechanisms of reactions. Thus, the two approaches are complementary and both are indispensable for the study of protein reactions.

Advances in measurement techniques (AFM, NMR, FTIR, FRAP; see various chapters in this book) have permitted the interpretation of protein structures at atomic resolution, enabling an examination of the reaction mechanism. Concurrently with such studies, developments have been made in techniques for kinetic studies of rapid protein reactions. The stopped flow method is now commercially available and commonly used in studies on the transient kinetics of protein reactions.

Steady state kinetic analysis has been systematized and widely used for multisubstrate reactions, and several studies have been published on this subject (Gutfreund 1972, 1995; Hiromi 1979; and general references). On the other hand, the study of transient kinetics of protein reactions is still an emerging field (Eccleston 1987; Goldmann and Geeves 1991). The objective of this chapter is to present the basic procedures used in transient kinetics with the goal of motivating the use of this technique in the field of cell and molecular biology.

8.2 Introduction to Protein Kinetics

This chapter concentrates on transient kinetics by the stopped flow method. The principles of steady state kinetics and the differences and significance of the two approaches will also be discussed.

Steady state kinetics is a powerful tool for distinguishing mechanisms, especially for multiprotein systems, as it may be used to demonstrate the preferred protein–protein binding sequence as well as the protein–protein dissociating sequence. Thus, this technique is limited in a reaction scheme in which protein molecules are chemically altered. Unfortunately, steady state kinetics yields no information regarding unimolecular processes, i.e. the

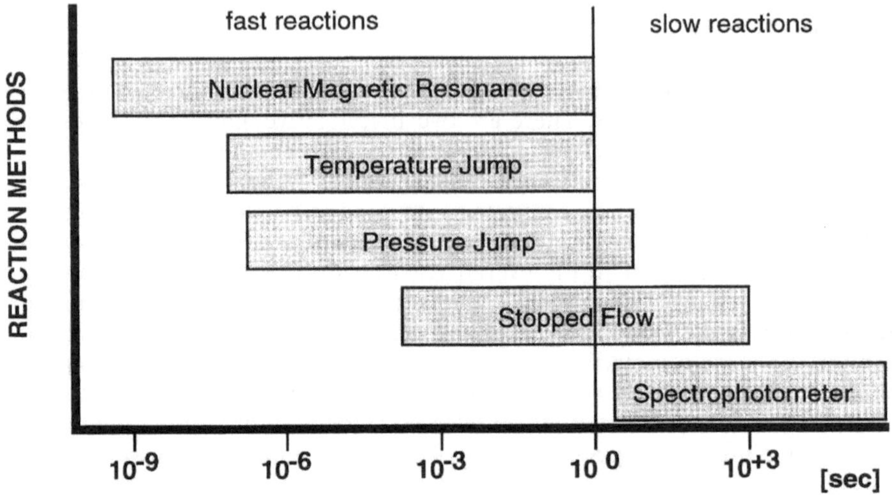

Fig. 8.1. Time ranges of fast reactions in solution and methods of study

"isomerization of complexes". The limitation of steady state kinetics resides in being unable to clarify the unique reaction mechanism amongst a plethora of possibilities which lead to the same overall rate equation. This may be removed by studying partial reactions or elementary steps directly instead of the overall reaction. For example, the spectral features of the protein–protein species involved in the isomerization process can be investigated and elementary steps can be determined. Transient kinetics, however, is a powerful tool as it permits observation of changes occurring in the molecule itself to clarify the elementary steps of the protein–protein reaction.

Techniques, which measure the rates of fast reactions in solution, are essential for transient kinetics since these steps usually occur within a few seconds. This can be accomplished by using the stopped flow method (Fig. 8.1). It is also important to examine the conformational change occurring in proteins during the reaction. The change in absorbance is most conveniently observed. Therefore, proteins having groups which absorb at specific wavelengths are best suited for transient kinetic studies. Over the years, devices have been developed which use fluorescence for detecting protein–protein interactions. Moreover, instrumentation and experimental techniques continue to be developed permitting studies with protein concentrations of as low as 10 nM. In this chapter, the use of the stopped flow method for determining protein reaction mechanisms is described.

8.3 A Manual to Measure Fast Reactions in Solution by the Stopped Flow Method

The stopped flow method involves three main components: (1) mixing two solutions rapidly to commence the reaction; (2) instantaneously stopping the two solution streams; (3) recording changes associated with the reaction in the observation cell. Despite the simplicity of these components, there are a few basic considerations when measuring rapid protein interactions correctly with a high degree of sensitivity:

1. The mixing of the solutions must be completed before detection commences. The efficiency of mixing depends on the structure of the mixer, the flow rate and viscosity of the solution.
2. The dead time (t_d) of the apparatus must be known, during which time the reaction cannot be observed. It is usually between 0.5–1.5 ms depending on flow velocity and dead volume (v_d) of the cell.
3. The rate constant of the reaction depends on the slit width of the monochromator and maximal signal to noise ratio (S/N).

The stopped flow device can be divided into two main parts: the fluid handling system and the detection system. The fluid handling system for a piston-driven apparatus SF 61 (Hi-Tech-Scientific, Salisbury. UK) is shown schematically in Fig. 8.2.

The two solutions in the reservoir syringes (B_1) are introduced through a valve into the driving syringes of 1 ml capacity each (B_2). The driving syringes are filled by manually drawing back the block (A). A valve is then turned to connect the syringes with the mixer (C) and the driving block (A) is moved by air pressure (4–8 atm). The two solutions are rapidly combined in the mixer, flow into the observation/reaction cell (D) and are finally displaced into waste (E_1). The flow is rapidly stopped by arresting the movement of the block (E_2) and activating the switch (F) for data collection. The observation/reaction cell has optical paths of 1.5 mm and 10 mm which can be used for absorbance, transmission or light scattering measurements. The

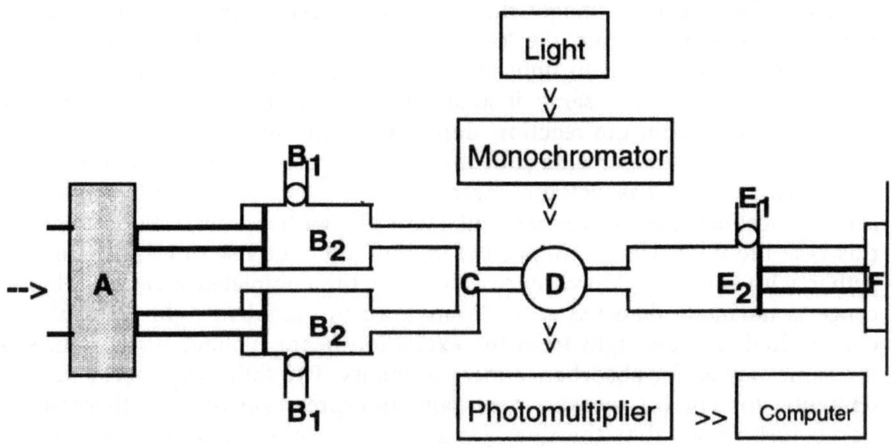

Fig. 8.2. Schematic diagram of the fluid handling system and detection system for the piston-driven SF 61 stopped flow apparatus (Hi-Tech-Scientific, Salisbury, UK): *A* pushing block; B_1 reservoir syringes; B_2 driving syringes; *C* mixer; *D* observation/reaction cell; E_1 waste; E_2 stopping block; *F* switch

dead volume (v_d) of the optical cell is 22.5 µl, and the minimum volume of the solutions needed for a single run is ≥100 µl.

The detection system consists of a light source, a monochromator, a photomultiplier, and an amplifier as well as a recording system. The light source is usually a 20 W deuterium lamp for the UV spectrum and a 75 W xenon lamp for the visible spectrum or a 100 W mercury lamp for selected peaks within the ultraviolet and visible spectrum. The reaction in the observation/reaction cell is recorded either by the change in transmittance or absorbance. The signal change detected by the photomultiplier is directly captured in digital memory after analog–digital conversion at a high sampling rate. This enables data analysis immediately after recording.

Prior to stopped flow experiments the following procedure should be adopted:

1. The magnitude of changes in e.g., absorbance, should be measured statically in a spectrophotometer. This allows the operator to set optimal detection ranges to carry out reliable kinetic measurements. A good method for testing the instrument is to mix water against water. The signal should be flat; any disturbance caused by turbulence, air bubble formation, mechanical vibration, cavitation etc. must be eliminated by optimizing the experimental protocol.
2. The reservoir syringes should be cleaned and filled with sample solutions. Three volumes of ~100 µl each should be pushed through the system to ensure equilibration of mixer and cell with the reaction solution.
3. The wavelength and slit width on the monochromator, voltage offset to compensate the voltage output and voltage of the photomultiplier should be set. The absorbance should be ideally <1 OD; the output of the photomultiplier increases with increasing slit width or voltage of the photomultiplier; and the S/N ratio should be as high as possible.

To detect fluorescence changes the total fluorescence intensity of the reaction mixture is measured from the 'on-off' voltage difference by using a shutter which cuts off the incident light. The change in fluorescence intensity due to the reaction under the same instrumental conditions can then be read in voltage directly from the reaction signal. The ratio of the 'on/off' voltage difference indicates the percentage of the total fluorescence intensity of the reaction mixture that has actually been observed as the reaction signal. The same procedure can be carried out statically with a spectrofluorometer. In this case the fluorescence change before and after the reaction is obtained together with the total fluorescence change of the completed reaction. Fluorescence is normally detected at a 90° angle to the incident light via a filter to cut out higher order light from the excitation beam. Voltage offset is used in the same way as for absorbance measurements. The following checks are fundamental for correct measurements and interpretation of data: there should be no turbulence after the flow has stopped; unsynchronized flow stopping; temperature differences between the cell and the solution; turbidity and dust; saturation of the photomultiplier; residual effects; and the optical density of the solutions should be the same.

8.4 Analysis of Fast Protein–Protein Reactions

Changes in optical properties of proteins are categorized as changes in the absorption or fluorescence spectrum. The differences in absorption when measuring light scattering is detected at a 90° angle from the incident light. The intensity of the light scatter signal increases with a decrease in wavelength ($1/\lambda^4$). Light scattering is normally measured at 350–380 nm to avoid the specific protein absorption at ~280 nm. The changes in light scatter are of the order of 0.5–2%. Since the fluorescence emission signal of the protein is more sensitive to environmental changes than the absorption, the relative change in fluorescence is larger than the relative change in absorption for the protein undergoing reaction. Therefore, fluorescent labels bound to proteins are used in stopped flow experiments. Certain fluorophores, which are attached to a protein covalently, i.e. pyrene, NBD etc. have been used to act as probes for reporting changes (Detmers et at. 1981; Kouyama and Mihashi 1981). When a protein which contains the indicator is mixed together with another protein in a stopped flow apparatus, fluorescence quenching upon protein–protein binding will normally occur. Examples of stopped flow traces of actin binding to filamin, alpha-actinin, vinculin and talin detected by light scatter are shown in Fig. 8.3 (see pp. 166, 167).

In the study of the transient phase of protein–protein reactions, experimental data sampled by fast reaction techniques are in most cases presented by a single exponential curve for the first order reaction, or by the superposition of more than one exponential curve. Analysis of these curves permits the apparent first order rate constant (k_{app}) or the relaxation time (τ), which is the reciprocal of k_{app}, for each curve to be evaluated. Generally, the rate constant is a function of the concentration of the reacting proteins. Determining the dependence of the rate constants upon protein concentration permits both the elucidation of the reaction mechanism in terms of its elementary step and the determination of the rate constants for each of these steps. The procedure adopted is essentially similar to that employed in steady state kinetics, in which the initial rate, instead of the rate constant, is used as an index of the rate of reaction. Convenient indices for the rate of reaction are the 'initial rate' in steady state kinetics, and the 'apparent first order rate constant' or 'relaxation time' in transient kinetics. The fundamental procedures for data treatment for relaxation methods of large perturbation, i.e. stopped flow, will be considered for (1) single step reactions and (2) two-step reactions in the Appendix.

8.5 Interpretation of Kinetic Data

As mentioned in the preceding sections, the first steps in protein kinetics are: to select some plausible kinetic schemes which represents the binding mechanism; and to show that the derived rate equations account for the experimentally observed kinetic behavior, i.e. the concentration dependence of relaxation times. The next important step is the molecular interpretation of

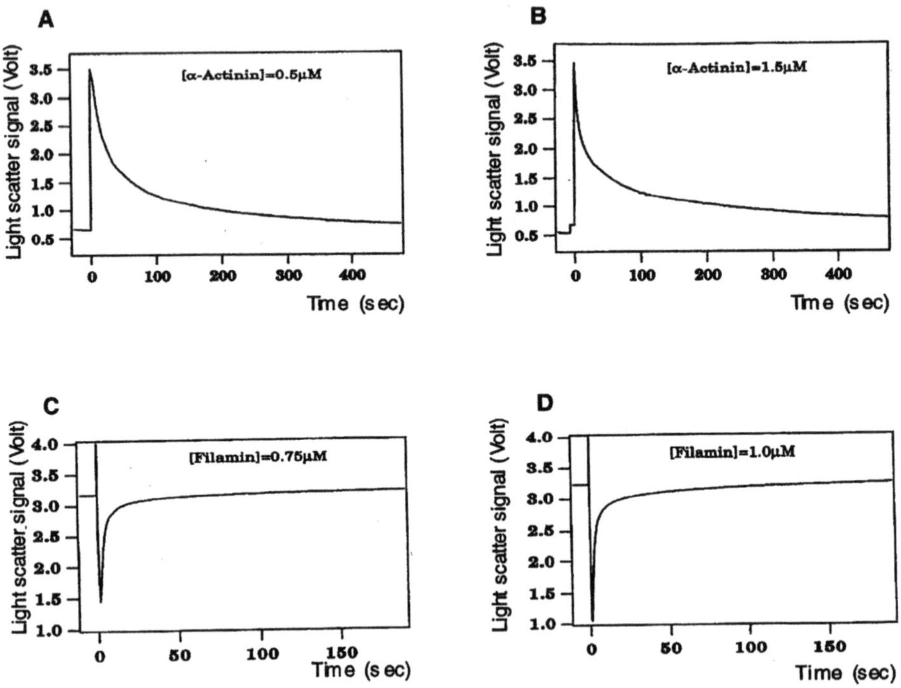

Fig. 8.3 A–H. Light scattering measurements at 90° angle to a 355 nm wavelength incident beam. All traces represent the average of five consecutive measurements in the stopped flow apparatus. Best fit to a double exponential is shown *superimposed*. A pretrigger was used to indicate the time of flow stopping and start of reaction (t=0). The observed reaction was the binding of F-actin (3 μM) to the following proteins: **A–B** Alpha-actinin. Buffer conditions: 100 mM KCl, 2 mM $MgCl_2$ 2 mM Tris HCl, pH 7.5, 0.2 mM $CaCl_2$, 0.5 mM ATP, 0.005% NaN_3; T=7 °C. Observed rates: **A** $k_1=0.06$ (s^{-1}); $k_3=0.005$ (s^{-1}); **B** $k_1=0.1$ (s^{-1}); $k_3=0.011$ (s^{-1}). **C–D** Filamin. Buffer conditions: 100 mM KCl, 2 mM $MgCl_2$, 2 mM Tris HCl, pH 7.5, 0.2 mM $CaCl_2$, 0.5 mM ATP, 0.005% NaN_3; T=15 °C. Observed rates: **C** $k_1=0.1$ (s^{-1}); $k_3=0.017$ (s^{-1}); **D** $k_1=0.18$ (s^{-1}); $k_3=0.017$ (s^{-1}).

the kinetic scheme, or characterization of the mechanisms at the molecular level. A careful analysis of kinetic data from various viewpoints provides useful information or indication regarding the molecular mechanisms of reactions. To gain a deeper understanding of molecular mechanism some relevant points are described briefly below.

For a consecutive reaction, the apparent rate constant of the overall reaction (k_{app}) may be related to the rate constant of each individual step, and the slowest step in a reaction series is called the 'rate limiting step;. In general the rate constant of 'n' elementary steps in the main path of the reaction cannot be smaller than the overall reaction rate constant, if so then the relevant process is not in the main path of the overall reaction but in the side path. The rate constant, k in a bimolecular reaction $A+B \leftrightarrow AB$ is expressed as follows: $k = A \times e^{-E_a/RT}$ where, A is the frequency factor and concerns the frequency of 'effective collisions' which may lead to the reaction, and E_a, is the Arrhenius activation energy which corresponds to the height of the en-

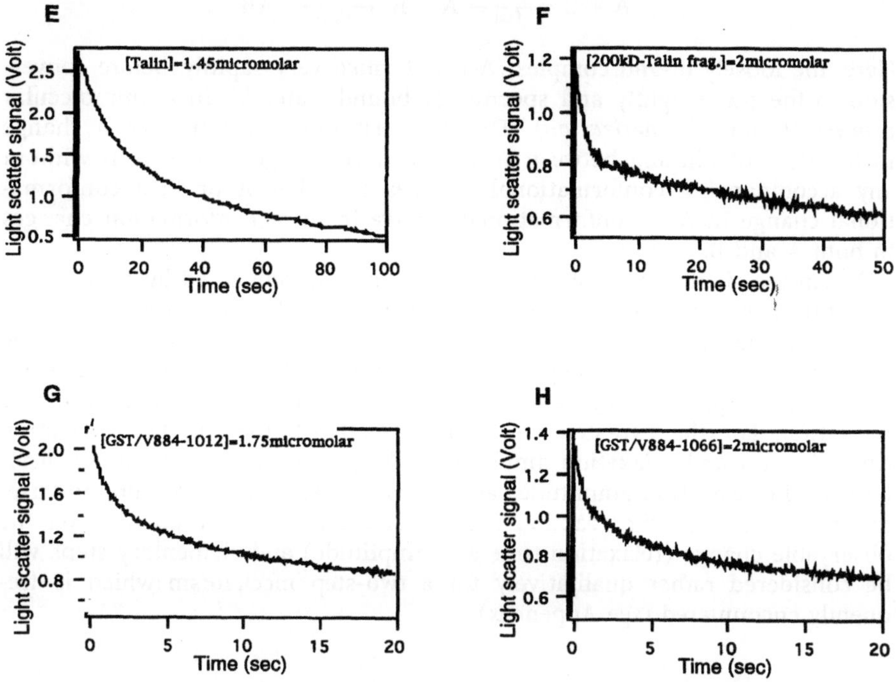

Fig. 8.3. **E** Talin. F-actin buffer: 2 mM $MgCl_2$, 10 mM Tris HCl, pH 7.0; 0.2 mM $CaCl_2$, 0.5 mM ATP, 0.2 mM DTT, 0.05 mM NaN_3; T = 20 °C. Talin buffer: 20 mM Tris HCl pH 7.0, 1 mM EGTA, 0.1 mM DTT. Observed rates: $k_1 = 0.115$ (s^{-1}); $k_2 = 0.017$ (s^{-1}). **F** 200 kDa Talin fragment. Buffer conditions same as Talin; T = 20 °C; observed rates: $k_1 = 0.74$ (s^{-1}); $k_2 = 0024$ (s^{-1}). **G** Vinculin fragment (GST/V884-1012). F-actin buffer: 100 mM KCl, 2 mM $MgCl_2$, 2 mM Tris HCl, pH 7.5, 0.2 mM $CaCl_2$, 0.5 mM ATP, 0.2 mM DTT, 0.05 mM NaN_3. Fragment buffer: 10 mM Tris HCl, pH 7.5, 150 mM NaCl, 1 mM EGTA, 0.3 mM NaN_3, 0.1 mM EDTA; T = 25 °C; observed rates: $k_1 = 1.62$ (s^{-1}); $k_3 = 0.133$ (s^{-1}). **H** Vinculin fragment (GST/V884-1066). Buffer conditions same as GST/V884-1066; T = 25 °C. Observed rates: $k_1 = 1.88$ (s^{-1}); $k_3 = 0.155$ (s^{-1})

ergy barrier to be surmounted for the reaction to occur. This equation implies that molecules which have an energy in excess of E_a, lead to the reaction. If the activation energy, E_a is zero the rate of reaction in solution will be determined simply by the frequency of effective collisions between A and B which is limited by the rate of diffusion in solution. It should be remembered that the rate constant of bimolecular association in solution cannot exceed the rate constant of a diffusion controlled reaction. Therefore, when an unreasonably high rate constant, i.e. $>10^{11}\ M^{-1} \times s^{-1}$ is obtained for a protein–protein reaction, the reaction scheme from which the rate constant was estimated should be rejected. Thus, the values of the rate constant itself could in some cases be a criterion for deciding the reaction mechanism.

Hyperbolic concentration dependence of $(1/\tau)$ is often observed in protein–protein interactions. This kinetic feature is consistent with that commonly expected for a two-step reaction mechanism in which a fast bimolecular association is followed by a slow unimolecular process (see Appendix).

$$A + B \underset{\text{fast}}{\longleftrightarrow} A - B \underset{\text{slow}}{\longleftrightarrow} AB$$

Here, the loosely bound complex, A-B is formed very rapidly, before conversion to the more tightly and specifically bound state, AB in a unimolecular process (termed *isomerization*). This isomerization could involve: a change in the state of binding between A and B or rearrangement of A–B without any accompanying conformational changes in either A or B; a conformational change in A; a conformational change in B; or conformation changes in both A and B.

When the single reversible step, A+B↔AB is accompanied by a change in the optical property of the protein the perturbation by the stopped flow will lead to a relaxation. This relaxation is accompanied by the relaxation time (τ) and a concentration change (Δ_c). The latter is observed through the molar optical property change and this change is termed the amplitude of the relaxation. For the two-step mechanism, A+B↔A–B↔AB, two relaxation times (τ_1; τ_2) and relaxation amplitudes (Δ_{AMP1}; Δ_{AMP2}) are observed. Analyses of the relaxation amplitudes as well as the relaxation times provide useful information on the reaction mechanism. The interrelation between the observable change (relaxation time and amplitude) and elementary steps will be considered rather qualitatively for a two-step mechanism which is frequently encountered (see Appendix).

8.6 Examples of Stopped Flow Analysis of Fast Reactions

This section describes the procedures for analyzing the transient kinetics in the case of fast protein reactions. Although several excellent examples are available in the literature only four protein–protein reactions including actin are considered here. The following is a brief description of these proteins (Isenberg 1995; Maciver 1995).

1. Actin is an abundant muscle and non-muscle cytoskeletal protein of ~42 kDa molecular mass. It consists of 374/375 amino acids and exists in various isoforms. The structure has been resolved at 2.8 Å resolution. Source: rabbit skeletal muscle (Pardee and Spudich 1982).
2. Filamin is a ~280 kDa homodimeric phosphoprotein that crosslinks actin filaments and binds to lipids. It has a molecular contour length of ~160 nm, consists of 2647 amino acids, and is ubiquitous in vertebrates. *Source*: turkey gizzard smooth muscle (Hartwig and Kwiatkowski 1991).
3. Alpha-actinin is a 94–103 kDa actin cross-linking and lipid-binding protein from muscle and non-muscle cells forming antiparallel homodimers of ~30–40 nm in length. Gen-Bank: X51753 (chicken), J03486 (smooth muscle), X15804 (non-muscle), Y00689 (*Dictyostelium*). Source: chicken or turkey gizzard smooth muscle (Craig et al. 1982).
4. Vinculin is a ~117 kDa actin and lipid-binding protein. A prominent linker protein of cellular junctions and focal adhesions. *Source*: chicken gizzard smooth muscle (Geiger 1979).

5. Talin is a ~270 kDa lipid- and actin-binding phosphoprotein that nucleates actin polymerization and links microfilaments to plasma membranes. It is a dumbbell-shaped homodimer of ~51 nm contour length and consists of 2541 amino acids (mouse). Gen-Bank: X56123 (mouse fibroblast). Source: human platelets (Burridge and Connell 1983).

The dependence of the overall reaction on the protein concentration enabled the evaluation of the dissociation constant (K_d) of protein–protein complexes by static measurements. The differences in absorbance or fluorescence signal were used for measuring the binding stoichiometry and dissociation constant (K_d) of a protein–protein complex. The kinetics of actin to alpha-actinin, filamin, vinculin and talin binding were studied by the stopped flow method detecting differences in absorbance. Results obtained from actin–alpha-actinin and actin–filamin binding are given in Fig. 3A–D. In both cases, a double exponential relaxation was observed. The binding of actin to full length talin and to talin fragment (200 kDa) as well as to vinculin fragment V884-1012 and V884-1066 attached to GST (Johnson and Craig 1995) is also consistent with a two-step mechanism in which a fast bimolecular association is followed by a slower unimolecular process resulting in cross-linking and/or bundling (Fig. 3E–H). The two conformational isomers A-B and AB of the general reaction scheme (A+B↔A–B↔AB) differ in their binding mode for these species. The observed relaxation time (τ_1) and (τ_2) represent the faster and the slower binding rate, respectively.

The preceding sections have shown that transient kinetics is an indispensable method for gaining a greater understanding of the mechanisms of protein–protein reactions. Thus, the applicability of transient kinetics is limited by two factors: (1) the availability of proteins and (2) the observability of the reaction under study. Fluorescence detection has advantages in both these respects. First, lower concentrations of proteins are required for fluorescence in comparison to detection of absorption. Second, the fluorescence is far more sensitive to environmental change than absorbance. In conclusion, improvements in the performance of the apparatus used for detecting fast reactions and the development of methods for monitoring reactions based on their physical properties should make transient kinetic studies of protein–protein interactions as easy as and more efficient than steady state kinetics.

Acknowledgement. This work was supported by the Deutsche Forschungsgemeinschaft and the American Cancer Society. We thank Dr. Nathan Busch for careful reading of this manuscript.

Appendix

1. *Summary of Symbols and Terms*

(AFM)	Atomic force microscopy
(NMR)	Nuclear Magnetic Resonance
(FTIR)	Fourier Transformed infrared Spectroscopy
(FRAP)	Fluorescence Recovery After Photobleaching
(NBD)	7-chloro-4-nitrobenzeno-2-oxa-1,3-diazole
(Pyrene)	N-(1-pyrenyl)iodoacetamide

($k_{app} = 1/\tau$)	Apparent first order rate constant
(τ)	Relaxation time
($K_d = k_2/k_1$)	Dissociation constant
(OD)	Optical density
(S/N)	Signal to noise ratio
(t_d)	Dead time
(v_d)	Dead volume
(Equilibrium binding studies)	Steady state kinetics
(Rapid binding studies)	Transient kinetics
(GST/V884-1012)	Vinculin fragment (amino acids 884–1012) attached to glutathione S-transferase fusion protein (GST).
(GST/V884-1066)	Vinculin fragment (amino acids 884–1066) attached to GST.
(200 kDa Talin)	Tail fragment of the talin molecule.

2. *Michaelis-Menton Equation*

$$E + S \underset{k_2}{\overset{k_1}{\longleftrightarrow}} ES \overset{k_3}{\longleftrightarrow} E + P$$

where E = enzyme; S = substrate; P = product.

- (E) associates with (S) to form an (ES) complex at a constant rate of k_1 assuming rapid equilibrium binding. The (ES) complex can now either dissociate at a constant rate of k_2 or proceed to form a product at a constant rate of k_3. The rate of the formation of (ES) = $k_1 \times [E][S]$; the rate of the breakdown of (ES) = $(k_2+k_3) \times [ES]$ and the *Michaelis constant*, $K_M = k_2+k_3/k_1$. In a steady state the concentrations of the [ES] stay the same whilst the concentrations of [E] and [P] are changing.

3. *Single-Step Binding*

$$A + B \underset{k_2}{\overset{k_1}{\longleftrightarrow}} AB$$

- This reaction can be expressed as follows: $1/\tau = k_1 \times ([A]+[B]) + k_2$. Plotting $1/\tau$ against ([A]+[B]) the rate constants for the forward and reverse process, k_1 and k_2 can be determined from the slope and the vertical intercept, respectively. τ is equal to the reciprocal of the apparent first-order rate constant, k_{app} and is related to t such that τ a $t_{1/2}/\ln 2$. (In a special case where [A]>>[B] the concentration of A can be neglected when calculating the rate constants). Note: $K_d = k_2/k_1$.

4. *Two-Step Binding*

$$A + B \underset{k_2}{\overset{k_1}{\longleftrightarrow}} A - B \underset{k_4}{\overset{k_3}{\longleftrightarrow}} AB$$

- The relaxation times can be expressed when the bimolecular process $A+B \leftrightarrow A-B$ is much faster than the unimolecular process $A\text{-}B \leftrightarrow AB$ or more exactly when $k_1 \times ([A]+[B]) + k_2 >> k_3+k_4$ as follows: $1/\tau_1 = k_1 \times ([A]+[B]) + k_2$ and $1/\tau_2 = k_3 \times ([A]+[B])/K_d + ([A]+[B]) + k_4$.

- In the case where the bimolecular process A+B↔A-B is much slower than the unimolecular process A-B↔AB or more exactly when $k_1 \times ([A]+[B])+k_2 << k_3+k_4$ the relaxation times can be expressed as follows: $1/\tau_1 = k_3+k_4$ and $1/\tau_2 = k_1 \times ([A]+[B])+k_2 \times k_4/k_3+k_4$. For further reading see Bernasconi (1976).

References

Bernasconi CF (1976) Relaxation kinetics. Academic Press, New York
Burridge K, Connell L (1983) A new protein of adhesion plaques and ruffling membranes. J Cell Biol 97:359–367
Craig SW, Lancashire CL, Cooper JA (1982) Methods Enzymol 85:316–335
Detmers P, Weber A, Elzinga M, Stephens RE (1981) 7-Chloro-4-nitrobenzeno-2-oxa-1,3-diazole actin as a probe for actin polymerization. J Biol Chem 256:99–104
Eccleston JF (1987) Spectrophotometry and spectrofluorimetry: a practical approach, chap 6. In: Bashford CL, Harris DA (eds) Stopped-flow spectrophotometric techniques. IRL Press, Oxford
Geiger B (1979) A 130 K protein from chicken gizzard: its localization at the termini of microfilament bundles in cultured chicken cells. Cell 18:193–205
Goldmann WH, Geeves MA (1991) A "slow" temperature jump apparatus built from a stopped-flow machine. Anal Biochem 192:55–58
Gutfreund H (1972) Enzymes: physical principles. Wiley-Interscience, London
Gutfreund H (1995) Kinetics for the Life Sciences. Cambridge University Press, New York
Hartwig JH, Kwiatkowski DJ (1991) Actin-binding proteins. Curr Opin Cell Biol 3:87–97
Hiromi K (1979) Kinetics of fast reactions. John Wiley, New York
Isenberg G (1995) Cytoskeleton proteins: a purification manual. Springer, Berlin Heidelberg New York
Johnson RP, Craig SW (1995) F-actin binding site masked by the intramolecular association of vinculin head and tail domains. Nature 371:261–264
Kouyama T, Mihashi K (1981) Fluorimetry study of N-(1-pyrenyl)-iodoacetamide labelled F-actin. Eur J Biochem 114:33–38
Maciver SK (1995) Microfilament organization and actin binding proteins. In: Hesketh JE, Pryme I (eds) The cytoskeleton, vol 1. Structure and assembly. JAI Press, Greenwich, Connecticut
Pardee JD, Spudich JA (1982) Purification of muscle actin. In: Frederiksen DW, Cunningham LW (eds) Methods in Enzymology, vol 85. Academic Press, New York, pp 164–181

General Reading

Kurstin K (ed) (1969) Methods in enzymology, vol XVI. Fast reactions. Academic Press, New York
Nordlie RC (1982) Kinetic examination of enzyme mechanisms involving branched reaction pathways. In: Purich DL (ed) Methods in enzymology, vol 87. Academic Press, New York, pp 319–548

CHAPTER 9

Biomolecular-Interaction Analysis (BIA-Technology)

A Universal Biosensor-Based Technology for Biochemical Research and Development

Francis Markey[1] · Franz Schindler[2*]

Contents

[1] Biacore AB, Uppsala, Rapsgatan 7, S-754 50 Uppsala, Sweden
[2] Biacore AB, Niederlassung Deutschland, Jechtinger Str. 8, 79111 Freiburg, Germany
[*] Corresponding address: Biacore AB, Niederlassung Deutschland, Jechtinger Str. 8, 79111 Freiburg, Germany, Tel.: +49-(0)761-4705-0, Fax: +49-(0)761-4705-14, e-mail: franz.schindler@eu.biacore.com

9.1 Introduction

BIA is a young technology in the biomolecular research field, first introduced only 6 years ago. The technology has, however, fast become established as an essential tool in many laboratories for screening applications as well as advanced research work. The ability to monitor molecular interactions in real time without the use of labels saves time and effort, and opens new horizons in the study of biological mechanisms and in the development of new pharmaceutical products.

Understanding of biological processes at the molecular level develops through a combination of the essentially separate areas of functional description and detailed structural investigation. A technique that has made major contributions to functional studies in a wide range of fields is real-time biomolecular interaction analysis (BIA), developed and marketed by Biacore AB (formerly Pharmacia Biosensor AB), Uppsala, Sweden. There are currently three instruments available for BIA from Biacore AB, ranging from the semi-automatic BIACORE X through BIACORE 1000 to BIACORE 2000 with fully automated sample handling, multi-channel analysis and sample recovery facilities.

BIA is a biosensor technology for monitoring interactions, in real time without the use of labels, between two or more molecules such as proteins and peptides, nucleic acids or pharmaceuticals, as well as interactions involving lipid vesicles, bacteria, viruses and mammalian cells. Questions that can be addressed include:

1. Which components interact, and under what conditions?
2. How fast do they bind and dissociate?
3. How strongly do they interact?
4. How much *active* interactant do I have in my sample?
5. How is the interaction influenced by effectors, cofactors, molecular modifications etc?
6. Do different components influence each other's binding to a common interactant, for example by steric, allosteric or cooperative effects?

The detection principle relies on the optical phenomenon of surface plasmon resonance (SPR), which detects changes in the refractive index of the solution close to the surface of the *sensor chip* (Jönsson and Malmqvist 1992). This is in turn directly related to the concentration of solute in the surface layer. The response resulting from an interaction is related to the change in mass concentration in the surface layer and is largely independent of the nature of the interacting species (Stenberg and Persson 1991).

BIA is a continuous flow technology which utilizes precision microfluidics for delivery of sample to the sensor surface. To perform a BIA analysis, one interactant is immobilized on the sensor chip surface, which forms one wall of a micro-flow cell. Sample containing the other interactant(s) is then injected over the surface in a controlled flow. Any change in surface concentration resulting from interaction is detected as an SPR signal, expressed in *resonance units* (RU). At the end of the injection, the sample is replaced by buf-

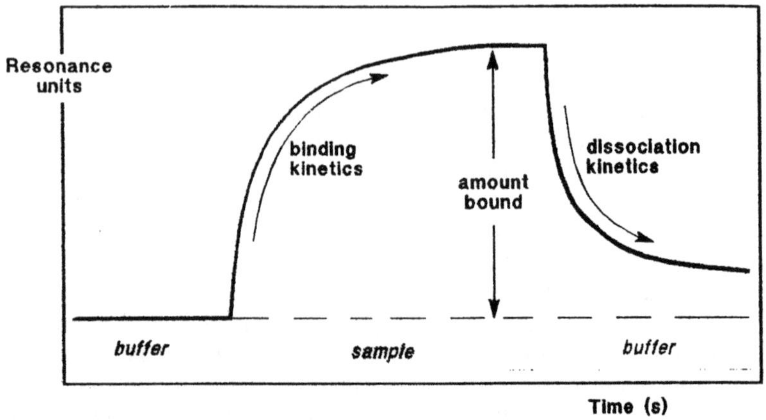

Fig. 9.1. The SPR angle is sensitive to the mass concentration of molecules close to the sensor chip surface. As this concentration changes, the SPR angle shifts and produces a response. The progress of an interaction is monitored in BIA as a sensorgram. Analyte binds to the surface-attached ligand during sample injection, resulting in an increase in signal. At the end of the injection, the sample is regulated by a continuous flow of buffer and the decrease in signal now reflects dissociation of interactant from the surface-bound complex

fer flow and the change in signal now reflects dissociation of interactant from the surface-bound complex. The continuous display of RU as a function of time, referred to as a *sensorgram*, thus provides a complete record of the progress of association and dissociation (Fig. 9.1). When analysis of one interaction cycle is complete, the surface can be regenerated by treatment with conditions that remove all bound analyte without affecting the activity of the immobilized ligand. A surface with for example immobilized antibody can typically be used for 50–100 analysis cycles depending on the stability of the antibody.

Advantages of BIA over other techniques for detecting and measuring interaction include:

1. User-controlled immobilization of the ligand gives full freedom in tailoring the specificity of the analysis.
2. Label-free detection means that practically any interactant can be studied, often without having to purify the interactant in advance.
3. Every step in a multi-step analysis protocol is monitored in the same way, providing a new dimension in experimental control information.
4. Real-time monitoring follows association and dissociation processes with a time resolution down to 0.1 s, providing a kinetic description of the interaction which is seldom available from other techniques.

To date, BIA has been applied to kinetic analysis, concentration measurement and binding site analysis in a wide range of application areas including signal transduction mechanisms, study of cell adhesion molecules, antibody characterization, nucleic acid research, protein engineering, drug design and screening. The technology is also finding increasing application in the fields of molecular biology and food and environment studies.

9.1.1 Surface Plasmon Resonance

Real-time BIA from Biacore AB is a label-free technology for monitoring biomolecular interactions as they occur. The detection principle relies on surface plasmon resonance (SPR), an optical phenomenon that arises when light illuminates thin conducting films under specific conditions. The resonance is a result of interaction between electromagnetic vectors in the incident light and free electron constellations called surface plasmons in the conductor. This phenomenon was initially investigated by Turbadar (1959) although it was the work by Otto (1968a, b) and Kretschmann and Raether (1968) that brought versatility to the technique. For reviews about SPR the reader is referred to Welford (1991) and Raether (1977).

To begin describing SPR, it is helpful to start with the phenomenon of total internal reflection (TIR). TIR occurs at an interface between media of different refractive index when light traveling from the denser to the less dense medium meets the interface at an angle above a critical angle (see Hirschfield 1967). When TIR occurs, an electromagnetic evanescent wave propagates away from the interface into the lower refractive index medium (Fig. 9.2).

If the interface is coated with a thin layer of a conducting material (gold is used in real-time BIA), the phenomenon of SPR can arise. A resonant coupling between the incident light energy and surface plasmons in the conducting film occurs at a specific angle of incident light, absorbing the light energy and causing a characteristic drop in the reflected light intensity at that angle. If the incident light is focused on the surface in a wedge, the drop in intensity at the resonance angle appears as a "shadow" in the reflected light wedge.

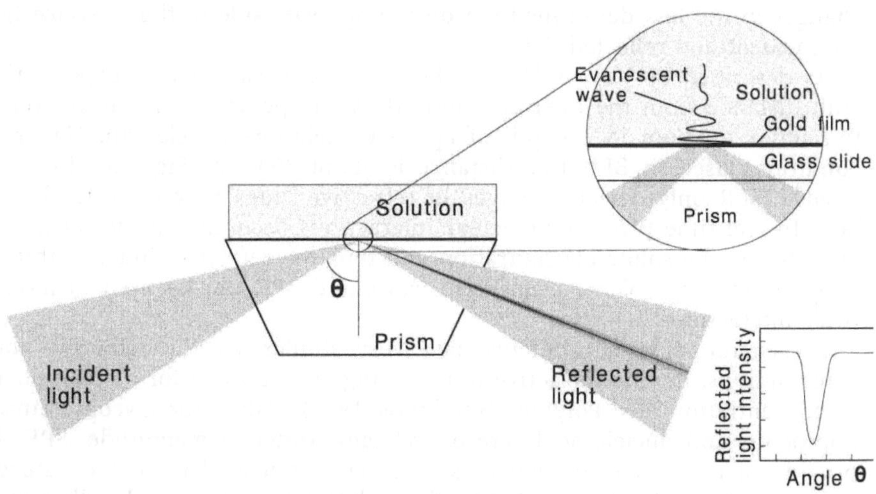

Fig. 9.2. SPR arises when light is totally internally reflected from a metal-coated interface between two media of different refractive index (a glass prism and solution in BIA). SPR is observed as a reduction in the intensity of the reflected light at a specific angle of reflection

Fig. 9.3. The evanescent wave decays exponentially with increasing distance from the reflecting interface

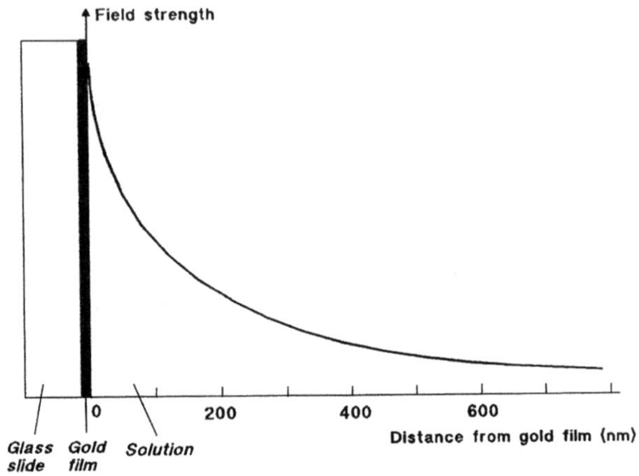

The resonance angle is sensitive to a number of factors including the wavelength of the incident light and the nature and thickness of the conducting film. Most importantly for BIA, the angle depends on the refractive index of the medium into which the evanescent wave propagates. When other factors are kept constant, the resonance angle is a direct measure of the refractive index of the medium. The dielectric equations describing this dependence and the application of the approach are discussed in detail by Kretschmann (1971), Liedberg et al. (1983) and Jönsson (1991).

To summarize the phenomenon, light reflected at an interface between media of two refractive indices separated by a thin film of conducting material will resonate at a specific angle and result in a reduction in intensity of the reflected light at that angle. The angle is very sensitive to refractive index changes in the less dense medium on the opposite side of the interface from the incident and reflected light.

As described by Kovacs (1982), the evanescent wave decays exponentially with distance from the interface, and effectively penetrates the lower refractive index medium to a depth of approximately one wavelength. Under the conditions used in BIA, this distance is about 300 nm (Fig. 9.3). In consequence, SPR only detects changes in refractive index very close to the surface. In real-time BIA, biomolecular interactions occurring at the sensor surface change the solute concentration and thus the refractive index within the penetration range of the evanescent wave, and SPR can be used to monitor these interactions.

In general, different proteins have very similar specific refractive index contributions, i.e. the refractive index change is the same for a given change in concentration (see Polymer Handbook 1989). Values for glycoproteins, lipoproteins and nucleic acids are of the same order of magnitude. SPR thus provides a mass detector which is essentially independent of the nature of the interactants. Most importantly, the technique requires no labeling of the interacting components, and the possibility of a mass detector is realized. Sensitivity considerations for SPR have been discussed by Kooyman et al.

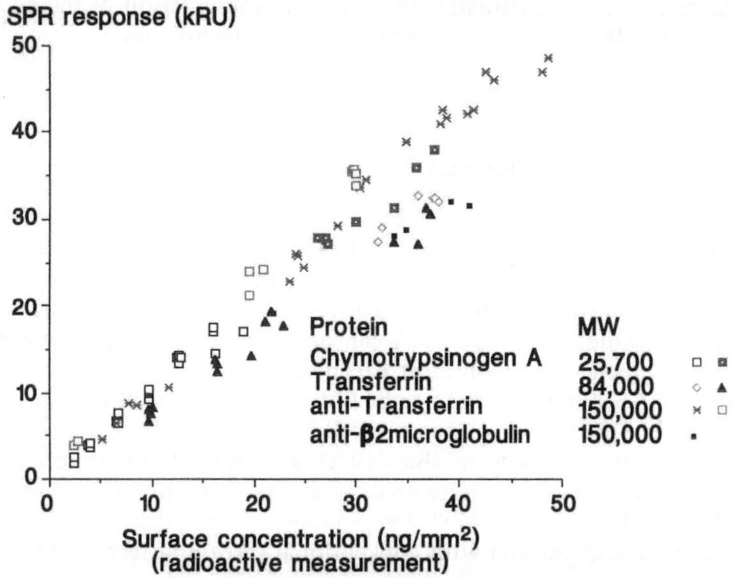

Fig. 9.4. The resonance signal obtained from BIA is a measure of the concentration of biomolecules on the surface. Shown here are absolute concentration measurements for four different proteins performed using radio labeled proteins as a comparison standard. The response is linear with the concentration on the surface. A response of 1000 RU corresponds to a change in surface concentration of 1 ng/mm^2. This correlation can be applied for proteins in general, independently of the amino acid composition

(1988), and linear correlation between resonance angle shift and protein surface concentration has been shown by Stenberg et al. (1991; Fig. 9.4).

9.1.2 Tailoring the Specificity

For BIA investigations, one interactant (generally referred to as the *ligand*) is immobilized on the sensor chip surface. It is the identity and biological activity of this ligand that determines the specificity of the interaction analysis. The ligand may be immobilized either by covalent attachment (exploiting intrinsic groups such as amine groups) or by ligand capturing, where a capturing molecule (e.g. streptavidin or a specific antibody) is immobilized and binds ligand (e.g. biotinylated nucleic acid or a specific antigen) from the solution (Johnsson and Löfås 1991). Sensor Chip CM5 is coated with a layer of carboxymethylated dextran that provides a matrix in which the ligand can be immobilized as well as preventing direct adsorption of proteins on to the gold surface and providing a hydrophilic environment for the interaction. Sensor Chip HPA provides a hydrophobic surface on which lipid monolayers can be adsorbed for studies of membrane-associated interactions.

The flexible immobilization procedure gives a high degree of freedom in designing BIA experiments. In particular when specific capturing techniques or site-directed chemistry are used, the presentation of ligand to analyte can be controlled by choice of immobilization technique.

9.1.3
Identifying Binding Partners

One major application area for BIA is screening for binding partners to target molecules of interest (e.g. hybridoma cultures for monoclonal antibody production, natural product extracts and combinatorial libraries for potential new therapeutics). Since BIA can detect binding in real time and without labeling, molecules of interest can be identified rapidly in crude preparations. Figure 9.5 illustrates the use of BIA in screening cell culture supernatants for ligands which bind to the extracellular domain of the "orphan" receptor ECK protein-tyrosine kinase (Bartley et al. 1994). With automated sample handling facilities for unattended analysis of up to 192 samples in BIACORE 1000 and BIACORE 2000, the savings in both time and labour can be dramatic in comparison with conventional screening techniques.

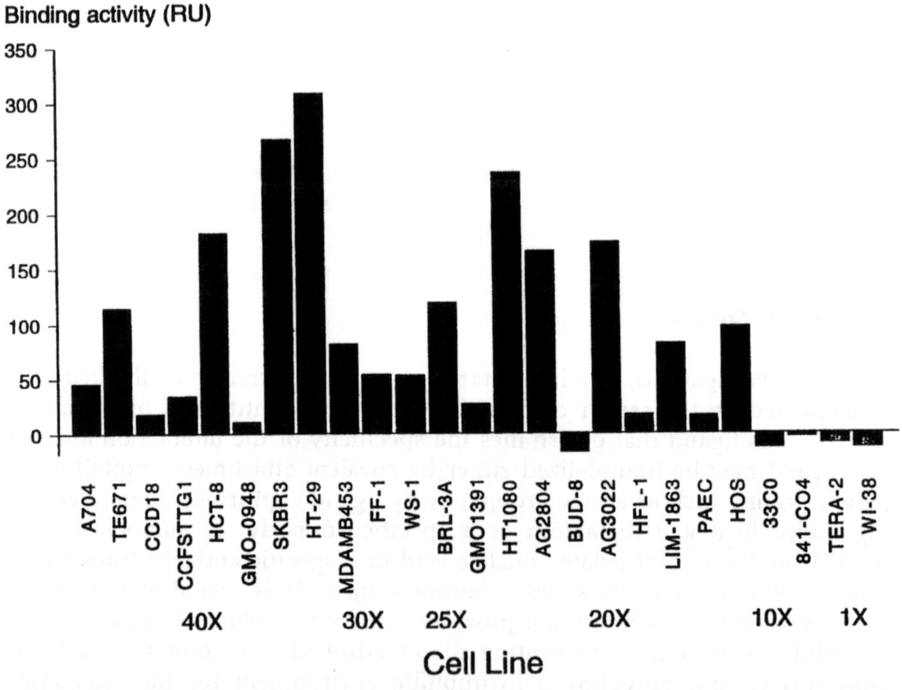

Fig. 9.5. Screening cell culture supernatants for potential ligands to the extracellular domain (ECK-X) of the "orphan" receptor ECK protein-tyrosine kinase. ECK-X is immobilized on the sensor chip surface, and binding activity in serum-depleted cell culture supernatants is measured as the response after a 25-s injection of supernatant. Supernatants were diluted between one- and 40-fold as appropriate. (Bartley et al. 1994)

9.1.4 Determining Binding Kinetics

Molecular interactions may be fast or slow, strong or weak, and the relevant question is often *how* fast or *how* strongly a component binds. A sensorgram monitors both the rates of association during sample injection and dissociation after the injection is completed, providing detailed kinetic information about the interaction.

Semi-quantitative kinetic and affinity ranking can often be established simply from the relative appearance of sensorgrams. With appropriate experimental design and data analysis methods, however, BIA can also be used for determination of apparent rate and equilibrium constants and for testing the validity of proposed interaction mechanisms (Karlsson et al. 1994).

9.1.5 Relating Structure to Function

By giving detailed information on the kinetics of interaction, BIA can probe the identity and role of residues involved in the interaction by monitoring the effect of changing single residues in the interactant. Since the interaction can be studied without labels and without rigorous purification of the inter-

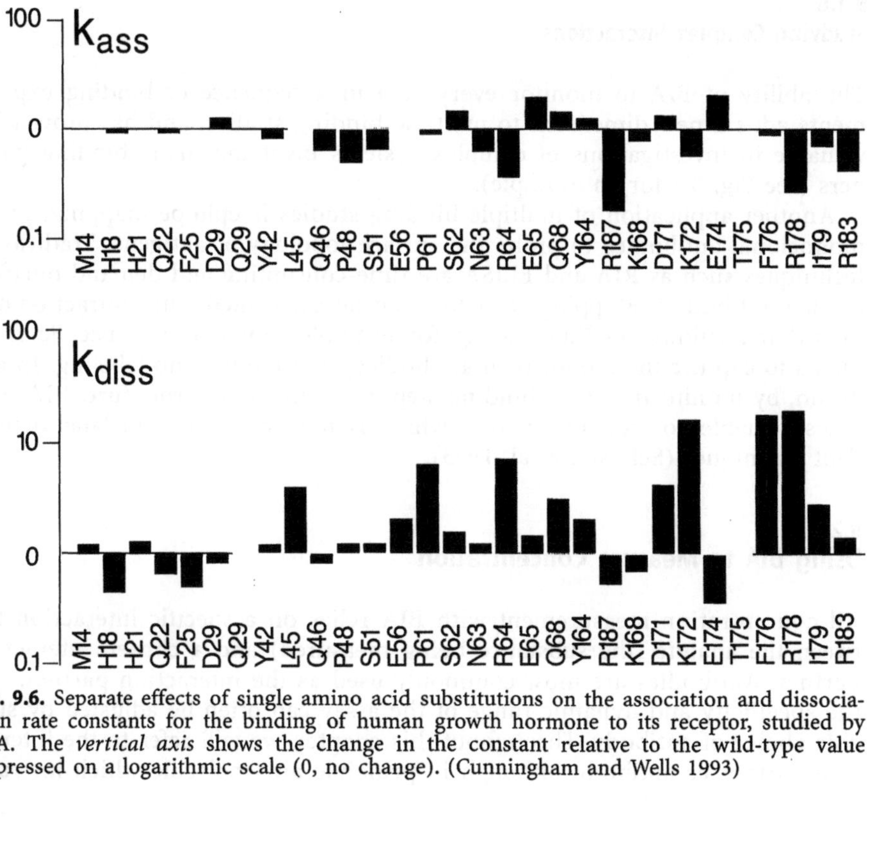

Fig. 9.6. Separate effects of single amino acid substitutions on the association and dissociation rate constants for the binding of human growth hormone to its receptor, studied by BIA. The *vertical axis* shows the change in the constant relative to the wild-type value expressed on a logarithmic scale (0, no change). (Cunningham and Wells 1993)

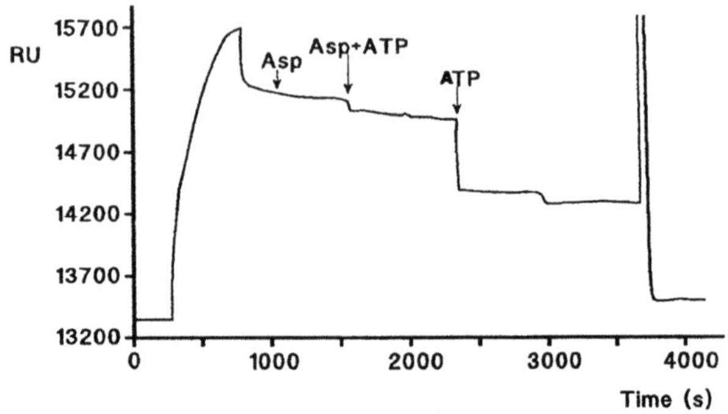

Fig. 9.7. A quaternary signalling complex from the chemotactic system of *E. coli* can be formed on the sensor chip surface and responds to physiological signals. Addition of aspartic acid alone or with ATP had little effect, but dissociation of the complex could be induced by addition of ATP alone. This corresponds to the physiological effects of aspartic acid and ATP on chemotaxis. (Schuster et al. 1993)

acting components, BIA is ideal for investigating the consequences of large numbers of separate modifications. Much of drug design in the pharmaceutical industry, as well as basic research into structure–function relationships in biomolecules, relies on this approach. Figure 9.6 gives an example, taken from the work of Cunningham and Wells (Cunningham and Wells 1993) who analyzed the importance of 30 amino acid residues in the interaction of human growth hormone and the extracellular domain of its receptor.

9.1.6 Studying Complex Interactions

The ability of BIA to monitor every step in a sequence of binding experiments adds a new dimension to multiple binding studies, and has proved invaluable in investigations of complex systems involving many binding partners (see Fig. 9.7 for an example).

Another application of multiple binding studies is epitope mapping, using monoclonal antibodies to probe the surface topology of antigens. Traditional techniques such as RIA and ELISA are time-consuming and demand purified labeled antibodies. Mapping with BIA can be performed with unfractionated hybridoma culture medium (using for example rabbit anti-mouse Fc fragments to capture the monoclonal antibodies) and requires no labeling. In addition, by monitoring every binding step in the mapping procedure, BIA provides valuable control information which is not obtained from label-dependent techniques (Schuster et al. 1993).

9.2 Using BIA to Measure Concentration

All concentration measurement with BIA relies on a specific interaction between the analyte (the substance being measured) and a chosen interaction partner. Antibodies are most commonly used as the interaction partner, and the specificity and dynamic range of the assay can often be adjusted by suitable choice of antibody. Throughout this chapter, we will refer to the interaction partner as antibody, although in principle any molecule which interacts

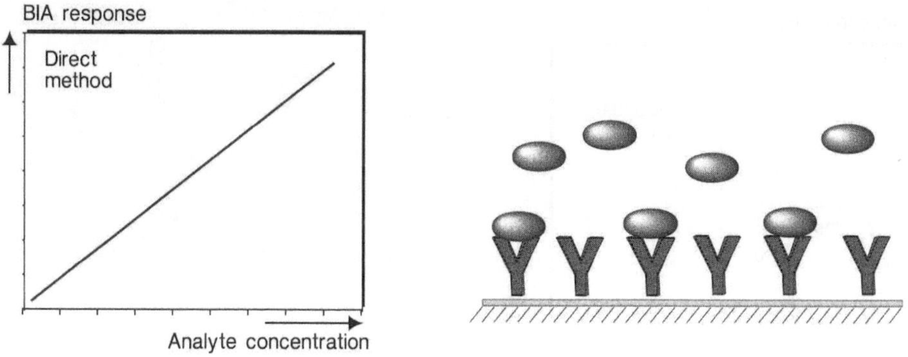

Fig. 9.8. Direct methods rely on the response obtained when the analyte binds directly to the immobilized ligand. The response increases with increasing analyte concentration

specifically with the analyte may be used. Macromolecular analytes can often be assayed directly. For low molecular weight analytes that give an intrinsically low response in BIA, competition (inhibition) methods are more appropriate. The BIA assays perform better than conventional methods in terms of speed, simplicity and in some cases reliability.

9.2.1
Direct Methods

When antibody is immobilized on the sensor chip surface and sample containing analyte is injected, binding of analyte to the surface is seen as a response on the sensorgram (Fig. 9.8). The response in a BIA assay is generally measured before the interaction has reached a steady state, so that a longer sample injection will increase the sensitivity of the assay (give a higher response for the same analyte concentration). Other factors influencing the sensitivity are the analyte size (larger molecules give a higher response per mole) and the amount of immobilized antibody (higher levels give a higher response).

When the level of response is used to determine analyte concentration, the measured level usually has to be corrected for the bulk refractive index of the sample. This can be done either by subtracting a blank response or by measuring the level of response shortly after the end of the sample injection. The second alternative is simpler, but may be less reliable if the dissociation rate constant for the interaction is significant.

Correction for bulk refractive index can be eliminated entirely by measuring the initial rate of analyte binding instead of the amount bound after a fixed time. This rate is determined by the association rate constant for the interaction and is directly proportional to the analyte concentration. Moreover, if measurements are made under conditions where mass transport of analyte to the surface is limiting, the binding rate is constant over time and independent of the antibody affinity altogether. Limiting mass transport is favored by a high association rate constant and a high level of immobilized

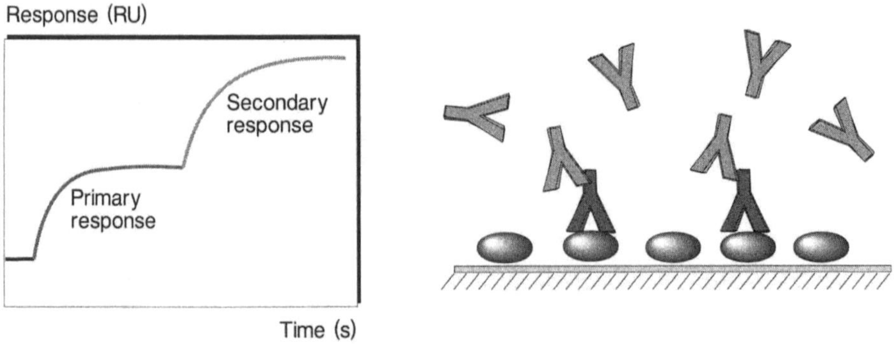

Fig. 9.9. With response enhancement techniques, the primary analyte response is amplified by binding a second interactant. This can also be exploited to enhance the specificity of the assay

antibody (these factors are discussed fully in the BIAtechnology Handbook 1994).

9.2.2 Response Enhancement

The response obtained from binding of analyte to immobilized antibody can be enhanced by injection of a second antibody or other interactant to improve the assay sensitivity, similar to sandwich techniques in ELISA and other methods. This technique uses a second interactant which gives a higher response than the analyte, either because it is a larger molecule or (less commonly) because it binds in a many-to-one ratio (Fig. 9.9). As an example, a secondary interactant of molecular weight 150000 binding in a 1:1 ratio will give potentially five times the primary response obtained with an analyte of molecular weight 30000.

Sandwich techniques can also improve the selectivity of an assay, since the measured analyte is identified by two separate interactions instead of only one.

For sandwich techniques to be successful, the primary interaction between analyte and immobilized antibody should be relatively stable so that dissociation of bound analyte is low during injection of the secondary interactant. It is also an advantage if the secondary interaction has a high association rate constant and a high affinity, so that more or less quantitative binding is achieved in a short time.

9.2.3 Indirect Methods

For small analytes which are difficult to measure with direct methods, inhibition or competition techniques are useful. For inhibition methods, a high molecular weight interactant such as an antibody is added to the sample and allowed to reach equilibrium binding with the analyte. With analyte or an

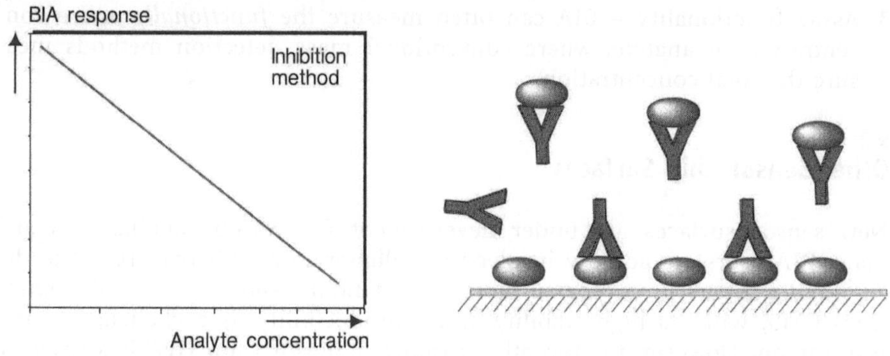

Fig. 9.10. With indirect and competition methods, the analyte interacts in solution with a secondary interactant and BIA is used to determine the remaining free concentration of secondary interactant in the mixture. The response is inversely related to the analyte concentration (more analyte leaves less free secondary interactant)

analog immobilized on the sensor surface, BIA is then used to determine the amount of free antibody remaining in solution (Fig. 9.10). The total concentration of analyte can then be derived from this value (most conveniently by reading off from a standard curve prepared with known analyte solutions). Note that with this technique the response is inversely related to the analyte concentration.

For competition methods, the analyte and a high molecular weight analog (e.g. analyte conjugated to a high molecular weight carrier) are allowed to compete for binding to antibody immobilized on the sensor surface. Again, the response (which is attributable almost entirely to the high molecular weight analog) is inversely related to the analyte concentration.

With both inhibition and competition methods, the sensitivity and working range of the assay are determined to a large extent by the affinity of the antibody for analyte. By a suitable choice of antibody, the performance of the assay can be adjusted to suit the particular requirements.

9.2.4
BIA Assays Versus Conventional Methods

Several articles describe the use of BIA for concentration measurement in food and environment studies. In each case, the BIA assay performed better than conventional assays, for reasons specific to the particular application – faster answers, less interference from other sample constituents, higher sensitivity etc. In addition, BIA frequently provides advantages over conventional assay methods in terms of:

1. Ease of operation – results are obtained directly with minimal pre- and post-assay processing.
2. Flexibility – by choice of antibody (or other interactant), the specificity and performance of the assay can be adjusted to the particular requirements at hand.

3. Assay functionality – BIA can often measure the *functionally active* concentration of analyte, where conventional mass detection methods measure the total concentration.

9.3
Other Sensor Chip Surfaces

New sensor surfaces are under development for various applications and many BIA users are actively involved in collaboration with Biacore AB to design and evaluate prototype surfaces. The standard and very versatile Sensor Chip CM5, with its high stability and reproducibility, is a challenge to improve upon. However, the recently introduced Sensor Chip HPA is a result of the efforts to supply users of BIA with new powerful tools and it enables researchers to study the membrane environment from a new angle. In addition to product-related material, references (Mayer et al. 1985; MacDonald et al. 1991; Kalb et al. 1992; Stelzte et al. 1993; Terrettaz et al. 1993; Plant et al. 1995) may be useful in the planning of experiments on Sensor Chip HPA. Sensor Chip NTA (QIAGEN GmbH, Hoffmann-LaRoche Ltd.) is used for convenient capture of histidine-tagged ligands.

9.4
Sensor Chip HPA

9.4.1
Approaching the Membrane

How membranes and membrane-bound molecules perform their function by interacting with other biomolecules is a central issue in biological research. Apart from using various techniques to monitor their function in vivo, many experimental situations necessitate a design where a membrane component is isolated from its native environment. This can be in order to achieve proper control of the binding specificity in e.g. a part of a signal transduction pathway or simply to accumulate concentrations of a particular molecule high enough to allow detailed analysis. A multitude of techniques and experimental designs for the isolation of a membrane functionality has been developed to produce membrane fragments, detergent-solubilized membrane receptors, subunits or truncated versions of receptors that are soluble in aqueous buffer, or even artificial membranes such as liposomes that have a precisely defined composition. Supported polar lipid monolayers that are deposited onto a solid surface fit into this context by providing a chemically and structurally defined and stable lipid environment that resembles a cell membrane in certain aspects. They can be used for the study of interactions involving the surface of the monolayer itself, or they can harbor selected molecules that have a part of their structure embedded in the lipid monolayer. Such a design, where ligands are anchored in a membrane-like environment, may be of significance for maintaining the integrity of the ligand as well as providing membrane-related secondary binding sites for analytes that interact with the ligand.

9.4.2 Surface Design and Methodology

Long-chain alkanethiol molecules form a flat hydrophobic quasi-crystalline layer on Sensor Chip HPA, designed to mediate *h*ydro*p*hobic *a*dsorption of polar lipids. Liposomes fuse spontaneously onto the surface to form a supported monolayer, suitable for biomolecular analysis in an aqueous environment. The operation profile is basically a five-step procedure:

1. Preparation of reagents includes the making of detergent-free degassed eluent buffer and user-defined liposomes. The molecular composition of the liposomes will determine the binding specificity of Sensor Chip HPA.
2. Liposomes are injected over Sensor Chip HPA and will fuse spontaneously with its hydrophobic surface. Coating the entire surface typically takes 0.5–3 h depending on liposome type and experimental circumstances. The process is allowed to continue until the sensorgram reading begins to level out. This will often occur at a signal which is of the order of 1000 RU or more.
3. The surface may initially carry multiple lipid layers and partially fused liposomes. Such loosely bound structures can be washed out by briefly increasing the flow rate. A brief exposure to alkaline pH will often produce an additional decrease in the sensorgram reading down to a level which will remain as a stable baseline.
5. With the lipid monolayer covering the surface, subsequent experimental strategies and data interpretation will be essentially the same as in experiments using Sensor Chip CM5. This includes running an application-specific negative control or a standard protein such as BSA to ensure that the observed binding is relevant.

Analytes that have bound to the lipid monolayer or to ligands anchored in the monolayer may be dissociated with variations in e.g., pH and ionic strength, leaving the lipid monolayer intact for additional cycles of interaction. The lipid layer is typically stable in aqueous, detergent-free buffers over a broad range of pH and ionic strength.

9.4.3 Liposome Preparation

Small unilamellar vesicles (SUVs) made from DMPC (dimyristoylphosphatidylcholine) or POPC (palmitoyloleoylphosphatidylcholine) have been used to develop the methodology for Sensor Chip HPA. They can be formed by extrusion through a polycarbonate filter (pore size 50 nm) in an aqueous buffer such as PBS or HBS, pH 7.4 which is a convenient way to reproducibly make various types of liposomes.

The physical and chemical properties of liposomes can be varied extensively and depend on their chemical composition, liposome formation technique, size, size distribution and lamellarity. Further, the concentration of ligands or binding sites built into the liposome membrane and transferred to the sensor surface will determine the signal that results from analyte binding.

The optimal target for the concentration of ligand that is built into the liposome membrane can be estimated from a simple calculation based on analyte molecular weight and desired response level. For more precise predictions of which surface concentration will be required for a certain application, consideration should be given to the rate constants, analyte concentration and contact time, since those values will determine how close to saturation a particular interaction will proceed. BIA simulation, software available from Biacore AB, enables the user to conveniently explore the effect of different combinations of these parameters to quickly choose an experimental design with regard to the binding capacity of the surface.

9.4.4 Coating the Surface with a Lipid Layer

An uncoated surface has a strong tendency to bind proteins non-specifically through hydrophobic interactions. A layer of polar lipid which generates a signal of around 1000 RU typically corresponds to a coating that drastically reduces or even eliminates non-specific binding. Using 0.1 mg/ml of BSA in PBS as a negative control on such a surface, binding is reduced to well below 100 RU, which should be compared to the more than 1000 RU of BSA that bind to an uncoated surface.

Coating with lipid is obtained by simply injecting liposomes over the surface in a flow of detergent-free buffer, typically at a physiological pH and ionic strength. Liposomes will spontaneously fuse with the hydrophobic surface. It has been postulated that the fusion will proceed faster for liposomes with a small diameter and at a temperature above the phase transition temperature (T_c). Both factors are assumed to increase the probability for the liposome to open up its spherical structure and expose the non-polar parts of the membrane to the chip surface. The rate of coating also depends on the liposome concentration. The flow rate for the liposome injection can be varied freely within the instrument specifications. The duration of the liposome injection that is required to obtain a sufficient reduction in non-specific hydrophobic binding, typically ranges from a few minutes to several hours, depending on the type of liposome used as well as application-specific demands.

9.4.5 Stability of the Lipid Layer

The coating procedure may result in a surface that has multiple lipid layers and carries partially fused liposomes. Such structures are only loosely bound and can easily be washed out by a brief increase of the flow rate to 100 µl/min. The remaining lipid layer is stable and can withstand wide variations in ionic strength and pH, as well as 10% solutions of ethanol and DMSO.

However, when exposed to the alkaline pH of a 10–100 mM NaOH solution, a drop in the baseline may occur down to a level which will then remain as a very stable baseline, even with repeated injections of the NaOH solution. The current interpretation of this event is that the high pH triggers an additional rearrangement of the lipid layer which removes loop structures and multilayer formations that resisted the wash with a high flow rate.

9.4.6
Interaction Analysis

The nature of the user-defined binding sites and their tolerance to various conditions will make each application unique. But the stability of the lipid layer indicates a general possibility to use an experimental design where analytes that have bound to the lipid layer, or to ligands anchored in the lipid layer, can be dissociated to yield a regenerated surface for additional cycles of interactions to occur. In this respect applications on Sensor Chip HPA typically follow applications on Sensor Chip CM5 both in design and stability to regeneration agents such as extremes of pH and ionic strength. Experiments with e.g. 10 mM NaOH as regeneration agent has shown very high cycle-to-cycle reproducibility on DMPC or POPC layers, suggesting that quantitative assays of e.g. kinetics can be carried out conveniently and with high accuracy.

9.5
Sensor Chip NTA

On the dextran matrix NTA is pre-immobilized. The chip is designed to bind histidine-tagged molecules via chelated nickel for subsequent analysis of analyte binding. Immobilization of ligands via histidine-tags provides a valuable and convenient methodological interface between BIA analysis and other laboratory techniques that rely on His-tags, e.g. chelating chromatography. Immobilization via a His-tag also has the advantage of orienting the ligand molecules in a homogeneous way and allowing the immobilization to be carried out without significantly changing the pH or ionic strength during the coupling procedure.

9.5.1
Integrating BIA with Other Technologies

A rapid increase in both the number of users and the range of application fields bore witness to the value of BIA as a general investigatory tool in both research and industrial development work, and after the BIA symposia in 1994 it was clear that the dominant question was no longer whether the technology works but how it can help to advance each particular field of research. In parallel with this acceptance was a growing understanding of the finer points of BIA technology, with for example careful use of oriented ligand capturing in place of random immobilization, and sophisticated approaches to the problem of how to interpret kinetic data (Markey 1995).

During 1995, the number of fields where BIA was used continued to grow, with important contributions in molecular biology as well as food and environment research. The continued maturation of BIA as a research technology is further supported by its integration with other technologies, a recurrent theme in the BIA symposia in September and October 1995. From being a new technology which generates its own questions and answers, BIA is beginning to attain the status of one among several resources in the laboratory, each with its own strengths and weaknesses and each with its own kind of information to provide. Among the techniques to which BIA has provided complementary information are analytical ultracentrifugation, microcalorimetry, NMR, light scattering, mass spectrometry, micropreparative HPLC and scintillation proximity assays.

9.5.1.1
Screening Applications

Drug screening at both primary and secondary levels must deal with large numbers of samples, asking essentially "yes/no" questions in automated analysis protocols. Although BIA instrumentation today is not optimized for this kind of handling, two articles describe the potential of BIA in this area (BIACORE 2000 with multichannel analysis facilities is particularly valuable here). However, BIA automatically provides much more than just a yes/no answer even in screening applications. Taremi et al. (1996) have used BIA as a primary screening tool and exploited the additional information in follow-up investigations on the positive results of the primary screen. Other workers have used BIA as a complementary technology to enhance the information obtained from conventional screening.

Todd Leff and coworkers at Parke-Davis Co. and the University of Michigan, Ann Arbor, for example, used a scintillation proximity assay (where a signal is detected only when the radioactively labeled probe associates closely with the surface of a scintillation bead) as a primary screen for compounds affecting the *myc-max* cell proliferation control system. This assay is easy to perform and provides a simple yes/no answer: BIA was then used to examine the detailed kinetic effects of identified positive compounds. The *myc-max* system, like many other regulatory systems in the cell, is a delicately controlled balance of signals, and the relative kinetics and affinity of effector binding is more important than the simplistic question of whether an effector binds or not. In such situations, complementary investigations that can provide detailed information early in the screening procedure are particularly valuable, since many of the candidates tentatively identified in the primary yes/no screening can be eliminated at an early stage.

9.5.1.2
Molecular Analysis

Researchers see BIA now as one of several approaches to the characterization of macromolecular interactions in solution. One such laboratory is in the Department of Macromolecular Sciences at SmithKline Beecham Pharmaceuticals, where Preston Hensley exploits an impressive range of biophysical techniques including BIA, ultracentrifugation, microcalorimetry, NMR and light scattering. Each of these techniques allows interaction processes to be visualized and quantitated from a different perspective, providing both complementary and confirmatory information.

Hensley illustrated this multi-disciplinary approach at the BIA symposium in San Diego with the characterization of the interaction between interleukin-5 (IL-5) and its receptor in solution. The receptor contains an α- and a β-subunit, required for IL-5 binding and signal transduction, respectively. A central question in the mechanism of IL-5 binding is whether the dimeric ligand binds between two α-subunits, promoting receptor dimerization, or between the α- and β-subunits of one receptor heterodimer. Ultracentrifugation analysis together with advanced mathematical techniques for analysis of sedimentation profiles of heterogeneous mixtures measures the apparent molecular weights in solution and indicates that IL-5 forms a 1:1 complex with solubilized receptor α-subunit. Mass spectrometry complements this information with exact molecular weights of the polypeptide chains, confirming that the major species seen in solution are monomeric. With BIA, detailed kinetic analysis can be performed for the interaction, potentially distinguishing between different binding models involving dimerization and/or conformational changes in the binding partners. Isothermal microcalorimetry measures the heat of binding associated with the interaction, which can be related both to the molar ratio in the complex and to the number of residues undergoing a significant change in microenvironment. The picture of the interaction obtained from these approaches together is significantly more complete than that provided by any one technique alone. In addition, using one approach may help to interpret the results of another – for example, indications of a conformational change obtained from microcalorimetry can help to choose a suitable kinetic model for analysing BIA data.

Another research team that has used analytical ultracentrifugation to complement BIA studies is at the University of Colorado in Denver, where Charles McHenry's group has studied the complex question of *E. coli* DNA polymerase holoenzyme structure. The holoenzyme consists of ten different subunits ($\alpha\tau\gamma\beta\delta\delta'\varepsilon\chi\psi\theta$), all of which have been cloned, over-expressed and purified in large quantities. BIA is used to identify which subunits can bind to each other, investigate the effects on complex formation of deletion mutations in selected subunits, measure kinetic parameters for the interactions and test the sequential formation of multi-subunit complexes. Analytical ultra-centrifugation provides complementary information on the apparent solution molecular weight of complexes and hydrodynamic properties related to molecular shape parameters. Together, these studies have led to refined models of holoenzyme structure, in particular with respect to the stoichiometry and binding role of the τ and γ subunits.

Mass spectrometry is another technique which finds increasing use in the study of biological macromolecules, giving precise determinations of molecular weights and correspondingly allowing identification and quantitation of molecular species. Randall W. Nelson and colleagues (Nelsson et al. 1995) at Arizona State University showed that MALDI (matrix-assisted laser desorption/ionization) time-of flight mass spectrometry can be integrated with BIA to identify molecules bound to the sensor chip surface. The sensor chip itself can be prepared directly in the solid phase sample matrix for MALDI-MS, making the interface with BIA particularly easy. Östen Jansson at Biacore AB, working with Carsten Sönksen at the University of Odense in Denmark, has investigated the alternative approach of using MS to identify molecules eluted from the sensor chip surface, and has shown that both MALDI and ESI (electrospray ionization) MS can be used for this purpose. The development of MS analysis of molecules eluted from the sensor chip opens new potential for on-line integration of the two techniques.

9.5.1.3
BIA Complements Chromatography

Edouard Nice and coworkers at the Ludwig Institute for Cancer Research in Melbourne, Australia have described synergies between BIA and micropreparative HPLC. These synergies work in both directions: HPLC is a valuable technique for purifying ligands before immobilization on the sensor chip and for desalting and buffer exchange into immobilization conditions, while BIA can be used to advantage for monitoring fractions obtained during chromatographic purification of biomolecules. This application of BIA is particularly useful when the biomolecule in question has no known activity other than interaction with another substance, so that there is no other simple way of identifying the molecule in chromatographic fractions. The work of Nice et al. [31] exploits this approach in searching for "unknown ligands" such as specific antigens for monoclonal antibodies raised against a complex mixture or new members of receptor families (e.g. structure–function studies on EGF and the EGF receptor)

9.6
Summary

Since BIA was first introduced in 1990, the technique has become an established and respected tool in both academic and industrial research. Benefits provided by BIA include both reduced sample consumption and higher throughput in comparison with many conventional techniques. The major gains are however often in the type of information obtained: in many cases, workers use BIA to ask questions and obtain data that cannot reasonably be approached by other techniques. An increasing number of laboratories are coming to regard BIA as one among several technologies available for studying macromolecular function in solution, and are using BIA together with other techniques to gain greater insight into their research questions. In this perspective, the dynamic interaction studies offered by BIA complement other methods for investigating both structure and function. The integrated approach bears witness to the acceptance of BIA technology by the scientific community.

References

Bartley TD, Hunt RW, Welcher AA et al. (1994) B61 is a ligand for the ECK receptor protein-tyrosine kinase. Nature 368:558–560
BIAtechnology Handbook obtainable from Biacore AB (formerly Pharmacia Biosensor AB) (1994) Code no BR-1001-84, Biacore AB, Uppsala
Cunningham BC, Wells JA (1993) Comparison of a structural and functional epitope. J Mol Biol 234:554–563
Hirschfield T (1967) Appl Optics 6:715
Johnsson B, Löfås S (1991) Immobilization of proteins to a carboxymethyldatran modified gold surface for biospecific interaction analysis in surface plasmon resonance. Analyt Biochem 198:268–277
Jönsson U (1991) Real time biospecific interaction analysis using surface plasmon resonance and sensor chip technology. BioTechniques 11(5):620
Jönsson U, Malmqvist M (1992) Real time biospecific interaction analysis. The integration of surface plasmon resonance detection, general biospecific interface chemistry and microfluidics into one analytical system. Adv Biosensors 2:291–336
Kalb E, Frey S, Tamm LK (1992) Formation of supported planar bilayer by fusion of vesicles to support phospholipid monolayers. Biochim Biophys Acta 1103:307–316
Karlsson R, Roos H, Fägerstam L, Persson B (1994) Kinetic and concentration analysis using BIA technology methods. Methods: a companion to Methods Enzymol 6:99–110
Kooyman RPH, Kolkman H, van Gent J, Greve J (1988) Surface plasmon resonance immunosensors: sensitivity considerations. Anal Chim Acta 213:35
Kovacs G (1982) Optical excitation of surface plasmon-polaritons in layered media. In: Boardman AD (ed) Electromagnetic surface modes. Wiley, New York, p 143
Kretschmann E (1971) Die Bestimmung optischer Konstanten von Metallen durch Anregung von Oberflächenplasmaschwingungen. Z Naturforsch B241:313
Kretschmann E, Raether H (1968) Radiative decay of nonradiative surface plasmons excited by light. Z Naturforsch 230a:2135
Liedberg B, Nylander C, Lundström I (1983) Surface plasmon resonance for gas detection and biosensing. Sensors Actuators 4:299
MacDonald RC, MacDonald RI, Menco BP, Takeshita K, Subbarao NK, Hu LR (1991) Small volume extrusion apparatus for preparation of large, unilamellar vesicles. Biochim Biophys Acta 1061:297–303
Markey F (1995) Interpreting kinetic data. BIAjournal 2(1):18
Mayer LD, Hope MJ, Janoff AS (1985) Solute distributions and trapping efficiencies observed in freeze-thawed multilamellar vesicles. Biochim Biophys Acta 817:193–196
Nelsson RW, Krone JR, Dogruel D, Williams P, Granzow R (1995) Interfacing mass spectrometric immunoassays with BIA. BIAjournal 3(1):17
Nice E, Lackmann M; Smyth F, Fabri L, Burgess AW (1994) Synergics between micropreparative high-performance liquid chromatography and an instrumental optical biosensor. J Chromatogr 660:169–185
Otto A (1968a) Excitation of nonradiative surface plasma waves in silver by the method of frustrated total reflection. Z Phys 216:398
Otto A (1968b) Phys Stat Solidi 26:199
Plant AL, Brigham-Burke B, Petrella EC, O'Shannessy DJO (1995) Phospholipid/alkanethiol bilayers for cell-surface receptor studies by surface plasmon resonance. Anal Biochem 226(2):342–348
Polymer handbook, 3rd edn. (1989) Wiley Interscience, New York, pp VII-469
Raether H (1977) Surface plasmon oscillations and their applications. Phys Thin Films 9:145
Schuster SC, Swanson RV, Alex LA, Bourret RB, Simon MI (1993) Assembly and function of a quarternary signal transduction complex monitored by surface plasmon resonance. Nature 365:343–346.
Stelzte M, Weismüller G, Sackmann E (1993) On the application of supported bilayers as receptor layers for biosensors with electrical detection. J Phys Chem 97:2974–2981
Stenberg E, Persson B (1991) J Colloid Interface Sci 143:513–526
Stenberg E, Persson B, Roos H, Urbaniczky C (1991) Quantitative determination of surface concentration of protein with surface plasmon resonance using radiolabeled proteins. J Colloid Interface Sci 143:513–526
Taremi SS, Prosisi WW, Rajan N, O'Donnell RA, Le HV (1996) Human interleukin 4 redeptor complex: neutralization effect of two monoclonal antibodies. Biochemistry 35:3222–2331

Terrettaz S, Stora T, Duschl C, Vogel H (1993) Protein binding to supported lipid membranes: investigation of the cholera toxin-ganglioside interaction by simultaneous impedance spectroscopy and surface plasmon resonance. Langmuir 9:1361–1369
Turbadar T (1959) Complete absorption of light by thin metal films. Proc Phys Soc (Lond) 73:40
Welford K (1991) Opt Quant Elect 23:1

CHAPTER 10

Measuring Cellular Traction Forces with Micromachined Substrates

Catherine G. Galbraith · Michael P. Sheetz*

Contents

Duke University Medical Center, Department of Cell Biology, Box 3709, Durham, North Carolina 27710, USA

*Corresponding address: Duke University Medical Center, Department of Cell Biology, Box 3709, Durham, North Carolina 27710, USA,
Tel.: +1 919 684 8091, Fax: +1 919 684 8592,
e-mail: m.sheetz@cellbio.duke.edu

10.1
Introduction to Microfabrication and Its Use in Biology

The ability to pattern silicon substrates revolutionized electronics by miniaturizing circuit components 40 years ago. The processes defined by this field have been extended within the past 25 years to make three dimensional silicon structures such as levers, diaphragms, valves, and gears which are the basis of a variety of miniature mechanical devices. These devices have a wide range of commercial applications, from accelerometers used for air bag release, to print heads for ink jet printers, pressure sensors for cardiovascular catheters, and even complete miniaturized gas chromatographs. More recently, micromachining has found applications in tissue engineering and fundamental studies on cell motility and guidance. This chapter will describe micromachining, its application in studying cell motility, and a new micromachined device that we are using to measure sub-cellular traction forces.

10.1.1
Definition of Micromachining

Micromachining is the process of directionally etching individual layers of silicon-based materials. Manufacturing of a micromachine begins with the use of photolithography to create patterns on the surface of a wafer (Fig. 10.1). The first step in the process is to grow a thin layer of silicon dioxide on a silicon wafer. The silicon oxide is then coated with photoresist, a polymer that is sensitive to ultraviolet radiation. A pattern is created on the photoresist by covering it with a photomask in which the desired pattern is made of metal to protect the underlying photoresist from irradiation. The exposed areas of the photoresist are weakened by the ultraviolet light and subsequently removed by a developing solution. Exposing the wafer to hydrofluoric acid etches only the silicon dioxide uncovered by the removal of the photoresist. The unexposed photoresist can now be removed with sulfuric acid, and the silicon dioxide acts as a mask for the underlying silicon during subsequent chemical etches. This basic process is used to create devices by repeated layering of material, and the use of etchants that act either isotropically, etching uniformly in all directions, or anisotropically, etching preferentially along a given axis, of the silicon crystal (Angell et al. 1983).

As a simple example of the steps involved in making a micromachined device, consider the method of making a cantilevered beam depicted in Fig. 10.2. A cantilever is a suspended beam that is attached at one end and free at the other, meaning that it is free from surrounding material on the top, bottom, and all but one end of the beam. This type of device is used in both the acceleration sensors that trigger the release of air bags and in our traction measuring device. A layer of silicon dioxide is grown on the surface of a silicon wafer. Holes are lithographically patterned and etched into the silicon dioxide to form anchor points for the beams. A layer of polycrystalline silicon is deposited over the silicon dioxide, and the pattern of the beams is lithographically patterned and etched into the polycrystalline silicon. Finally, the silicon oxide is etched away to leave the cantilevered beams.

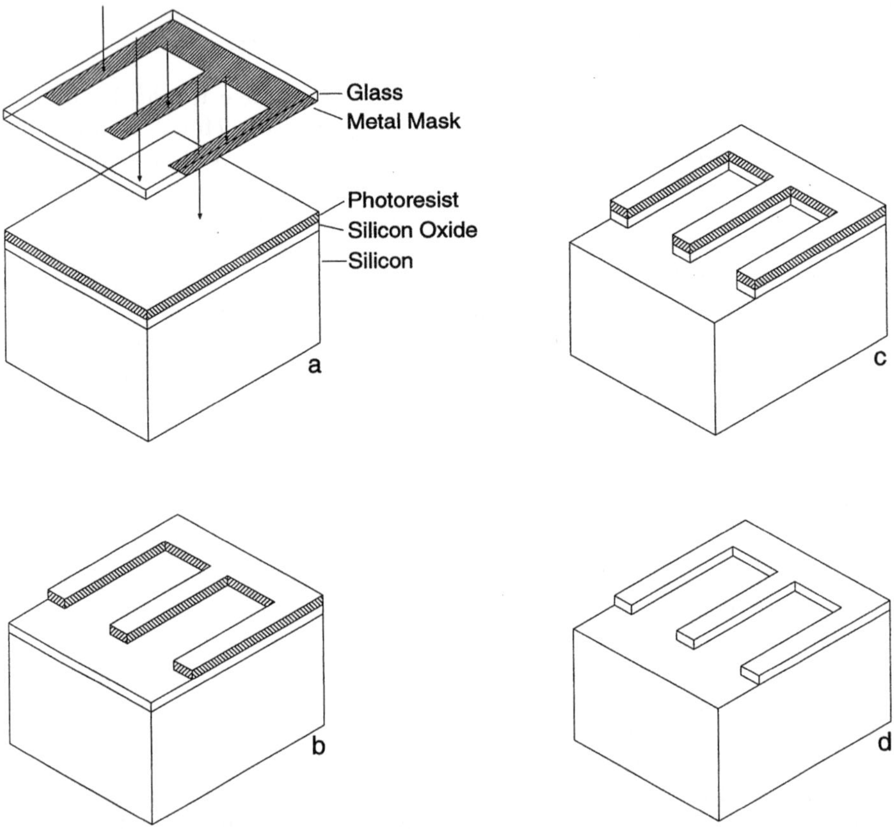

Fig. 10.1 a–d. Steps involved in photolithography. **a** A silicon wafer is coated sequentially with a silicon oxide and a photoresist, and a mask with the desired pattern made of metal is placed over the wafer and subjected to ultraviolet radiation. **b** The portion of photoresist which is covered by the metal in the mask protects the underlying polycrystalline silicon from ultraviolet radiation. **c** The silicon oxide not protected by the photoresist is etched away. **d** The photoresist is removed

10.1.2
Applications in Biology

Most applications of micromachining in biology do not involve devices with movable parts, but instead use directional etching to pattern the substrate or imbed microelectronics. One of the first biological applications of micromachining was the development of implantable two-dimensional microelectronic sensors for the study of signal processing in biological neural networks (Najafi et al. 1985). This approach has been extended to three-dimensional devices (Hoogerwerf and Wise 1994) in order to make recordings from an entire volume of brain tissue rather than a small number of nerves. Both of these devices have clinical applications within neural prostheses. Other types of implantable miniaturized silicon devices have been used to provide guid-

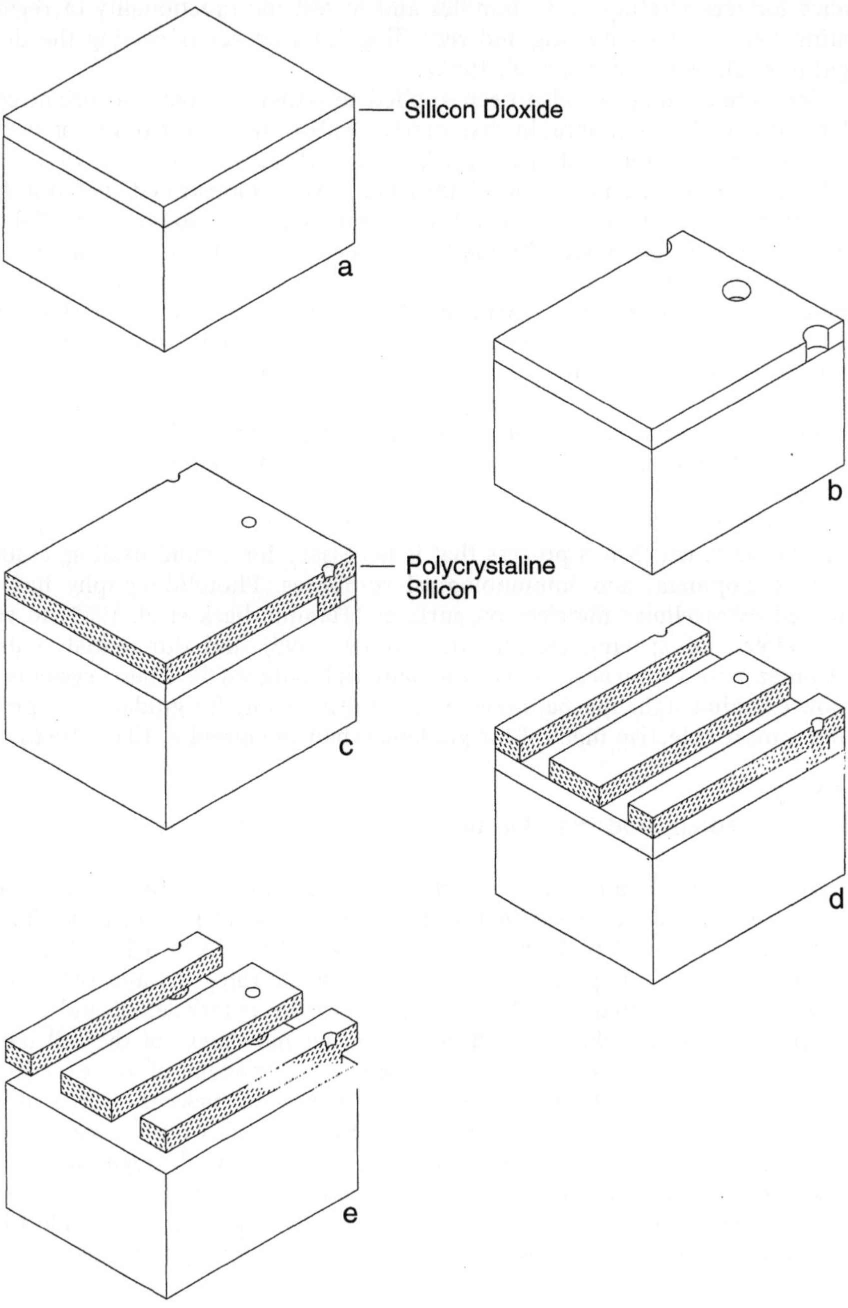

Fig. 10.2 a–e. Manufacture of a cantilevered beam. **a** A layer of silicon dioxide is grown on the surface of a silicon wafer. **b** Holes are lithographically patterned and etched into the silicon dioxide to form anchor points for the beams. **c** A layer of polycrystalline silicon is deposited over the silicon dioxide. **d** The shape of the beams is lithographically patterned and etched into the polycrystalline silicon. **e** The silicon dioxide is etched away to free the cantilevered beams

ance for regenerating nerve bundles and to test the functionality of regenerating nerves by stimulating and recording from nerves traversing the device (Akin et al. 1994; Kovacs et al. 1992).

Micromachining has also been applied to pattern surfaces to orient cellular growth. Photolithography has created substrata with grooves of various orientations, depths, and spacing (Brunette 1986a,b; Clark et al. 1987, 1990, 1991). Experiments using these substrata have demonstrated that fibroblast alignment depends upon the depth and spacing of the groove. Epithelial cells are more sensitive than fibroblasts to topography, aligning on all but the shallowest grooves and showing less dependence upon groove spacing, but neurons are less responsive than fibroblasts to topography (Clark et al. 1987, 1990, 1991). This type of information can then be applied to pattern the interface of prosthetics to increase directional tissue attachment and integration of the prosthetic to the body. For example, connective tissue attachment along the longitudinal axis of an implant can be enhanced if the implant has horizontal grooves which impede the down growth of epithelia (Chehroudi et al. 1990).

Micropatterning has also been used to study more fundamental aspects of cell motility, which is a process that is necessary for wound healing, embryonic development, and immunological responses. Photolithography has patterned extracellular matrices on surfaces (Hammarback et al. 1985) to study the effects of spacing (Kindt and Lander 1995) and differential matrices (Gomez and Letourneau 1994) on neuronal outgrowth. These systems can now examine signaling pathways that are responsible for guidance by providing a more selective method for guidance than topography (Clark 1996).

10.1.2 Traction Forces and the Measurement of Traction Forces

Cell migration requires cells to exert forces against their substrata in order to move forward. In two-dimensional systems, cells exert a force against the substrate in contact with their ventral surface. This force arises from many interactions, including: the adhesion between the integrins and the extracellular matrix, the coupling of the integrins to the cytoskeleton, and the contraction of the cytoskeleton. Although the net movement of the cell parallel to the surface indicates that the dominant component of force is in this plane, the number of components that combine to make the force indicate that the force also acts in other directions. Since we cannot measure all of these components and determine how they interrelate throughout an entire cell, we are developing a device to measure the force that the cell exerts at a single point contact, so that we can alter the linkages that are made to this contact and understand how they sum to yield net forward motion.

10.2.1 Definition of Traction

The net force exerted by the cell on the substrate can be defined mathematically by considering the interaction between two surfaces such as a cell and its substrate. The interaction between the cell and the substrate is due to two types of forces, the body forces which include forces due to gravity and the surface forces. The surface forces are the forces we are interested in using to describe cell motility. They can be expressed by considering a small region of the substrate of area ΔS which has a unit vector υ that is normal to ΔS. The cell exerts a force ΔF on the substrate which depends upon the size of the area ΔS and the orientation of the normal vector. The traction in the direction of the unit normal, T, can be defined as $\Delta F/\Delta S$ as ΔS approaches a definite limit (Fung 1969). Therefore, traction is a vector which has units of force per unit area.

10.2.2
Previous Methods Used to Measure Traction

Previous experiments have measured the force between the cell and its substrate over the entire area of the cell. These experiments cross-linked a 1 µm layer of polydimethyl siloxane (60 000 cp) over uncross-linked fluid which acted as a lubricant (Harris et al. 1980). Chick heart fibroblasts moving along the surfaces produced compression wrinkles in the substrate which were perpendicular to the long axis of the cell. These wrinkles developed 5–25 µm behind the leading edge of the cell, where adhesions with the substrate were seen to develop when visualized with interference reflection microscopy. Using calibrated microneedles it was determined that the cells exerted forces against the substrate in excess of 10 nN/µm of advancing margin (Harris et al. 1980). This method was also used to compare the traction forces generated by different cell types (Harris et al. 1981). The strongest tractions were exerted by fibroblasts from embryonic chick heart, skeletal muscle and tendon, dermis of the eye and BALB/3T3 mouse cells. Glial cells from dorsal root ganglia and platelets from human blood also exerted strong tractions, but epithelial cells were considerably weaker. Still weaker tractions were generated by L-929, CHO and SV40 transformed 3T3 cells. The weakest tractions were exerted by macrophages, polymorphonuclear leukocytes, and nerve growth cones. The weakness was not just a reflection of size, since dozens of nerve growth cones arranged side by side could not distort the substrata. These data demonstrate that the most mobile cells exert the weakest traction (Harris et al. 1981).

A second generation of experiments improved the technique of Harris by embedding marker beads in the substratum to provide more accurate information about the location and orientation of the in-plane component of traction forces (Lee et al. 1994). These experiments illustrated that for rapidly moving keratocytes, the largest forces were forward directed under the rear half of the cell, and forces along the lateral margin were directed inward, toward the nucleus. These experiments had several limitations: if the substrate was too compliant, about one half of the beads did not recover to their initial position after the cell passed over them, the temporal resolution of the assay was limited by the relaxation time of the film, and the displacement of an individual bead did not measure the displacement of a point contact with the substrata, but rather, a displacement which was influenced by all of the contacts that the cell and its neighbors had with the surface. For example, if contact A is in the immediate vicinity of contact B, and contact B exerts a strong force which applies a tension on the substrate, then the amount of bead displacement generated by contact A will be less than if contact B was not present.

The displacements measured using the elastic substrata were used to calculate traction forces (Oliver et al. 1995; Dembo et al. 1996). These calculations were limited to the measurements which were taken from the elastic substrata which were calibrated by applying localized displacements with calibrated microneedles. The displacements in these experiments were on the order of 1–2 μm, the relaxation time was on the order of a second, and the stiffness of the films were on the order of 20 nN/μm. The displacement of beads on the periphery of the cell and the motion of the cell centroid were used to predict the traction forces generated at discrete mesh points beneath the cell surface.

These experiments have provided an estimate of the force generated against the substrate by keratocytes as 2.5×10^{-8} N (Lee et al. 1994), while the force generated by fibroblasts is estimated at 20×10^{-8} N (Harris et al. 1980). Comparing these forces with other forces that cells exert, locomoting leukocytes generate a protrusive force of approximately 3×10^{-8} N (Usami et al. 1992), and advancing neurites pull with a force greater than 6×10^{-9} N (Lamoureaux et al. 1989; see Oliver et al. 1994 for other examples). These forces provide design criteria for new devices.

10.3 Design of a New Device to Measure Traction Forces Generated by Single Adhesive Contacts

As mentioned above, the previous methods used to measure tractions measured forces over the entire area of the cell. We have applied the technology of micromachining to design a device where the cell exerts force on only a small movable region of the substrate. This allows us to measure the force generated by local regions of contact between the cell and the substrate.

10.3.1 Design Objectives

We would like to measure the tractions generated by single adhesive contacts. The ability to measure the forces generated by single adhesive contacts will then allow us to alter components of the contact such as the extracellular matrix or cytoskeletal contractility to see how this affects tractions. To perform these experiments, it is necessary that measurements be made locally over the area of the cell which makes a single adhesive contact. This approach eliminates some of the difficulties of earlier techniques because surrounding cells cannot influence the measurement, and the measurement is made at a local region, not over the whole cell. By making the experimental device elastic, it returns to its original position when unloaded, and the measurements can be made at video rates. Developing such a device represents the first application where cells are used to displace micromachined pieces in order to measure forces.

10.3.2
How the Device Makes Traction Measurements

To measure the forces that cells exert on the substrate as they move, we have developed a TRaC, traction recording chip, in collaboration with MCNC (Research Triangle Park, North Carolina) which is based on micromachined cantilevered levers. The modifications for our experiments were to place another stationary layer on top of the beams, covering all but a small portion of the movable end of the beam (Fig. 10.3). A pedestal with a square island, or pad, was placed on the end of the beam so that the pads were planar with the top stationary layer. A cut-out in the surface surrounds each pad, allowing it to be displaced by a cell. This design is shown schematically for a single beam in Fig. 10.3 where the different layers have been "peeled apart" to illustrate the salient features of the device.

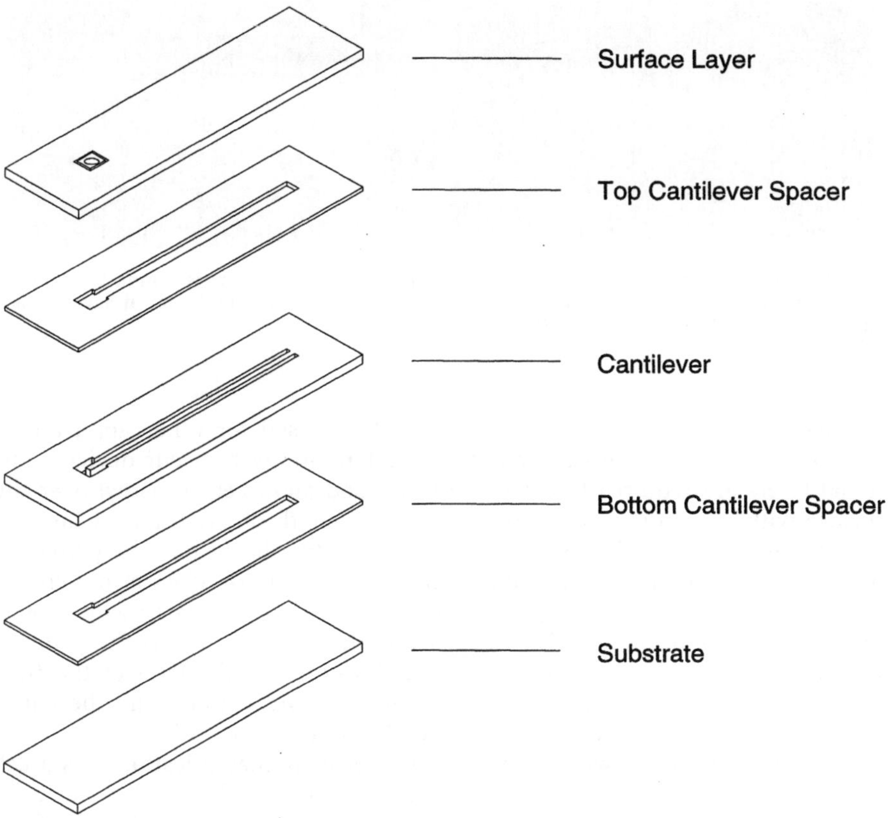

Fig. 10.3. Isometric of beam. Slices taken through the cantilevered beam used in the TRaC. The square pad is planar with the top surface and attached to the cantilever by a pedestal. The cantilever spacer allows the bottom of the beam to be free

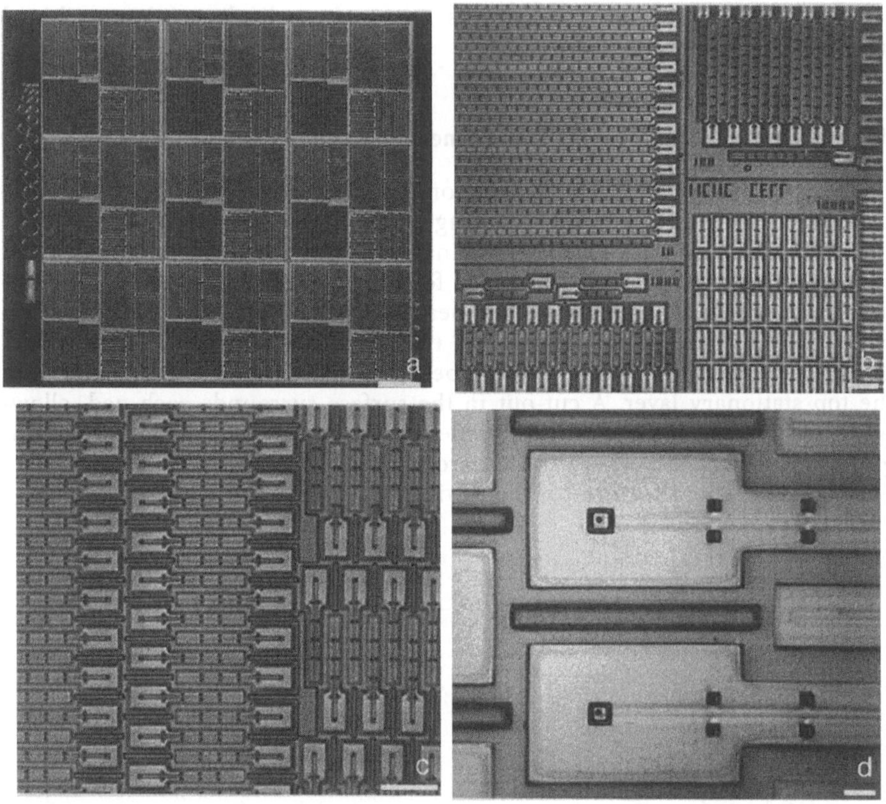

Fig. 10.4 a–d. Photographs of a wafer and chip. **a** A single TRaC. *Bar* is 1 mm. **b** Beams of various lengths. *Bar* is 0.1 mm. **c** Beams of 0.18 mm length. *Bar* is 0.1 mm. **d** Several pads on 0.18 mm length beams. *Bar* is 10 μ

Micromachined cantilever devices have been used for other applications and have been described earlier in the chapter. However, in our device additional layers are required to cover the beam and allow cells to crawl over the surface without coming into contact with the length of the beam. A layer of spin-on glass is placed above the polysilicon after the beams are patterned by lithography. A small hole is made through the spin-on glass at the end of each beam. This new top surface of spin-on glass is then coated with a plasma-deposited amorphous silicon which fills the hole and forms the pedestal which anchors the pad to the beam and elevates it to the level of the final surface. The top layer is etched to form the pad and cut-out, and the entire lever assembly is then etched free with hydrofluoric acid.

A single TRaC is shown in Fig. 10.4 at increasing magnifications. As a cell moves across the surface of the device, it contacts a pad and displaces a lever. The device has 5904 cantilevered levers ranging in length from 0.086 mm to 0.86 mm, and the pads range in size from 4 to 25 μm^2. The displacement of the lever is proportional to the force that the cell generates

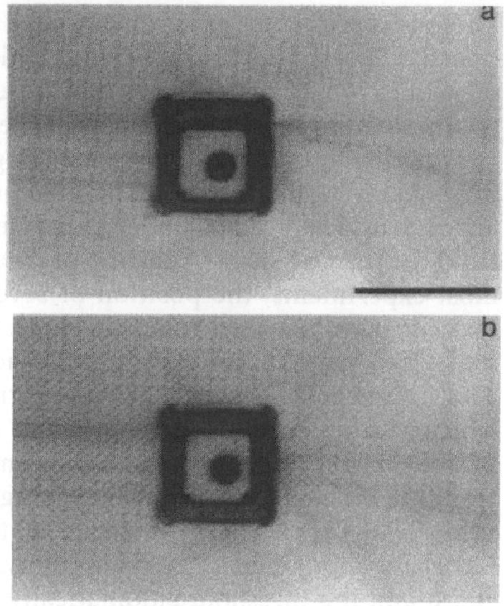

Fig. 10.5 a–b. Needle bending beam. **a** A calibrated microneedle is bending the pad downward. *Bar* is 10 μ. **b** The needle has been elevated and refocused, and the pad has returned to its original position

against the surface. The proportionality constant is the beam stiffness which can be calculated according to linear beam theory as $4Eh^4/l^3$ where l is the length of the beam, E is the modulus of elasticity, and h is the width and thickness of the beam. In designing the chip, we chose four different-length beams to span four orders of magnitude of beam stiffness, ranging from 970 nN/μm to 0.97 nN/μm, which covers the levels of forces expected to be generated by a variety of cell types. In the experiments reported here, we have only used the 0.18 mm beams.

The beams have been independently calibrated with glass microneedles. Using the method of Dennerll (Dennerll et al. 1988), an 8 mm length of 25 μm diameter chromel wire was mounted in a piece of capillary glass and attached to the stage of a small upright observation microscope which was turned on its side so that its optical axis was parallel to the table surface. The wire was calibrated by hanging weights made of different lengths of wire on its free end and noting the deflection with a reticule. These calibrations gave a stiffness for the wire which agreed with the values reported by Dennerll to within 1%. The same procedure was used to calibrate glass microneedles which had been pulled from solid-core glass and had uniform diameters for lengths greater than 10 diameters from the tip.

The calibrated microneedles were then used to calibrate the levers. Using a Zeiss Axiophot, a 40×0.75 NA water immersion objective, and reflection optics, a needle was positioned into the cut-out surrounding the pad using a DeFonbrunne manipulator. The needle was displaced with the manipulator so that its end caused the pad to move. Next, the needle was lifted, allowing the pad to bounce back to its original position, and the needle was brought back into focus (Fig. 10.5). The relative displacement of the pad and the needle were used to calculate the beam stiffness from the needle calibration.

These experiments gave an average beam stiffness of 75.8±11.4 nN/μm (mean ±SD, 41 trials on 6 beams) which agrees with the theoretical stiffness to within 20%. This stiffness can be used to calculate forces from displacements when the force is directed perpendicular to the long axis of the beam.

Experiments measuring the forces generated by cells are performed on a Zeiss Axiovert microscope. The attachment point of the pedestal to the lever appears optically dark under polarized reflection microscopy, and its centroid is tracked as a measure of beam deflection during experiments. In most experiments, the position of several cells is tracked by cycling the motorized stage between them at regular intervals. Since the accuracy of the motorized stage is less than the displacements which can be measured by the TRaC, the centroid of the pad is subtracted from the centroid of the cut-out which surrounds the pad in order to eliminate any rigid-body displacement of the entire TRaC. These measurements are obtained by capturing data from the video with a Scion LG3 image board in a Macintosh 7100 computer using NIH Image 1.60 (developed at the US NIH and available from the Internet by anonymous FTP from zippy.nimh.nih.gov). The images are density sliced and the threshold automatically determined by NIH Image.

10.3.3 Design Concerns

Developing a micromachined device for a biological application involves both the design and fabrication of the device as well as its biological compatibility with the cells or tissue of interest. During the development of the device described in this chapter problems in both of these areas were encountered. In the first generation of the TRaCs, the sacrificial layer that was applied in order to "bury" the beams beneath the substrate could not be completely removed, and the beams were fused to the well in which they were contained. This problem was corrected in the next design, but this generation of chip lost most of its pads during the final etch in hydrofluoric acid. By performing a titration curve of etch times, we were able to pick a time which released the beams and maintained the integrity of the pads. However, cells did not grow well on these mechanically functional chips. If the TRaC was placed in a petri dish that was then seeded with cells, cells on or near the chip often died, despite the fact that cells grew well on unetched substrata used as controls. We suspect that some residual hydrofluoric acid was probably trapped in the spaces beneath the surface. Accordingly, chips that were soaked in a buffer for longer periods of time after release had more favorable cell growth. We found that the majority of cell viability problems were alleviated by soaking the chip for several hours in a 25% solution of $MgSO_4$ to sequester any free Fl^- and then rinsing and coating with extracellular matrix. Some of the design limitations that are still present include the two longer beams, 0.36 and 0.86 mm, possibly exhibiting "stiction" which results in the beams adhering to the walls of the channel so that they rest beneath the surface of the chip. Many of these beams are "pinned" upon release. We suspect that residual internal stresses or stress gradients between the layers cause the top layer of amorphous silicon to "peel" and eventually decay the chip surface over time.

10.4 Examples of Measurements Made with the TRaC

Figure 10.6a shows a fish keratocyte migrating across a 25 μm² pad. Scales were isolated as described previously (Kucik et al. 1991), but grown on TRaCs in 70% fish Ringer's (Cooper and Schliwa 1985), 30% Phenol-red free DMEM supplemented with 10% fetal bovine serum and 20 mM HEPES. Chips were placed in a sealed chamber made of a glass slide and coverslip

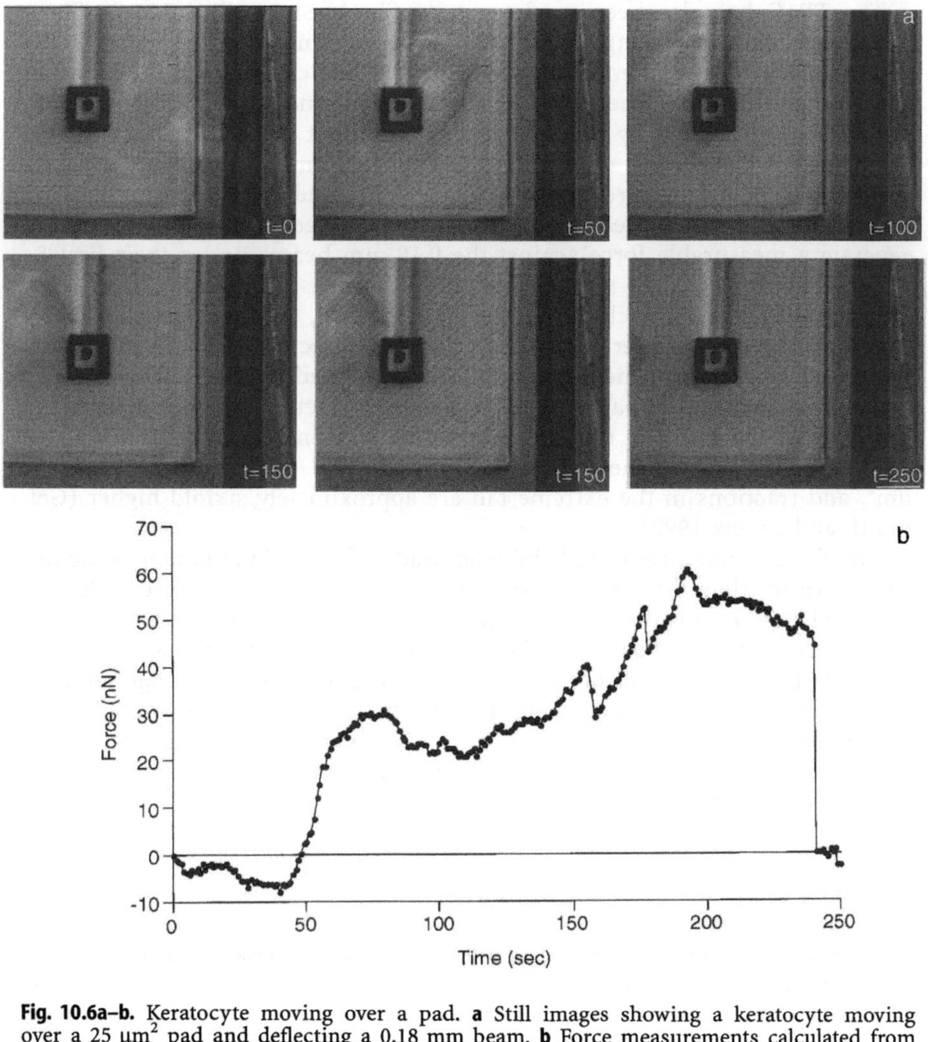

Fig. 10.6a–b. Keratocyte moving over a pad. **a** Still images showing a keratocyte moving over a 25 μm² pad and deflecting a 0.18 mm beam. **b** Force measurements calculated from displacement of pad. Positive force indicates movement of the beam in the direction of cell migration

that was mounted on the motorized stage of the microscope. The TRaC was visualized with light limited to 640 nm $<\lambda<$760 nm. An additional polarizer and an analyzer were placed in the optical path to obtain more cell contrast. The entire experiment was recorded on SVHS tape and individual frames were digitized and analyzed as previously described. Figure 6b shows that the front of the keratocyte pulls the beam underneath the cell toward its center, but the rear of the cell pulls the beam forward in the direction of cell migration. These directions agree with those determined from the silicon sheet experiments, and the value of these forces are on the same order of magnitude (Lee et al. 1994). These results demonstrate the ability of the TRaC to measure the forces generated by motile cells and also show the changes in force as a function of time.

The TRaC has also been used to study fibroblast motility. Motile fibroblasts have a preferred angle of orientation, so the measured force can easily be corrected for any attenuation made by the cell crossing the beam at an angle other than 90°. Therefore, all displacement measurements made with fibroblasts are divided by the sine of the angle that the cell makes with the beam. Forces are plotted according to the convention that positive (+) values are oriented along the direction of cell migration and (–) values are oriented against the direction of cell migration. The leading edge of fibroblasts do not generate a measurable force against the 0.18 mm beams under their first 5–10 μm. The force is then negative, meaning that it is against the direction of cell migration. As the nuclear region passes the pad, the force switches direction, and the force under the rear is positive, directed with cell migration. From each experiment, the maximum force generated by that cell is recorded and normalized by the pad area to determine a traction (force/unit area). In the front of the cell, the traction is negative and on the order of 1–1.4 nN/μm^2; tractions behind the nucleus are positive and on the order of 0.6 nN/μm^2, and tractions in the extreme tail are approximately sixfold higher (Galbraith and Sheetz 1997).

The force is not steady, and the magnitude of some fluctuations is significantly greater than that of measurement noise. Measurement noise is defined as any change in position of the centroid of the pad in the direction parallel to the long axis of the beam. Fluctuations twofold greater than the noise occurred in less than half of the cells analyzed and have an average duration of several minutes, suggestive of non-steady force generation or adhesion remodeling. Labeling these CEFs with an antibody to beta-1 integrin illustrates that only a few adhesive contacts generate the forces measured by each pad (Galbraith and Sheetz 1997).

10.5 Summary

The experiments outlined above demonstrate that the TRaC can measure the forces generated by several adhesive contacts in motile cells. In the future we would like to extend these studies by measuring the forces of individual adhesive contacts. Obviously, the TRaC lends itself to further substrate patterning to investigate the role of various extracellular matrices on the forces generated through these adhesive contacts. Future generations of the TRaC will eliminate the need to use polar cells to obtain accurate force measurements by allowing movement in multiple axes.

This work demonstrates that the use of micromachining can enable us to obtain answers to questions which were previously unavailable because of technological limitations. Unfortunately, its widespread use in cell biology may be limited at this time because of the high cost of developing and manufacturing limited numbers of devices.

Acknowledgments. The staff at MCNC was instrumental in fabricating the TRaC, in particular Dave Koester, Vijay Dhuler, Richard Faire, Scott Goodwin-Johansson, and Karen Markus. We thank Bruce Nicklas, members of his lab, Dahong Zhang and Suzy Ward, and Fred Siedenburg of Sutter Instruments for their assistance in making microneedles and calibrating the levers. Ken Yamada provided the hybridomas for the ES66 antibody, and Jim Galbraith provided suggestions and assistance with the figures. Funding was provided by MCNC, a NIH grant (MPS), the Geller Fund (MPS), and the Whitaker Foundation (Duke Center for Cellular and Biosurface Engineering).

References

Akin T, Najafi K, Smoke R, Bradley R (1994) A micromachined silicon sieve electrode for nerve regeneration applications. IEEE Transact Biomed Eng 41:305–313

Angell J, Terry S, Barth P (1983) Silicon micromechanical devices. Sci Am 248:44–55

Brunette D (1986a) Fibroblasts on micromachined substrata orient hierarchically to grooves of different dimensions. Exp Cell Res 164:11–26

Brunette D (1986b) Spreading and orientation of epithelial cells on grooved substrata. Exp Cell Res 167:203–217

Chehroudi B, Gould T, Brunette D (1990) Titanium-coated micromachined grooves of different dimensions affect epithelial and connective-tissue cells differently in vivo. J Biomed Material Res 24:1203–1219

Clark P (1996) Cell and neuron growth cone behavior on micropatterned surfaces. In: Hoch H, Jelinski L, Craighead H (eds) Nanofabrication and biosystems. Cambridge University Press, New York, pp 356–366

Clark P, Connolly P, Curtis A, Dow J, Wilkinson C (1987) Topographical control of cell behavior: I. Simple step cues. Development 99:439–448

Clark P, Connolly P, Curtis A, Dow J, Wilkinson C (1990) Topographical control of cell behavior. II. Multiple grooved substrata. Development 108:635–644

Clark P, Connolly P, Curtis A, Dow J, Wilkinson C (1991) Cell guidance by ultrafine topography in vitro. J Cell Sci 99:73–77

Cooper M, Schliwa M (1985) Electrical and ionic controls of tissue cell locomotion in DC electric fields. J Neurosci Res 13:223–244

Dembo M, Oliver T, Ishihara A, Jacobson K (1996) Imaging the traction stresses exerted by locomoting cells with the elastic substratum method. Biophys J 70:2008–2022

Dennerll T, Joshi H, Steel V, Buxbaum R, Heideman S (1988) Tension and compression in the cytoskeleton of PC-12 neurites II: quantitative measurements. J Cell Biol 107:665–674

Fung Y (1969) A first course in continuum mechanics. Prentice Hall, Englewood Cliffs, 340 pp

Galbraith C, Sheetz M (1997) A micromachined device provides a new bend on fibroblast traction forces. PNAS 11:9114–9118

Gomez T, Letourneau P (1994) Filopodia initiate choices made by sensory neuron growth cones at laminin/fibronectin borders in vitro. J Neurosci 14:5959–5972

Hammarback J, Palm S, Furcht L, Letourneau P (1985) Guidance of neurite outgrowth by pathways of substratum adsorbed laminin. J Neurosci Res 13:212–220

Harris A, Wild P, Stopak D (1980) Silicone rubber substrata: a new wrinkle in the study of cell locomotion. Science 208:177–179

Harris A, Stopak D, Wild P (1981) Fibroblast traction as a mechanism for collagen morphogenesis. Nature 290:249–251

Hoogerwerf A, Wise K (1994) A three-dimensional microelectrode array for chronic neural recording. IEEE Transact Biomed Eng 41:1136–1146

Kindt R, Lander A (1995) Pertussis toxin specifically inhibits growth cone guidance by a mechanism independent of direct g protein inactivation. Neuron 15:79–88
Kovacs G, Storment C, Rosen J (1992) Regeneration microelectrode array for peripheral nerve recording and stimulation. IEEE Transact Biomed Eng 39:893–902
Kucik D, Kuo S, Elson E, Sheetz M (1991) Preferential attachment of membrane glycoproteins to the cytoskeleton at the leading edge of lamella. J Cell Biol 114:1029–1036
Lamoureaux P, Buxbaum R, Heideman S (1989). Direct evidence that growth cones pull. Nature 340:159–162
Lee J, Leonard M, Oliver T, Ishihara A, Jacobson K (1994) Traction forces generated by locomoting keratocytes. J Cell Biol 127:1957–1964
Najafi K, Wise K, Mochizuki T (1985) A high-yield IC compatible multi-channel recording array. IEEE Transact Electronic Devices 32:1206–1211
Oliver T, Dembo M, Jacobson K (1995) Traction forces in locomoting cells. Cell Motility Cytoskeleton 31:225–240
Oliver T, Lee J, Jacobson K (1994) Forces exerted by locomoting cells. Sem Cell Biol 5:139–147
Usami S, Wung S, Skierczynski B, Skalak R, Chien S (1992) Locomotion forces generated by a polymorphonuclear leukocyte. Biophys J 63:1663–1666

CHAPTER 11

Viscoelasticity, Rheology and Molecular Conformational Dynamics of Entangled and Cross-Linked Actin Networks

E. Sackmann

Contents

Physics Department, Biophysics Laboratory, Technische Universität München, 85747 Garching, Germany, Tel.: +49 89 289–12471, Fax: +49 89 289–12469, e-mail: sackmann@physik.tu-muenchen.de

11.1 Introduction

This chapter deals with the viscoelastic properties of purely entangled and cross-linked networks of actin. It concentrates on experimental techniques, physical concepts and biological implications. In particular the question of correlations between the phenomenological viscoelastic parameters (measured for instance in a rheometer) and the conformational dynamics or the bending elasticity of single polymer filaments is addressed. Another central aim is to show that high precision measurements of viscoelastic properties provide a very useful tool to detect and analyse (even subtle) effects of the manifold of actin manipulation proteins on the structural and dynamic properties of the actin-based cytoskeleton.

Last not least I attempt to show that actin is highly interesting from the point of view of pure physics. It is a prototype of a semiflexible macromolecule enabling studies of fundamental physical properties of macromolecular fluids and gels on a micrometer scale, thus yielding direct insight into correlations between phenomenological and molecular properties. One prominent example is the direct visualization and analysis of the self-diffusion of actin filaments in networks which yielded a first direct proof of the beautiful concept of diffusion by reptation introduced by de Gennes (Käs et al. 1994; see also J. Käs, this Vol.).

New scaling laws characteristic for semiflexible macromolecules have been established by quasielastic light scattering studies (Schmidt et al. 1989; Janmey et al. 1995) by rheology (Zaner and Stossel 1983; Janmey et al. 1990; Müller et al. 1991) or by direct visualization of actin filaments using microfluorescence (cf. Käs et al. 1996 for an overview). This has greatly promoted the development of new theories of semiflexible macromolecular systems (MacKintosh et al. 1995; Isambert and Maggs 1996; Kroy and Frey 1996) which revealed the remarkable differences between networks made up of semiflexible and flexible macromolecules. In fact this is a beautiful example of the stimulation of progress in physics by biological material research. On the other side these theories are absolutely necessary in order to relate phenomenological physical parameters as measured by rheology to molecular properties.

Depending on their flexibility macromolecules are divided into three classes: flexible, semiflexible and stiff chains. The cell cytoskeleton is made-up of all three classes. Spectrin, dystrophin or fodrin (the major elements forming the membrane-associated fishnet-like intracellular scaffolds) belong to the flexible class although they consist of rather large (triple-helical) rigid domains of about 100 Å diameter (forming the segments) which are interconnected by highly flexible hinges. Actin and intermediate filaments belong to the class of semiflexible filaments while microtubili can be considered as rigid rods.

Actin appears to be the most interesting candidate of the three subsystems of the cytoskeleton for the following reasons:

1. It is a living polymer which exists in a dynamic equilibrium state owing to the fact that it exhibits a fast growing and a slowly growing end (cf.

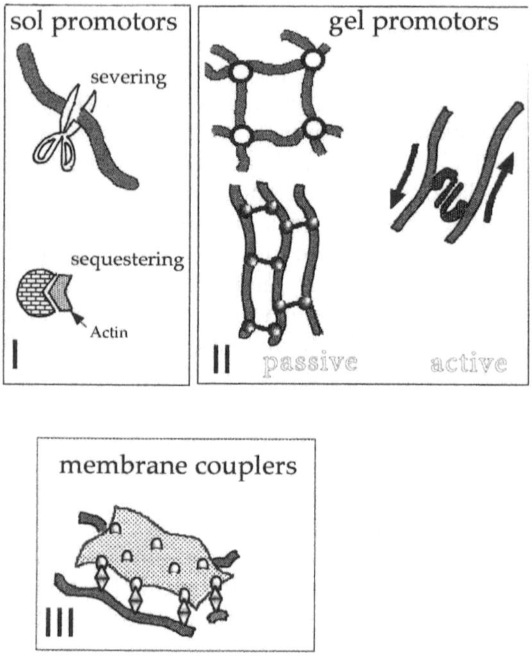

Fig. 11.1. Schematic view of exceptional features of actin-based cytoskeleton. Schematic view of three major sub-systems of actin-manipulating proteins. These may be classified: **I** as polymers controlling the filament length (comprising severing and sequestering proteins); **II** as cross-linking proteins (such as α-actinin and filamin) but which include also myosin II; **III** as actin-membrane-coupling proteins such as talin and hisactophilin.The activity of the manipulating proteins can in addition be controlled by second messengers

Korn et al. 1987 for a review). Therefore the chain length can be controlled most effectively by variation of the concentration of polymerizable (non-sequestered) actin monomers or by blocking the end with capping molecules.

2. Nature provides a whole phalanx of actin-manipulating proteins (AMP's). These include (cf. Fig. 11.1):
   - Firstly, molecules which can control the chain length such as sequestering molecules (e. g. profilin), severing proteins (e. g. gelsolin) or cappers (e. g. gelsolin). These molecules can simultaneously act as nucleators and can thus in principle both promote and impede actin filament growth.
   - Secondly, molecules which can cross-link actin networks. Depending on their structure or flexibility the cross-linkers tend to form random networks or bundles. Below we will argue that the structure of the cross-linked network depends on the concentration of both actin and cross-linker and bundling appears to be always favoured at high cross-linker concentrations independently of the structure of the cross-linker (cf. Chap. 8). To the class of cross-linkers one may also count myosin II.
   - Thirdly, actin-membrane coupling molecules. Examples are talin or hisactophilin. The latter is a pH-dependent actin-binding protein which may act simultaneously as cross-linker and nucleator (cf. Behrisch et al 1995).
3. The activity of the actin-manipulation proteins may be controlled by second messenger or by phoshorylation. A prominent examples of the former

control mechanism is gelsolin which may be activated by Ca^{2+} (which is thus promoting gel–sol transition) or inhibited by PIP_2 (thus promoting gelation; Janmey 1995). An example of the second control pathway are bundles of myosin II which may be decomposed (and therefore effectively inactivated) by phosphorylation of a tyrosine side group at the heavy chain (Pasternak et al. 1989).

The outstanding features of the actin-based cytoskeleton are responsible for the highly complex elastic and rheological properties of cells. The overall cellular deformability is determined by the interplay of all three partially interpenetrating subsystems of the cytoskeleton but also by the intracellular compartments and vesicles. The latter may interact with the macromolecular networks through direct coupling and by competing for the available free volume enclosed by the plasma membrane.

Cells may resist astonishingly high shear forces varying from 10^2–10^4 Pa (Evans 1993). However, the deformability is not a well defined quantity since cells can change it by orders of magnitude in response to an external perturbation. Moreover, it varies locally. Judged from our present knowledge this large variability of the cell elasticity is largely determined by the actin moiety of the cytoskeleton and its coupling strength to the lipid–protein bilayer of the plasma membrane. Indeed, the composite shell comprising the bilayer and the associated actin–myosin cortex forms the machine responsible for cell locomotion.

A successful strategy to come to an understanding of the physicochemical basis of the self-organization and function of biological material is based on systematic and parallel studies of natural systems and reconstituted model systems of growing complexity. This strategy proved very successful in the field of membrane research (cf. Lipowsky and Sackmann 1995). In the same way systematic studies of the viscoelastic properties of in vitro models of the actin-based cytoskeleton will help to come to a better understanding of the physical basis of the control of chemo-mechanical processes of cells by the actin cortex.

The field has gained further impetus by the hypothesis that the mechanical coupling between the plasma membrane and the nucleus may play a role in the regulation of the cell composition and function by signal cascades (cf. Forgacs 1995 for a review).

In this review we intend to summarize the basic physical principles of the viscoelasticity of macromolecular fluids and gels and in particular point out the distinct features of networks composed of semiflexible macromolecules. Special emphasis is put on the universality of elastic properties of entangled and cross-linked networks.

It is further emphasized that high precision measurements of viscoelastic parameters by rheology provide a powerful tool to study the modulation of the stuctural and dynamic properties of both single filaments or networks by actin-binding proteins.

11.2 Definition of Chain Flexibility and Geometrical Characterization of Macromolecular Networks of Semiflexible Filaments

11.2.1 Bending Elastic Modulus of Wormlike Chains (Fig. 11.2)

The flexural rigidity of semiflexible macromolecules may be characterized by two equivalent stiffness parameters: the bending elastic modulus B (in units of energy times length) and the persistence length L_p. These measures are connected by the universal relationship

$$B = k_B T \cdot L_p \,, \tag{1}$$

where k_B is the Boltzmann constant and T the absolute temperature.
B is defined through the elastic energy associated with the local bending of the filament which is given by

$$G_{Ela} = \frac{1}{2} B \int_0^L \left(\frac{\partial^2 R}{\partial s^2} \right)^2 ds \,, \tag{2a}$$

where the integral extends over the length L of the contour. R(s) denotes the local deflection of the chain from an original straight line and $\partial^2 R/\partial s^2$ is the inverse of the local radius of curvature $\rho(s)$ (cf. Fig. 11.2). Equation (1) is Hooks law for a weakly bent (free) rod, where the strain is equal to the local curvature $\partial^2 \vec{R}(s)/\partial s^2$. Sometimes the bending energy is expressed in terms of the local variation of the tangent to the filament $\vec{t} = (\partial \vec{R}/\partial s)^2$ (cf. Käs et al. 1993).

$$G_{Ela} = \frac{1}{2} B \int_0^L \left(\frac{\partial \vec{t}}{\partial s} \right)^2 ds \tag{2b}$$

The persistence length, L_p, is defined by considering the randomness of the orientation of the local tangent to the filament $\overrightarrow{t(s)} = \partial \vec{R}/\partial s$. Any segment can perform local Brownian motions (rotational and translational diffu-

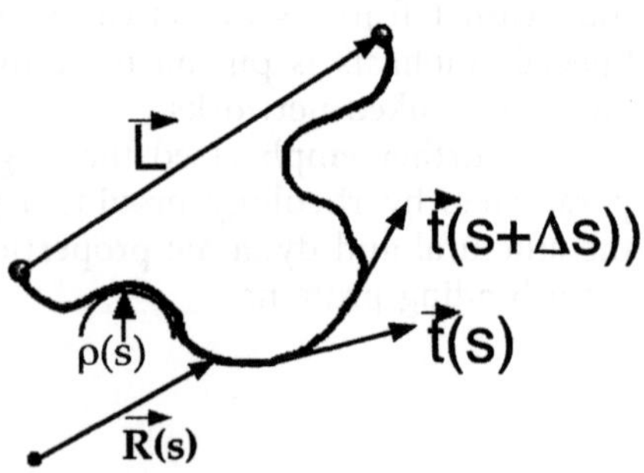

Fig. 11.2. Momentary conformation of semiflexible (wormlike) polymer chain of contour length L and definition (1) of radius vector of chain segment ($\vec{R}(s)$) at contour length s, (2) of local radius of curvature ($\rho(s) = (\partial^2 \vec{R}(s)/\partial s)^{-1}$) and (3) of local tangent $\vec{t}(s) = \partial \vec{R}(s)/\partial s$. $\vec{L}$ is the vector characterizing the end-to-end distance

sion) owing to thermal fluctuation forces exerted by the environment. Owing to the finite stiffness of the filaments the diffusive motions of neighbouring segments (s and s′) are correlated and the correlation is expected to decrease with the distance $\Delta s = s - s'$ between the segments. For random processes obeying Gaussian statistics (such as the segment Brownian motion) an exponential law holds for the correlation between the orientation of two segments separated by a distance Δs along the contour.

$$\langle \vec{t}(0)\,\vec{t}(\Delta s)\rangle = \left\langle \int_{-\infty}^{+\infty} \vec{R}(s)\,\vec{R}(s+\Delta s)\,ds \right\rangle = \exp\{\Delta s/L_p\} \tag{3}$$

where $\langle ... \rangle$ denotes the statistical average (or time average) over all conformations. Note that the correlation function is only dependent on the absolute difference between the two segments.

The major value of this geometric definition of chain flexibility is that the chain stiffness (that is L_p) can be determined by measuring the end-to distance of the filaments for instance by analysis of microfluorescence images. This distance can be calculated by classical methods of statistical mechanics (Chap. 8.8 in Doi and Edwards) and is

$$\langle L^2\rangle = \langle |\vec{R}(\vec{L}) - \overrightarrow{R(0)}|^2\rangle = 2\,L_p^2\left[\frac{L}{L_p} - (1 - \exp(-L/L_p))\right]. \tag{4}$$

There are two limiting situations:

1. The random coil: $L_p \ll L$

$$\langle L^2\rangle = 2\,LL_p = 2\left(\frac{L}{L_p}\right)L_p^2 \tag{5}$$

which corresponds to an ideal random coil made-up of L/L_p segments of diameter L_p.

2. The rigid rod: $L_p \gg L$

$$\langle L^2\rangle = L^2\,. \tag{6}$$

11.2.2 Entanglement regimes of polymer solutions

Solutions of semiflexible and stiff macromolecules exhibit four concentration regimes with well distinguished viscoelastaic behaviour: dilute, semidilute, concentrated and nematic liquid crystalline. The crossover between the regimes depends critically on the chain length and may exhibit further subtleties (cf. Fig. 11.3 below).

In the first regime the filaments are isolated and only the viscosity of the fluid solution is increased owing to the impeded rotational and lateral diffusion.

When the average distance between the filaments becomes comparable to the average filament length $\langle L\rangle$ the molecules become entangled and the so-

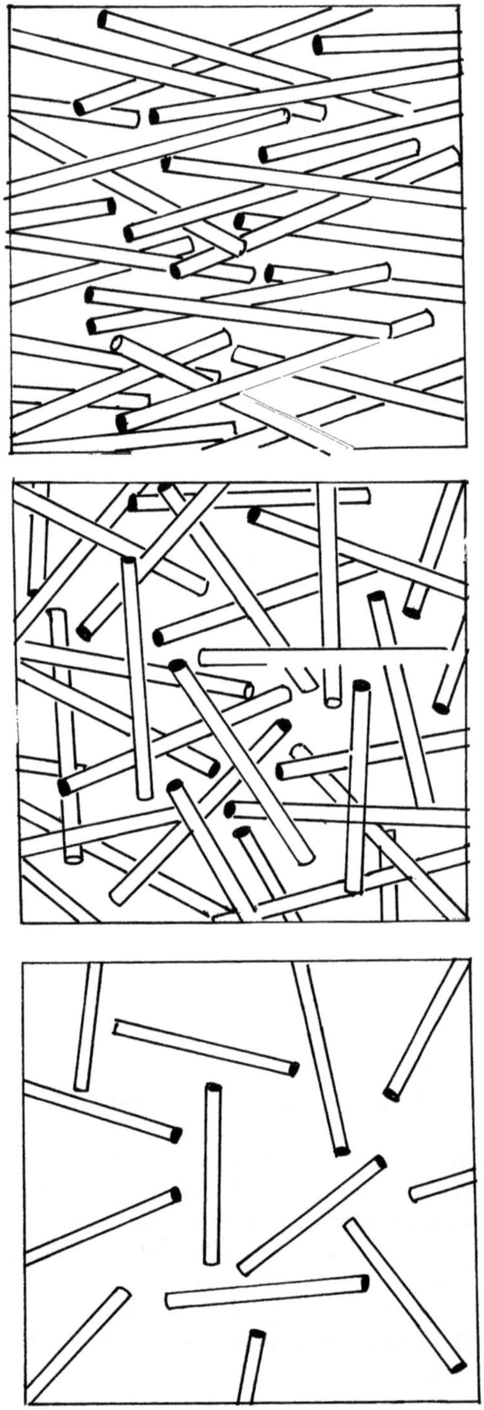

Fig. 11.3 a–c. The three concentration regimes of semiflexible or rod-like polymer filaments. **a** Dilute solution. **b** Semidilute solution to concentrated isotropic solution. **c** Liquid crystalline solution. The concentration can be given in terms of (1) monomer number density, n, (2) the actin monomer concentration c_A (Mol/ltr) or (3) the fraction of the volume occupied by the polymer, φ_P. The structure of semidilute and concentrated isotropic solutions is characterized by the geometrical mesh size ξ

lution exhibits elastic behaviour which is, however, strongly dependent on the chain length. The dilute-to-semidilute transition occurs at a monomer concentration (measured in units of monomers/m^3) of

$$n/\bar{N} = \langle \bar{L} \rangle^{-3} \tag{7}$$

where $\bar{N}$ is the average number of monomers per filament. Note that $n/\bar{N}$ can be expressed as $\rho N_L/\bar{M}$ where N_L is Avogadros number and ρ is the polymer mass density (mass per unit volume) and $\bar{M}$ is the average molecular weight per filament.

In the concentrated regime the average distance between the filaments is small compared to the filament length and the viscoelastic properties become only dependent on a single geometric parameter: the effective mesh size ξ (= average distance between filaments, cf. Fig. 11.3) but is independent of chain length (cf., however, Sect. 5.5 for a more subtle consideration of this point).

The transition to the fourth regime, the ordered fluid, is a consequence or the finite diameter a of the filaments. A freely rotating molecule requires a free volume of the order $a \times L^2$ and (for a given concentration) the repulsion between the molecules increases strongly with increasing length. Similar to the behaviour of Van der Waals gases this repulsion leads to a condensation of the solution with parallel orientation of the molecules (thus reducing the entropy associated with the molecular rotation). As shown in a seminal paper by Onsager (1994), the transition to the anisotropic (in general nematic) phase for rigid rods occurs if the polymer number density $n/\bar{N}$ (filaments/cm^3) becomes of the order of aL^2. Simultaneously, the volume fraction, φ_P, of the polymer (volume occupied by polymer to volume of solution) increases abruptly. The exact change depends on the ratio of the length to the diameter of the filament and the theory predicts $\varphi_P^{iso} \approx 3.3\ L/a$ for the isotropic and $\varphi_P^{aniso} \approx 4.5\ L/a$ for the nematic phase. For the semiflexible macromolecules the isotropic–nematic transition occurs at higher polymer volume fractions and is shifted to higher values with increasing chain flexibility (cf. Käs et al. 1996 for references).

11.3
Basic Principles of Viscoelasticity

11.3.1
The Elastic Modulus Is a Time-Dependent Quantity

Viscoelasticity is a very broad field since it has to deal with many different types of matter ranging from geological materials to body tissues or bones. Despite its conceptional simplicity the field is very complex since it involves continuum mechanics, hydrodynamics and molecular dynamics of materials. Since we are only concerned with soft actin networks of low density we restrict ourselves to a presentation of the basic principles and methods of measurement of viscoelastic parameters by rheological methods adapted to such systems. For further reading on phenomenological principles the book by Ferry (1980) on viscoelasticity of polymers is highly recommended.

Fig. 11.4. Definition of viscoelastic parameters for simple case of shear tension σ_{zx} (force per unit area) applied paralled to the horizontal plane at height $z=\Delta z$ in direction x. $\gamma(t)$ is the time dependent tilt angle. Note that on average a quadratic cross section is deformed into a parallelogram by stretching of one diagonal and compressing of the other without changing the volume of the body. The shear deformation may also be realized by two (equal) tensions of opposite sign applied in the direction of the diagonals of the square (cf. Fung (1981) for a clear and simple representation)

The most complex problem is to find connections between phenomenological viscoelastic parameters and the molecular structure and dynamics of the biomaterials which are of course of primary interest in life science. A comprehensive theoretical treatment of the viscoelasticity of flexible polymers is given in the book of Doi and Edwards (1986) which should be consulted by anybody who wants to go deeper into the field. We are, however, concerned with semiflexible macromolecules, for which the theoretical basis is just laid (cf. Sect. 11.5).

In order to introduce the concept of viscoelasticity we consider for simplicity a cubic body composed of a purely entangled polymer network. If we apply a pure shear tension (=stress) parallel to the top surface (while keeping the bottom one fixed) the body will be tilted as shown in Fig. 11.4. The deformation (=strain) is simply represented by the tilt angle γ.

It is intuitively clear that the deformation of a soft body after application of a sudden stress depends on time. At very short times ($<\mu s$) the macromolecules will be fixed and the deformation is associated with a distortion of the (random) lattice. Therefore a momentary elastic tension, σ_m, arises within the network which is related to the maximum deflection γ_m by Hook's law:

$$\sigma_m = G(0) \cdot \gamma_m \tag{8}$$

where G(0) is the elastic modulus at time t=0.

After some time (determined by the relaxation time of the filament bending fluctuations) the dynamically wrinkled filaments will stretch resulting in a decrease of the internal stress of the body and a further deformation of the lattice. This corresponds to a smaller shear elastic modulus G, that is G(t) decreases within the time scale required to pull out the wrinkles (elastic regime I).

Now filaments start to diffuse (by reptation) in the direction of the force. However, G(t) remains constant as long as the filaments remain entangled resulting in a plateau regime (II in Fig. 11.5).

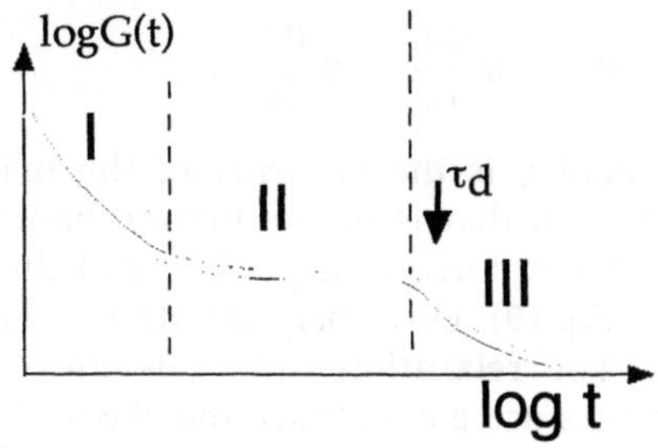

Fig. 11.5. Hypothetical time dependence of the shear elastic modulus of entangled polymer solution showing three regimes: Solid-like behaviour at short times (*III*), a rubber-like plateau (*II*) and a so-called terminal transition regime (*I*) leading to fluid-like behaviour. The latter vanishes after cross-linking the filaments (gelation)

It is intuitively clear that the filaments disentangle if they have moved by a distance of the order of the average end-to-end distance of the chains. The time $\tau_d \approx \langle L^2 \rangle / D$ required for disentanglement is called the terminal relaxation time since after this time the viscoelastic body starts to flow as a liquid (terminal regime III). Since the viscoelastic body flows during the relaxation, its behaviour is also determined by a viscosity $\eta(t)$. It is a measure for the local energy dissipation during the relaxation of mechanical stresses by the internal conformational changes. Cross-linked networks show the same behaviour as entangled solutions at short times, but do not exhibit the terminal regime.

The above consideration shows that dynamic studies of the elasticity provide highly valuable information on the dynamics of molecular conformational changes of the polymeric fluids and gels. A prerequisite is that theory provides the tools to connect the molecular structure and dynamics to the phenomenological viscoelastic parameters $G'(t)$ and $\eta(t)$. Since the elastic stress relaxes with time, $G(t)$ is called the relaxation modulus.

11.3.2
Viscoelastic Bodies May Behave as Solids and Fluids

In order to get a more vivid insight into the viscoelasticity it is helpful to write down the differential equation describing the dynamic elastic response within an entangled network following an enforced shear deformation $\gamma(t)$. The time dependence of the elastic tension within the body is described by the following differential equation:

$$\tau \frac{d\sigma_{xz}}{dt} + \sigma_{xz} = \eta \frac{d\gamma}{dt} \tag{9}$$

which can be rationalized as follows.

For times long compared to τ the first term can be neglected and the internal stress, σ_{xz}, induced by a shear flow is given by Newton's law of internal friction in fluids,

$$\sigma_{xz} = \eta \frac{dv_x}{dz} = \eta \frac{d\gamma}{dt}$$

where η is the viscosity of the material behaving as a fluid. At short times ($t \leq \tau$), that is before the mechanical internal stress starts to relax, the flow-induced force is negligible and the behaviour is determined by the left side of Eq. (9). Obviously the stress relaxes exponentially as $\sigma = \sigma_o \exp(-t/\tau)$.

The relaxation time τ is the time at which the viscous and the elastic forces are equal. Since the elastic force (per unit area) is $G d\gamma$ and the viscous force (per unit area) is $\eta d\gamma/dt$ it follows $\tau G = \eta$ or

$$\tau = \eta / G \tag{10}$$

In Section 11.4.3 we will consider periodic excitations of the viscoelastic body. It is then convenient to replace the viscosity η by a so-called loss modulus and write

$$G'' = 2\pi\eta/\tau = \omega\eta \tag{11}$$

where ω is the angular frequency of the excitation.

Equation (9) can be written as

$$\frac{\tau G'' d\gamma/dt}{2\pi} + G'\gamma = \sigma_{ext} \tag{12}$$

where σ in Eq. (9) is replaced by $\sigma = -G'\gamma$ and σ_{ext} now denotes an externally applied stress. G' is now called the storage modulus. It is easily verified that the two equations (9) and (12) are mechanically equivalent. Equation (12) describes the time dependence of the strain γ after application of a small external stress, while Eq. (9) describes the time dependence of the stress following an enforced shear deformation.

In the following we will show that Eqs. (9) and (12) can be represented by two simple mechanical equivalent circuits.

11.3.3 Mechanical Equivalent Circuits of Viscoelastic Bodies

In order to describe the mechanical behaviour of viscoelastic bodies Maxwell and Boltzmann introduced the concept of mechanical equivalent circuit. Such models are very helpful in order to get more intuitive insights into the viscoelastic responses following transient shear deformations or stresses. For the above reason and since we will make use of the models below we give a brief summary of such equivalent circuits.

The Maxwell Element

The behaviour of the body described by Eq. (9) is completely analagous to the response of a mechanical circuit composed of a spring (spring constant μ) and a dashpot (of viscosity η) in series (cf. Fig. 11.6a). If a step force is applied (at time $t=0$) the spring is immediately displaced by a distance $u=f/\mu$ and the dashpot starts to flow with a velocity $du/dt=f/\eta$. In order to establish the motional equation we consider the time dependence of the dis-

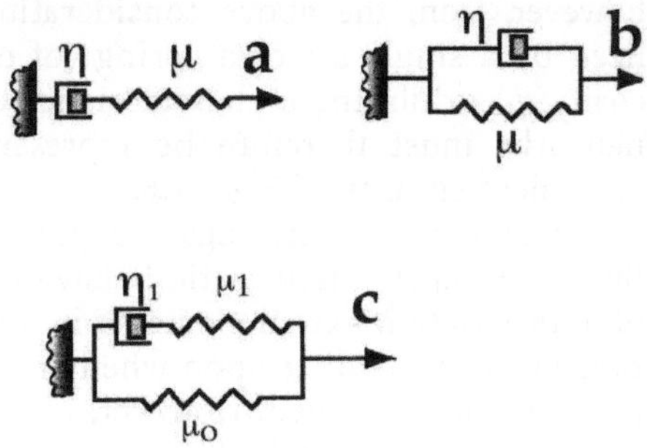

Fig. 11.6 a–c. Mechanical equivalent circuit of viscoelastic bodies. **a** Maxwell body consisting of spring (force constant μ) and dashpot (viscosity η) in series. **b** Voigt body consisting of parallel arrangement of a spring and a dashpot. **c** Kelvin- (or Zener-) circuit consisting of parallel arrangement of Maxwell body and additional spring of force constant μ_0 (cf. the book of Fung (1981) for a more extensive representation of the various equivalent circuits of viscoelastic bodies)

placement (du/dt) which is equal to the sum of the deflections of the spring and of the dashpot, respectively, and one can write

$$\eta\, du/dt = \tau\, df/dt + f \tag{13}$$

with $\tau = \eta/\mu$. This equation is completely analogous to Eq. (9). The meaning of τ becomes evident if we consider the force (stress) within the body after cessation of the flow (at time t=T). It relaxes according to

$$f = f_o \exp(-t/\tau)$$

and the dashpot remains deflected by a distance $u_o = P \cdot du/dt$.

The Voigt Element

Conceptually simpler is a parallel circuit of a spring and a dashpot (the so-called Voigt body shown in Fig. 11.6b). In this case it is convenient to consider the response to an external force f_{ex}. It is counterbalanced by the forces generated in the spring (μu) and in the dashpot (η du/dt), respectively. Thus

$$\eta\, du/dt + \mu u = f_{ex}\,. \tag{14}$$

It is easily verified that after removal of f_{ex} the system relaxes according to $u = u_o \exp\{-t/\tau\}$ (the relaxation time $\tau = \eta/\mu$ is identical to that of the Maxwell body). The deformation response following a step force is $u = u_o$ $(1 - \exp\{-t/\tau\})$ (with $u_0 = f_{ex}/\mu$).

We will show below that the parallel circuit of the Voigt body is best suited to describe the so-called oscillatory experiment while the Maxwell body is best suited to evaluate the so-called creep experiments of entangled networks.

It is clearly evident (and will become more so in Sect. 11.5) that none of the above simple circuits is suited to account for the contributions of the dynamic conformational changes (wrinkling) of the filaments to the viscoelasticity since they are characterized by a single relaxation time. As follows, however, from the above considerations any relaxation process can be realized by a single set of a spring (of elastic constant μ_i) and a dashpot (viscosity η_i) exhibiting a characteristic relaxation time $\tau_i=\eta_i/\mu_i$. Macromolecular networks must therefore be represented by a combination of the simple equivalent circuits of Fig. 11.6.

Let us consider first again a purely entangled network. Since it must be fluid at $t>t_d$ its viscoelastic behaviour can be described by an arrangement of appropriate Maxwell elements in series or of Voigt elements in parallel arrangement depending upon whether a creep-response or a force-response experiment is performed. However, in order to account for the sol-like terminal regime, the Voigt body must be augmented by a dashpot in series with the parallel arrangement of Voigt elements (cf. Fig. 11.12b).

The Kelvin Element

In order to realize the behaviour of a cross-linked network (devoid of the terminal relaxation regime) in a creep experiment a purely elastic element has to be introduced forming a parallel circuit with one or several Maxwell elements. This is called a Kelvin body which is shown in Fig. 11.6c. A somewhat cumbersome consideration of the force equilibrium yields (cf. Fung (1981) for a detailled derivation):

$$f + \tau_1 \frac{df}{dt} = \mu_0 \left[u + \tau_2 \, du/dt\right] \tag{15a}$$

$$\tau_1 = \eta_1/\mu_1 ; \quad \tau_2 = \left(\frac{\eta_1}{\mu_1} + \frac{\eta_1}{\mu_0}\right) . \tag{15b}$$

τ_1 characterizes the relaxation of the force if u is constant, while τ_2 is the time of strain relaxation at constant force.

In summary, the above considerations show:

1. That any viscoelastic body can be represented by a mechanical equivalent circuit;
2. That each relaxation process characterized by a single relaxation time can be represented by a spring–dashpot pair;
3. That, depending on the type of experiment, parallel or in-series arrangements of the pairs of elements should be considered.

It is also essential to point out that the representation of viscoelastic bodies by mechanical circuits is arbitrary since there exist indeed an infinite number of circuits exhibiting the same viscoelastic response curves (cf. Ferry 1980).

11.4 Rheological Techniques

Owing to the time dependence of the elastic modulus, the stress generated at any instant within a viscoelastic body depends on the history of the strain generated by a strong external force. Due to this viscoelastic memory the change of the internal stress $d\sigma(t)$ at time t depends on the strain $d\gamma(t - t')$

at time t. Therefore Hook's law has to be written in the form of a convolution integral

$$\sigma(t) = \int_{-\infty}^{T} G(t-t') \frac{d\gamma(t')}{dt'} dt' \tag{16}$$

Since the memory of materials composed of large polymers of cross-linked networks can last for hours, reproducable results are often difficult to obtain.

Fortunately, the situation may be greatly simplified by performing two types of experiments: namely the observation of the mechanical response after either stepwise or oscillatory excitations. These are briefly summarized below together with experimental techniques enabling macroscopic and microscopic measurements of viscoelastic moduli of bio-macromolecular solutions and gels.

11.4.1
The Oscillatory Experiment (Impedance Spectroscopy)

The simplest approach is to start from the general equation for a purely entangled network (Eq. (9) and consider the shear stress generated by an oscillatory deformation $\gamma(t)$. This is most easily done by working in complex space by introducing the complex elastic impedance (cf. Fig. 11.7)

$$G^*(\omega) = G'(\omega) + iG''(\omega), \tag{17}$$

where the real part is called the storage modulus and the imaginary part the loss modulus. The periodic excitation is then expressed in terms of

$$\gamma^* = \gamma_0 \exp(i\omega t)$$

Inserting into Eq. (9) yields (with $d\gamma/dt = i\omega\gamma$)

$$\sigma^* = \frac{i\omega\eta\gamma^*}{1 + i\omega\tau}$$

by assuming $\sigma^* = \sigma_0 \exp(i\omega t)$. Multiplication of the denominator and the numerator by $(1-i\omega\tau)$ and by considering Eq. (11) yields

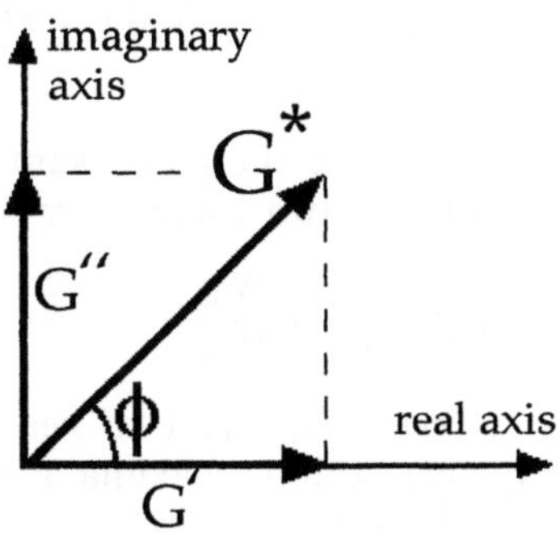

Fig. 11.7. Representation of complex impedance in complex coordinate system with G′ forming the real and G″ the imaginary axis. The complex elastic constant G* is simply given by the addition of the vectors $\vec{G}'$ and $\vec{G}''$ and $\tan\varphi = G''/G'$ the angle formed by the vector G* with the horizontal axis

$$\sigma^* = \left(\frac{\omega^2\,\tau^2}{1+\omega^2\,\tau^2} + \frac{\mathrm{i}\,\omega\tau}{1+\omega^2\,\tau^2}\right) G_0\,\gamma^* = G^* \cdot \gamma^* \tag{18}$$

Comparison with Eq. (16) shows that the first term in the bracket is equal to G'/G_0 and the second to G''/G_0.

As illustrated in Fig. 11.7 it is very helpful to represent the complex elastic modulus in a complex plane by plotting the loss modulus along the imaginary axis and the storage modulus along the real axis. The complex modulus is then determined by the sum of these two vectors.

Note: The reader not familiar with working in complex space can also solve the problem by a cosine type excitation $\gamma = \gamma_0 \cos \omega t$ and show that the differential equation (9) can be solved by

$$\sigma = \frac{\eta\omega\gamma_0}{1+\omega^2\,\tau^2}\,(\omega\tau\,\cos\,\omega t - \sin\,\omega t)\,,$$

which obviously consists of a component which is in phase with the excitation: $G' \propto G_0\,\tau^2\,\omega^2/(1+\omega^2\,\tau^2)$ and one which is out of phase by 90 degrees, $G' \propto G_0\,\tau\,\omega/(1+\omega^2\,\tau^2)$. The advantage of solving the differential equation in complex space is that it can be easily expanded to the more complex visco-elastic bodies.

The origin of the names storage and loss modulus for G' and G'' becomes clear when we consider the energy dissipated during the oscillatory experiment by averaging the energy over one cycle of the experiment.

Since the elastic energy associated with a small change in shear strain is $\delta W = \sigma \cdot d\gamma = G' \cdot \gamma \cdot d\gamma$. The total amount of energy (per unit volume) stored during a half cycle of period $T = \omega/2\pi$ is easily obtained by integration of δW. With $\gamma = \gamma_o \cos \omega t$ one obtains

$$W_{stor} = \frac{1}{T}\int_0^{T/2} G'\gamma\,d\gamma = \frac{1}{4}\,G'\,\gamma_0^2$$

The energy (per unit volume) dissipated due to hydrodynamic flow is equal to the viscous force $\eta\, d\gamma/dt$ times the change in shear, . One obviously obtains for the energy dissipated per cycle

$$W_{disp} = \frac{1}{T}\int_0^{T/2} \eta\,\frac{d\gamma}{dt}\,d\gamma = \pi\,G''\,\gamma_0^2\,.$$

In this image the phase shift angle obtains a simple meaning since it is related to the two energies by

$$\mathrm{tg}\,\varphi = \frac{G''}{G'} \propto \frac{W_{disp}}{W_{stor}}\,. \tag{19}$$

It is very important to keep in mind that the loss modulus G'' given by Eq. (17) exhibits maxima at frequencies $(\omega_i \sim \tau_i^{-1})$ which are associated with a relaxation processes.

11.4.2
The Creep Experiment

The second basic way of studying the viscoelastic response of viscoelastic bodies is based on the analysis of the deformation after application of a step-wise stress σ^* at time $t=0$. For a single step-like shear stress the strain response can be expressed as

$$\gamma(t) = J(t)\,\sigma(t)\,. \tag{20}$$

The elastic constant J(t) is called the compliance and is a measure for the softness of the viscoelastic body.

In the present review we will consider only the simple case of a single step and of a purely entangled network. In this case σ in the above equation represents the applied step-like stress: ($\sigma=0$ at $t\leq 0$, $\sigma=\sigma_0$ at $t\geq 0$). It is easily verified that for this scenario the shear is given by

$$\gamma(t) = J_{ss}\cdot\sigma_0 + \frac{t}{\eta_z}\,\sigma_0\,. \tag{21}$$

The first term represents the elastic response (corresponding to a step of $\gamma(t)$) immediately following the step-like force. J_{ss} is the stationary state compliance. The second term accounts for the fluid like flow. η_z is the limniting viscosity at $\omega\rightarrow 0$ the so-called zero shear viscosity (cf. Sect. 11.4.5). An example of such a creep experiment will be shown below in Fig. 11.11. Note that the network does not respond immediately but with a certain response time. This is a consequence of the above-mentioned dynamic wrinkling of the filaments the pulling-out of which occurs with a temporal delay.

In order to study the large variety of natural and synthetic viscoelastic bodies ranging from fluids to solids an impressive number of viscoelastic techniques has been developed many of which are described in the book by Ferry (1980). Most of these are mechanical methods which are especially suited to measurements in the 10^{-6} to 10^{6} Hz frequency region. An elegant and promising new technique is based on the analysis of shear-induced birefringence (Fuller and Mikkelsen 1988). In the following, techniques are summarized which in the author's view are best suited to study the viscoelasticity of in vitro models of the cytoskeleton.

11.4.3
Classical Rheometers

The classical method of rheology of macromolecular networks is based on the torsional instrument. The sample is contained between a fixed and a rotatable plate. The latter often exhibits a conical cross section. This geometry results in a constant shear in the tangential direction since the height, h, of the sample increases from the centre to the outside in the same way as the tangential velocity, v_t, and the shear rate v_t/h is thus constant. Very elaborate and high precision rheometers of this type are commercially available and belong to the basic equipment of any polymer laboratory.

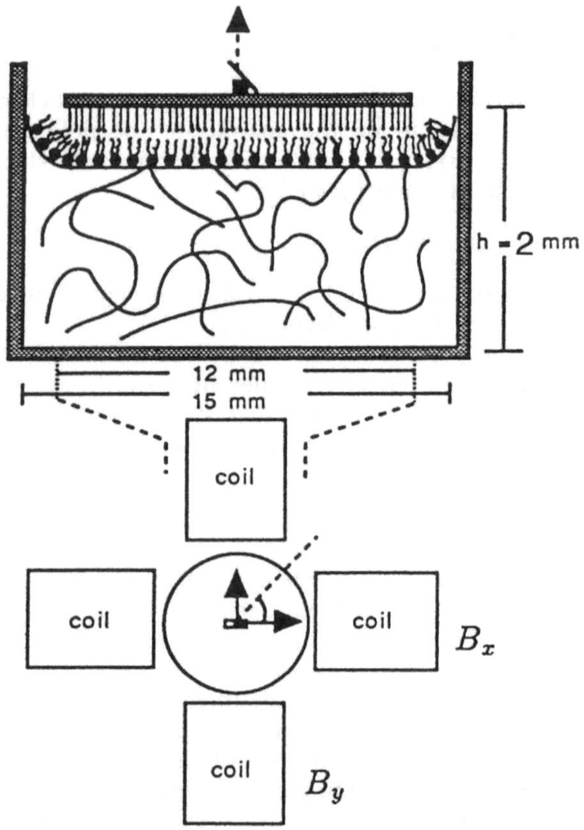

Fig. 11.8. Schematic view of magnetically driven torsional rheometer according to Müller et al. (1991). The inner dimensions of the cuvette are: radius 0.4 cm, height 0.2 cm. The monolayer of lipid on top of the actin solution is essential to prevent denaturing of actin at the air–water interface

Biological networks are often soft since (due to the great length and the semiflexible nature of the filaments) the volume fractions are in general very low: typically $\varphi \leq 10^{-4}$. Moreover contact of the proteins with metal surfaces or the air/water interfaces tend to promote gel formation due to protein denaturing. A simple but very reliable and precise parallel-plate rotational rheometer designed in the author's laboratory is shown in Fig. 11.8. The sample is contained in a cylindrical glass cell and the top plate is oscillated by magnetic forces. Denaturing of the monomeric actin at the air/water interface (and thus polymerization under non-polymerizing conditions) is avoided by covering the top with a lipid monolayer (Müller et al. 1991). The mechanical contact of the top plate with the lipid monolayer is achieved by deposition of a silane monolayer on the lower side of the plate. The force is applied as follows: a small rod-like magnet is fixed to the centre of the top plate and two pairs of coils allow the application of two crossed AC magnetic fields. One serves the definition of a resting orientation and the other the application of an oscillatory force. The precision is greatly improved by shielding the measuring device with a cage of μ-metal. Single domain magnets – such as whiskers – exhibiting very high saturation fields of 10^4 Gauss enable applications of shear fields of up to 10^3 Pa.

An elegant modification of a torsional rheometer was designed by the Harvard group (cf. Janmey et al. 1994). The rheological measurement is based on the analysis of the damped oscillation of a torsional pendulum placed on top of the macromolecular network following a stepwise angular deflection of the plate attached to the end of the torsional wire. The decay constant of the oscillation yields directly the ratio G''/G'.

11.4.4 Microrheometer

Biological macromolecular networks are in general heterogeneous and microrheometers enabling quantitative measurements on μm scales are highly desirable. One promising microrheometer based on the application of magnetic tweezers has been developed by Ziemann et al. (1995). Paramagnetic beads (such as Dynabeads) of μm-diameter are embedded into the polymeric network and observed by phase contrast or ultramicroscopy. The beads are locally deflected by an inhomogeneous magnetic field generated by magnetic coils.

The motion of the bead is analysed by dynamic image processing. As demonstrated below, both oscillatory and creep experiments can be performed (Ziemann et al. 1995). In the case of an oscillatory experiments the phase shift of the deflection of the bead with respect to the applied force is measured together with the deflection amplitude yielding both the storage and the loss modulus [$G'(\omega)$ and $G''(\omega)$]. By embedding non-magnetic colloidal probes the viscoelastic parameters may be simultaneously measured at several sites within the networks.

An important problem for quantitative measurement is that the shear field about a sphere is not homogeneous as in the case of the torsional rheometer. Nevertheless, according to Ziemann et al. (1995) absolute values of the viscoelastic parameters can be measured (cf. Fig. 11.9b). This has been attributed to the fortunate situation that the elastic strain field and the hydrodynamic flow field exhibit the same symmetry. However, more theoretical work is required to clarify this point.

Magnetic tweezers in combination with non-magnetic probes open the possibility of mapping strain fields within macromolecular networks and measuring viscoelastic response curves locally (cf. Schmidt et al. 1996). The method is illustrated in Fig. 11.10. A small number of magnetic beads and a larger number of non-magnetic colloidal probes are embedded into the network. An area containing one magnetic tweezer and an assembly of probe beads is selected for the measurement. As shown in Fig. 11.10, a deflection of the magnetic bead by a step-force is followed by the displacement of the probe beads.

The pure shear experiment enables local measurements of the Young modulus, $E(r)$ and the Poisson ratio, $\delta(r)$. From these parameters the shear modulus $\mu = E/2\,(1+\sigma)$ as well as local viscosities $\eta\,\vec{n}$ can be obtained. A basic assumption is that the force field within the network is determined by the elastic strain field and that the latter can be approximated by the field generated

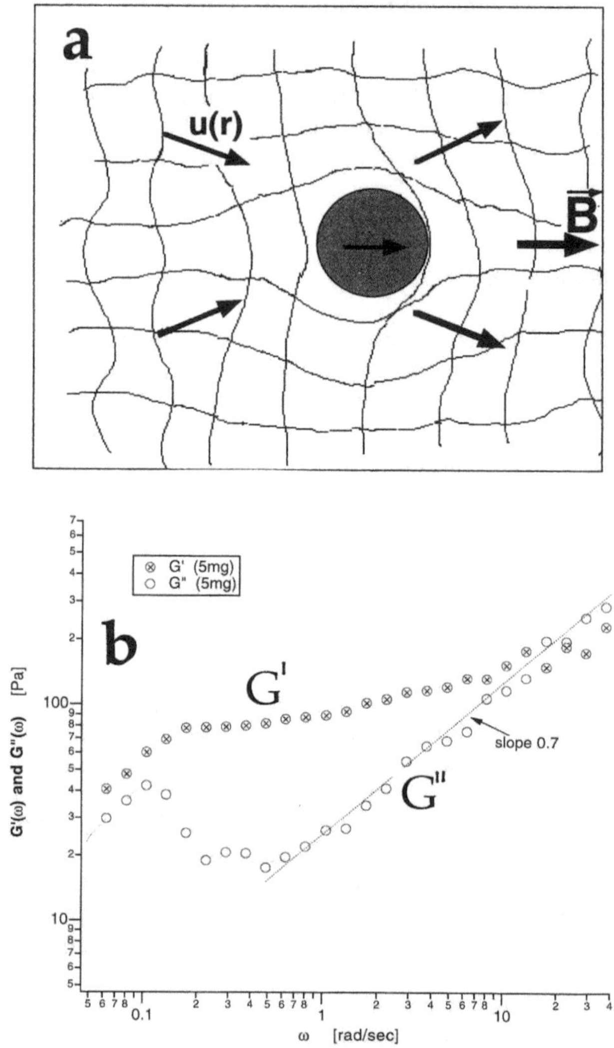

Fig. 11.9. **a** Schematic view of deformation of entangled network by magnetic bead. **b** Frequency dependence of storage modulus of entangled actin network ($C_A = 2.2\ \mu M$, or $\xi \sim 1\ \mu m$) measured with magnetic bead of 4.8 μm diameter. Note the similarity with the equivalent plots obtained by torsional rheometers shown in Fig. 11.12

by a local point force. According to a general postulate of elasticity, the traction on a surface of a spherical cavity (that is the surface bound by the magnetic bead) is equivalent to a force at a point applied at the origin. The deformation field $\vec{u}(\vec{r})$ of a point force is

$$\vec{u}\,\vec{r} = \frac{1+\sigma}{8\pi E(1-\sigma)} \quad \frac{(3-4\sigma)\vec{f} + \vec{n}(\vec{n}\vec{f})}{r} \tag{22}$$

where r is the distance from the point force and $\vec{n}$ is a normal in the direction of the radius vector $\vec{r}$.

Since the local stresses $\sigma(r)$ can be determined by the strain field mapping experiment it is possible to measure the local viscoelastic parameters

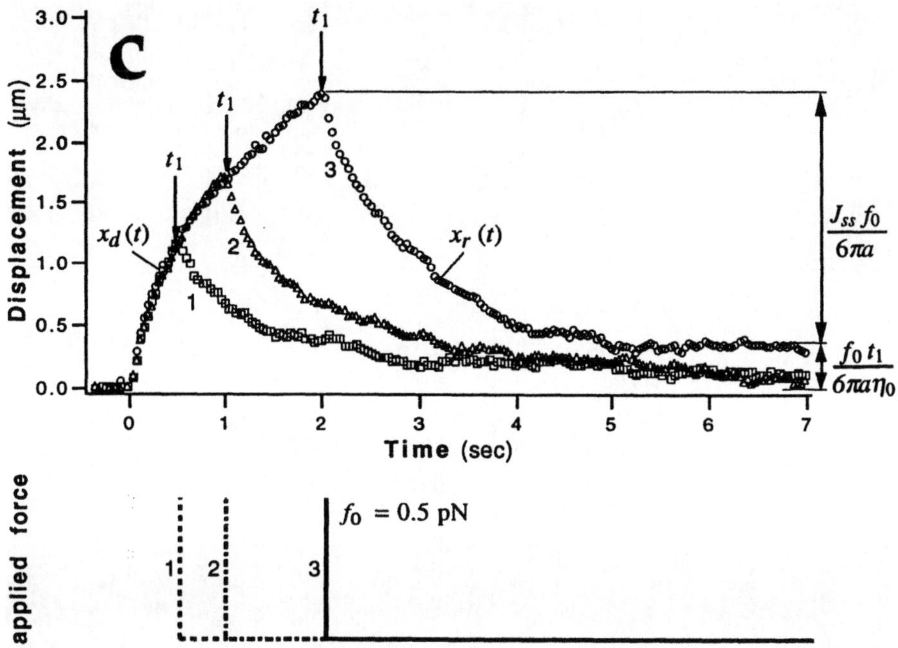

Fig. 11.9. c Creep experiment performed by repeated application of step forces of increasing amplitudes (steps 1–3 indicated by *numbers*

quantitatively by performing creep response or stress relaxation experiments. Fig. 11.10b shows an example of the former. It is seen that the trajectories of the colloidal probes in the direction parallel and perpendicular to the force direction can be analysed in terms of a creep response curve

$$\pi a^2 u(t)/\sigma = J_{ss} + t/\eta , \tag{23}$$

where σ is the local stress and u(t) is the deflection in the parallel and perpendicular direction. J_{ss} is the component of the stationary compliance in the two directions. A major benefit of the microrheological technique based on magnetic tweezers is that it allows the mapping of the local viscoelastic parameters of heterogeneous actin networks (Schmidt and Sackmann, unpublished results).

The Young modulus and Poisson ratio are determined locally by measuring the ratio u_x/u_y of the deflection of the colloidal probes parallel (u_x) and perpendicular, u_y, to the force f and the absolute value $u = \sqrt{u_x^2 + u_y^2}$ of the displacement. For the case of Fig. 11.10 a Poisson ratio $\sigma = 0.4$ was found which is close to the upper bound $\sigma \approx \frac{1}{2}$ for an ideally incompressible body. The shear modulus agrees within a factor of 1.3 with the value measured by the torsional rheometer, showing that the basic assumptions of the strain field mapping are fulfilled.

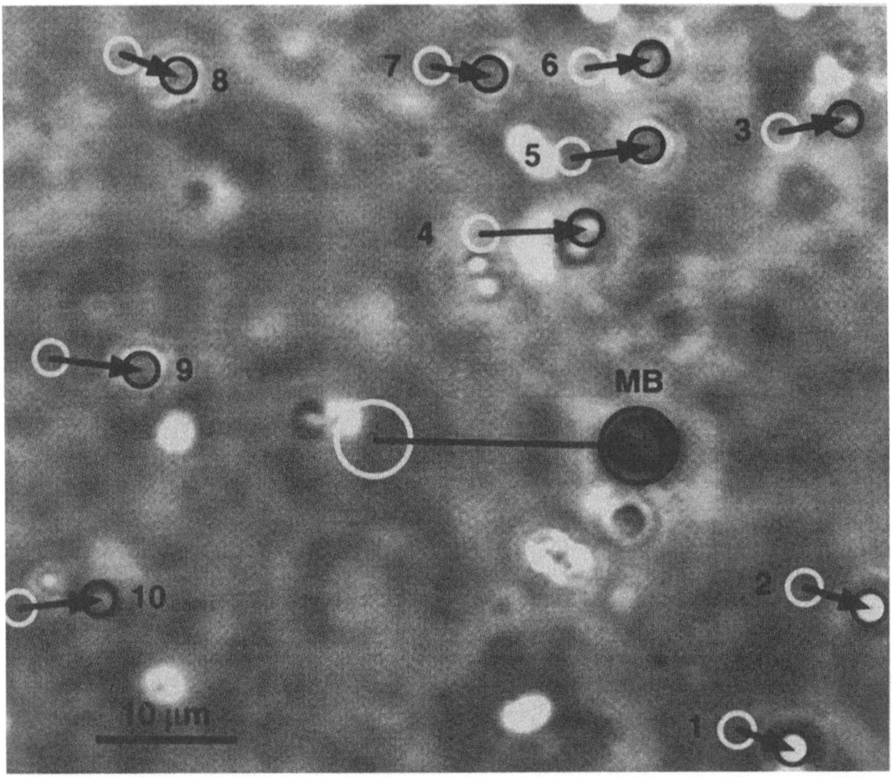

Fig. 11.10. a Visualization of strain field within cross-linked actin network. A paramagnetic 5 μm diameter bead (*MB*) together with an assembly of non-magnetic (2 μm) beads (numbered *1* to *10*) is embedded in the macromolecular network (here partially cross-linked actin of mesh size $\zeta \sim 1$ μm). Local deflection of magnetic bead by a force pulse induces displacement of the colloidal probe as shown by marking the initial and final positions of beads by *bright* and *dark contours*, respectively

11.4.5 Zero Shear Viscosity

The creep experiment is the method of choice to measure the static viscosity, the so-called zero shear viscosity

$$\eta_z = \lim_{\omega \to 0} \{G''(\omega)/\omega\}. \tag{24}$$

The zero shear viscosity is a highly important parameter since it gives insight into the degree of entanglement at the transition from the dilute to the semidilute solution. Owing to the large filament length ($L \sim 25$ μm) η_z is difficult to measure for actin solutions by oscillatory experiments. The method of choice is the creep experiment. It is customary to measure η_z by plotting J(t)/t and m J(t)/t (with m=dlog J(t)/dt) as a function of time and to extrapolate these curves to t=0. Of course only the terminal regime of the J(t)-versus t curves can be used for this procedure. Fortunately, the creep compli-

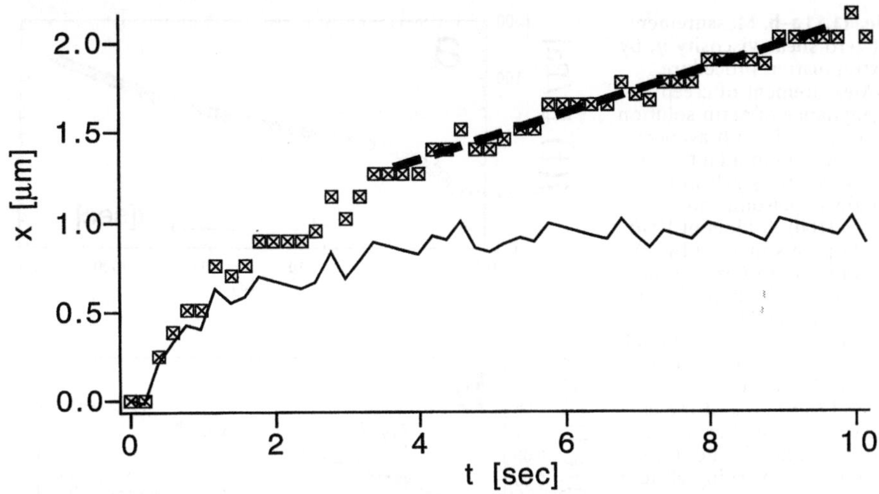

Fig. 11.10. b Local measurement of elastic constant and viscosity by analysis of creep response curve in terms of an elastic response (showing saturation) and a continuous deflection (straight line regime). Both directions (parallel and perpendicular to force) can be evaluated (Schmidt et al. 1996)

ance J(t) of actin solutions can be well described by the superposition of a linear term and a stretched exponential according to

$$J(t) = J_{ss} + t/\eta_z + C_1 \exp\{-(t/C_2)^{\beta}\}. \tag{25}$$

In order to measure η_z the above function is fitted to the plots of J(t)/t and m J(t)/t versus 1/t by varying the free parameters J_e, C1, C2, and β. η_z is obtained from the extrapolated value of the ordinate at t=0. The fitting of the experimental curve by the stretched exponential improves the accuracy of the η_z-measurement greatly. An example is shown in Fig. 11.11.

11.5 What Can We Learn from Rheology?

As pointed out several times, rheology provides a very valuable and sensitive tool to gain insight into structural and dynamic properties of actin networks and their manipulation by actin regulation proteins. Recent progress in the theory of semiflexible macromolecular solutions opened the possiblity of obtaining quantitative information on static and dynamic properties of the actin filaments. Moreover, one is in the fortunate situation that the rheological data can be directly related to the direct observations of the actin filament conformational dynamics and motions (reptations) by microfluorescence (treated in the article by J. Käs).

In this subsection we present some essential correlations between the frequency dependent viscoelastic moduli and topological parameters of entangled networks such as filament length or actin concentration (or mesh size). One follows the classical strategy of polymer research and tries to es-

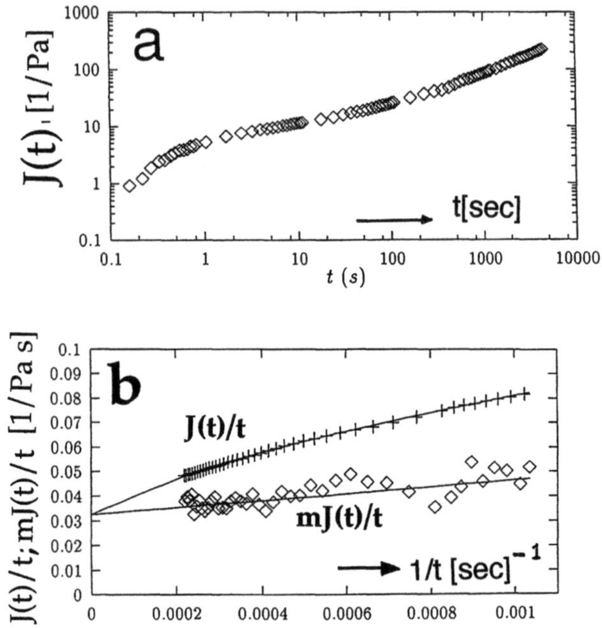

Fig. 11.11 a–b. Measurement of zero shear viscosity η_z by extrapolation procedure. **a** Measurement of creep compliance of actin solution ($C_A = 25$ μM) with average contour length adapted to $L = 5.25$ μm by gelsolin (actin-to-gelsolin ratio $r_{AG} = 2100$). **b** Plots of J(t)/t (data points marked by +) and of mJ(t)/t (data points marked by diamonds) (with m=d log J(t)/d log t) as a function of 1/t for terminal regime (III) of J(t)-versus-t curve. The *curves* correspond to the fits of the empirical stretched exponential curve defined by Eq. (25) obtained by varying all four parameters

tablish power laws relating the viscoelastic moduli to the above-mentioned parameters. Such power laws provide first tests for the validity of theoretical models. This aspect will be considered in the second part of this chapter.

11.5.1
Some Power Laws

Fig. 11.12 shows an example of a frequency dependent measurement of $G'(\omega)$ and $G''(\omega)$ by the torsional and magnetic bead rheometer for a semi-dilute entangled actin solution in the frequency regime from 4×10^{-5} to 10 Hz. The terminal regime, the plateau regime and the ascending branch at high frequencies which were defined in Fig. 11.5 are well pronounced. Systematic measurements of the viscoelastic moduli as function of the actin concentration (Müller et al. 1991; Janmey et al. 1994) and the chain length have led to the following series of power laws connecting these phenomenological parameters to dynamic and structural properties of the semiflexible actin filaments and the entangled networks.

1. Frequency Dependence of Viscoelastic Moduli. Both viscoelastic moduli exhibit a power law

$$G''(\omega) \propto G'(\omega) \propto \omega^{0.5} \tag{26}$$

at the high frequency regime ($\omega/2\pi \geq \tau_i$). This power law holds in particular for small actin concentrations ($C_A \leq 10$ μM) and the slope decreases somewhat with increasing concentration (at $C_A > 10$ μM). Very recently a detailed

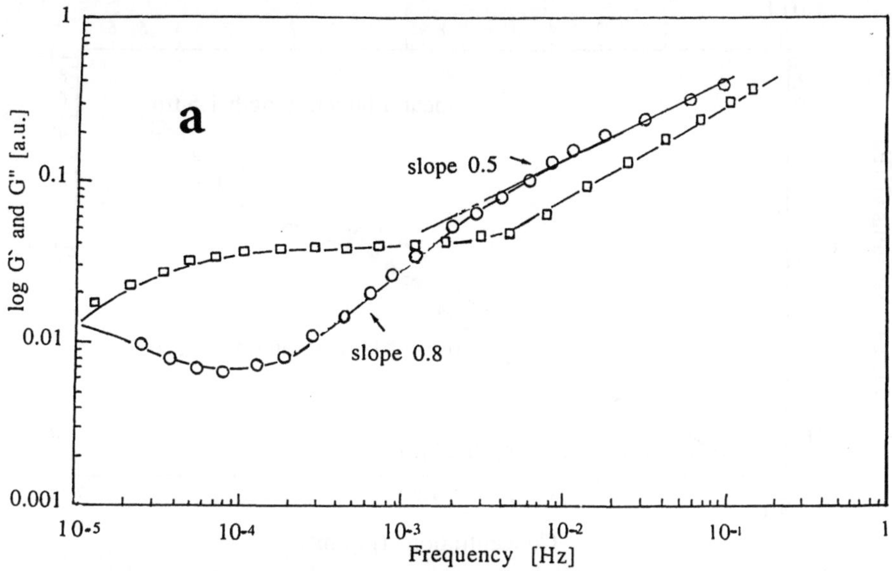

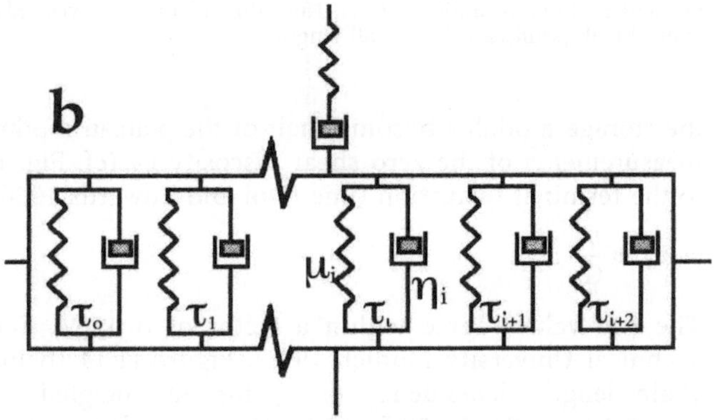

Fig. 11.12. **a** Frequency dependence of viscoelastic moduli of entangled network of actin of monomer concentration $C_A = 7\ \mu M$ (corresponding to mesh size $\zeta \approx 1\ \mu m$). Three regimes are clearly visible (cf. also Fig. 5 and note that $\omega \propto 1/t$): (1) a *terminal regime* where $G'(\omega)$ decays to zero and which corresponds to the transition to fluid state; (2) a *plateau regime*; and (3) an *ascending branch* with an apparent strain hardening. **b** Mechanical equivalent circuit of *acending branch* (regime III) of impedance frequency spectrum

correlation analysis of the Brownian motion of beads embedded in the cytoskeleton provided a scaling law $G'(\omega) \propto \omega^{0.75}$ (MacKintosh and Schmidt, pers. comm.) which agrees rather well with the result shown in Fig. 11.9. A theory of this behaviour is not available yet although an interesting model was proposed by Forgacs (1995).

2. Chain Length Dependence of Terminal Relaxation Time τ_d. The time τ_d was measured in two ways: firstly, from the value of the frequency at which

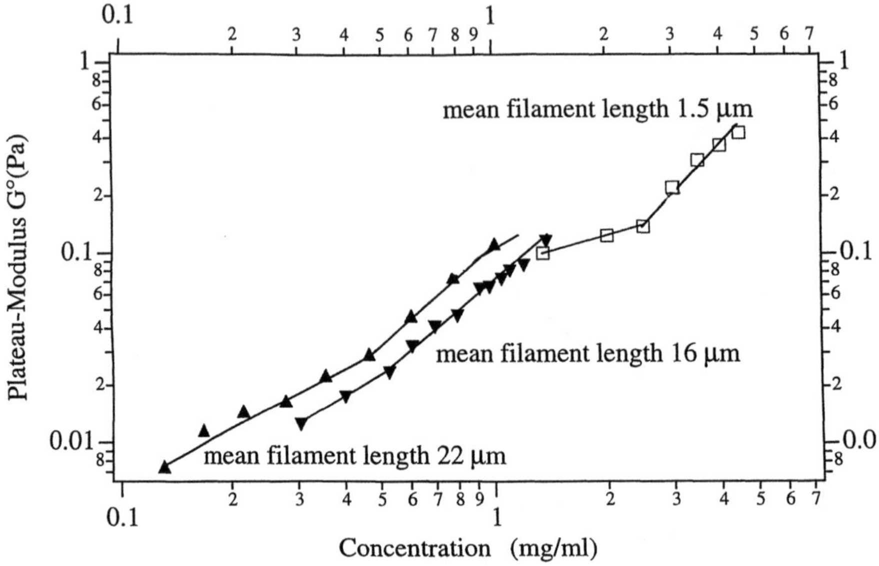

Fig. 11.13. Variation of plateau modulus of entangled actin solution with actin monomer concentration for average contour lengths L = 1.5 µm; L = 16 µm and L = 22 µm. (After Hinner and Sackmann, unpubl. data). Note that all three curves exhibit breaks and that slopes above break points are 1.7 for all lengths

the storage modulus becomes half of the plateau modulus and secondly, from measurements of the zero shear viscosity η_Z (cf. Fig. 11.11), which is related to the terminal relaxation time (Doi and Edwards 1986) by

$$\tau_d \approx \frac{\eta_z}{G'_0} \,. \tag{27}$$

The two values agree within a factor of two (M. Tempel, Doctoral Thesis, Technical University Munich 1996). Figure 11.14 (below) shows a plot of the chain length dependence of τ_d for an entagled network of $C_A = 25$ µM ($\xi \sim 0.3$ µm). For $L \geq 20\xi$ one finds a power law

$$\tau_d \propto L^{1.5} \,, \tag{28}$$

while for $L < 20\xi$, τ_d is proportional to $L^{2.5}$. We will see below that this behaviour can be understood in terms of the tube model [cf. Eq. (37)] and that measurements of τ_d as a function of L enbable measurements of the persistence length.

3. Concentration Dependence of Plateau Modulus G'_0. For networks of ideally flexible (Gaussian) chains the plateau modulus is a universal function of the mesh size and the monomer concentration

$$G'_0 = k_B T C_A / N_e \,, \tag{29}$$

where N_e is the number of monomers between two points of entanglement and C_A/N_e is therefore the number density of sides of the network.

For semiflexible chains the situation is more complex. Figure 11.13 shows the behaviour for three chain lengths. Similar behaviour was found by Müller et al. 1991 and Janmey et al. (cf. MacKintosh et al. 1995). One finds a transition between two well defined regimes of behaviour at a length dependent critical concentration C_A^*:

$$G_0' \propto C^{1.7\pm0.1} \quad \text{for} \quad C_A > C_A^* ,$$
$$G_0' \propto C_A^{1.1} \quad \text{for} \quad C_A < C_A^* . \tag{30}$$

This behaviour is found for different contour lengths (cf., however, Fig. 11.19).

11.5.2 Connection of G′(ω), G″(ω) to Reptation and Chain Conformational Dynamics

For flexible polymers the relaxation can be related to the self diffusion and the conformational dynamics of the single chains by application of the celebrated Rouse model (cf. Chaps. 4 and 7 in Doi and Edwards 1986). A rigorous theory has not been worked out yet for entangled solutions of semiflexible filaments.

The Terminal Regime. The transition to the fluid-like behaviour at $t \le \tau_d$ can be again explained in terms of the reptation model (by ignoring fluctuations of the tube). The relaxation of the internal stress following a sudden strain and leading to the decay of $G'(\omega)$ at $\omega/2\pi > \tau_d^{-1}$ is determined by the diffusion of the filament within the tube. If the tube length is L_T, the terminal relaxation time is

$$\tau_d = \frac{L_T^2}{2\,D_{rept}} , \tag{31}$$

where D_{rept} is the self diffusion coefficient along the tube. By ignoring the hydrodynamic screening (cf. Chap. 5 in Doi and Edwards 1986; Götter et al. 1996), D_{rept} can be approximated by

$$D_{rept} = \frac{k_B T}{2\,\pi\eta} \frac{\log(L/a)}{L} . \tag{31}$$

By considering hydrodynamic screening the term log (L/a) is replaced by log (Λ/a) where Λ is the so-called hydrodynamic screening length which is about equal to the mesh size ξ.

The limiting value of $G'(\omega)$ at $t > \tau_d$ is given by the value G_0' of the rubber plateau and therefore (cf. Chap. 9 in Doi and Edwards)

$$G_\omega' \cong G_0' \frac{\omega^2 \tau_d^2}{1 + \omega^2 \tau_d^2} . \tag{33}$$

The value of G_0' for semiflexible polymers will be considered below. Note that for flexible polymers it would be given by Eq. (29).

The high frequency relaxation regime. In analogy to the traditional model for flexible chains the ascending branch of $G^*(\omega)$ at $t \leq \tau_i$ is explained in terms of the stretching of the dynamically wrinkled chains. In this image the elastic relaxation modulus $G'(t)$ defined in Eq. (34) can be expressed as (cf. Doi and Edwards, Chap. 7)

$$G'(t) = \sum_p G_p \exp\{-t/\tau_p\}, \tag{34}$$

where τ_p are the relaxation times of the bending modes. In the frequency space this leads to

$$G'(\omega) = \sum_p \frac{G_p\,\omega^2\,\tau_p^2}{1+\omega^2\,\tau_p^2}\,; \quad G''(\omega) = \sum_p \frac{G_p\,\omega\,\tau_p}{1+\omega^2\,\tau_p^2}\,. \tag{35}$$

This behaviour can again be represented by a mechanical equivalent circuit consisting of a parallel array of Voigt bodies (cf. Fig. 11.6). Unfortunately, the situation is much more complex than for Rouse chains for which G_P is the universal function Eq. (29). The bending undulations are partially truncated due to the constraints imposed by the tube and the amplitude of the bending modes, G_P, are complex functions of the wavelength of the bending mode. Therefore a satisfactory microscopic model which can explain the frequency dependence of the viscoelastic moduli is not available yet despite a promising approach by Isambert and Maggs (1996) and by Forgacs (1995). It is hoped that such models become available soon since they would make rheology a powerful and sensitive tool for the evaluation of the structural and dynamic molecular properties of actin networks and their control by actin manipulation proteins.

11.5.3
Measurement of Persistence Length (=Bending Stiffness) from Zero Shear Viscosity

The terminal relaxation time is determined by the time required for a filament to diffuse such a distance that it has lost the memory of its original orientation. For semiflexible polymers, this length is not equal to the average filament length, L_T, but has to be augmented by the persistence length. It has been pointed out by K. Kroy (pers. comm.) that in order to account for this effect L_T in Eq. (31) is replaced by $\bar{L} + 2\,L_p - 3\,\xi$ and one obtains (Tempel et al. 1997)

$$\tau_d = \frac{12\,\eta_z}{\pi^2\,G_0'} = \frac{2\,\eta\,L\,(L + 2\,L_p - 3\xi)^3}{\pi\,k_B\,T\,\log(\xi/a)}\,. \tag{36}$$

The persistence length can then be determined by fitting the equation

$$\tau_d \propto \bar{L}\,(\bar{L} + 2\,L_p - 3\xi)^2 \tag{37}$$

to the measured τ_d-versus-$\bar{L}$ curves.

As demonstrated in Fig. 11.14 the power law is astonishingly well fulfilled over an order of magnitude of the length. The fitting procedure yields a value

$$L_p = 17 \pm 5\ \mu m\,.$$

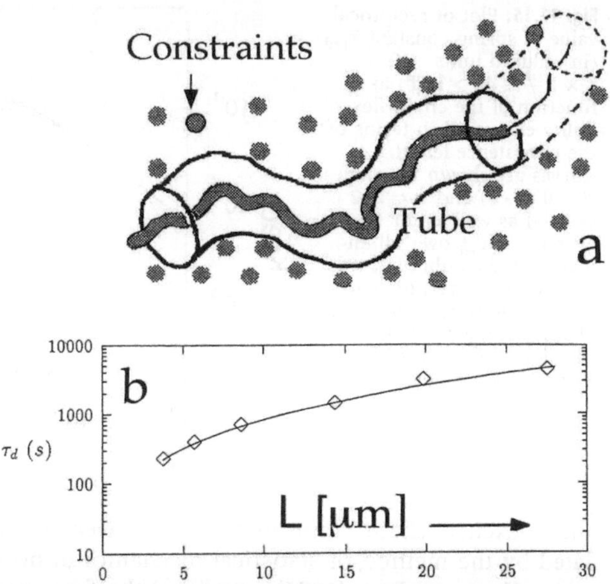

Fig. 11.14. a Tube model of network of semiflexible macromolecules. The constraints imposed on the motion of a filament singled out are represented by a tube of length L_T. For semidilute solutions, the width is about equal to the entanglement length introduced below whereas for concentrated solutions it is equal to the geometric mesh size. The *dashed piece* of the tube indicates a previous position and stresses the fact that the orientation of the newly formed tube segment is determined by the original orientation until it has moved over a distance equal to the persistence length L_P. **b** Plot of terminal relaxation time, τ_d, as function of the average chain length. τ_d was obtained by measuring the zero shear viscosity. The *curve* corresponds to the theoretical prediction and was obtained by fitting Eq. (37) to the data points (M. Tempel, Doctoral Thesis, Technical University Munich 1996)

This value agrees well with the value obtained by the analysis of the mean square amplitudes of the long wavelength (>10 μm) fluctuations of single filaments (cf. Käs, this Vol.).

11.5.4 Plateau Modulus and Filament Bending Stiffness

Following one major aim of this review to relate phenomenological viscoelastic parameters with molecular properties we establish first the relation between the storage modulus G_0' in the plateau regime and the filament bending modulus.

The stiffness of the semiflexible filaments is characterized by the force (or spring) constant k according to

$$\langle \delta L \rangle = k \cdot f_{\|} , \tag{38}$$

where $\langle \delta L \rangle = |\langle \vec{L} \rangle_f - \langle \vec{L} \rangle_o|$ is the average elongation of the chain and $f_{\|}$ is the component of the force vector $\vec{f}$ in the direction of the end-to-end vector of the chain. Since the chain elasticity is determined by the manifold of ther-

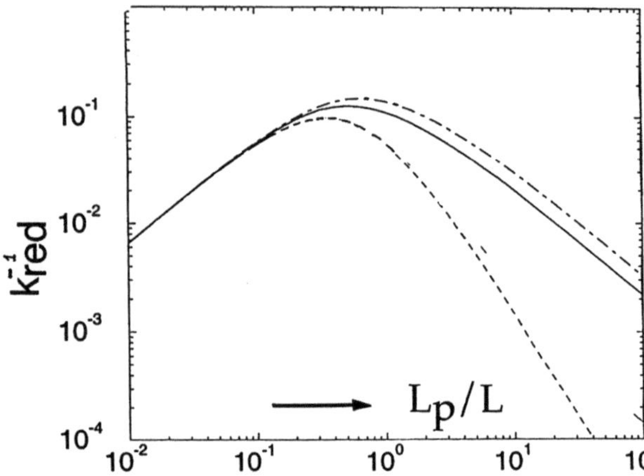

Fig. 11.15. Plot of reciprocal value of spring constant k_{red} (in reduced units $k \times L^2 / < \delta L > k_B T$) as a function of the chain flexibility expressed in terms of the persistence length L_P. Curves are shown for angles $\vartheta = 0$ (- - -) and $\vartheta = \pi/2$ (-·-·-) as well as for the average of k_{red}^{-1} over all angles (—). Note that left side of abscissa corresponds to the flexible (Rouse-like) chain and the right side to the rigid rod limit. (After Kory and Frey 1996)

mally excited chain conformations, the average elongation has to be calculated by the method of statistical mechanics in order to obtain the force constant. Kroy and Frey (1996) calculated the force constant for the whole range of flexibilities from flexible random coils to stiff filaments and as a function of the orientation of the force with respect to the average long axis of the filaments grafted at one end. The result is reproduced in Fig. 11.15 where the chain flexibility is given in terms of the ratio (L_p/L) of the persistance length to the contour length. The essential results are:

1. The force constant exhibits a minimum at $L_p = L$ (that is semiflexible filaments appear more flexible than flexible or stiff ones).
2. For the force parallel to the end-to-end vector the spring constant for a semiflexible chain is

$$k = \frac{B}{L^3} \tag{39}$$

3. For completely flexible chains the force constant is given by the familar law $k = k_B\, T / R_p^2$.

Since dissipative processes can be neglected in the plateau regime, the plateau modulus can be calculated following the classical theory of the elasticity of solids. Since this approach is very informative it will be briefly sketched below.

We consider the entangled actin solution as a cubic square lattice of lattice constant ξ_e. The elastic energy associated with a small (shear) deformation of the lattice can be expressed in therms of a Taylor expansion of the interaction potential V(R) of the lattice with respect to the lattice displacements from the equilibrium positions R_0:

$$G_{ela} \approx \frac{1}{2} \left| \frac{\partial^2 V(r)}{\partial r^2} \right|_{R_o} (\vec{R} - \vec{R}_0)^2 = \frac{1}{2}\, k\, (\vec{R} - \vec{R}_0)^2 \, . \tag{40}$$

By introducing the shear strain $(\vec{R}-\vec{R}_o)/\vec{R}_o)$ and the (shear) elastic constant G_0' the elastic energy may be expressed as

$$G_{ela} = \frac{1}{2}\, G_0'\, v\, (\vec{R}-\vec{R}_0)^2/R_0^2\,, \tag{41}$$

where v is the volume of the unit cell (note that G' has dimension of energy per volume). Comparison of the two equations yields finally

$$G_0' = c_e\, k\, \xi_e^2\,, \tag{42}$$

where c_e is the number of points of entanglement per unit volume which is about equal to the reciprocal volume of the unit cell of the network. Equation (42) thus relates the elastic modulus to the structure of the network and the flexural rigidity of single filaments. Equation (42) provides the basis for the application of rheological techniques to gain useful information on the structure and deformability of networks.

Since the force constant depends on the orientation of the chain segments with respect to the external force, the effective force constant is determined by the average over all orientations. Following Kroy and Frey (1996) the average force constant in the limit of semistiff chains ($L \ll L_p$) is given by

$$\bar{k} = B/L_e^3\,, \tag{43}$$

where L_e is the contour length stretching between two points of entanglement.

In the following it will be shown that G_0' may be related to the bending modulus and the actin concentration. However, the result depends on the ratio $\bar{L}/\xi$ of the filament length to the geometric mesh size:

1. For values $\bar{L}/\xi$ of the order of one, ξ_e is an arbitrary parameter which is considerably larger than the geometric mesh size. This situation will be considered below (Sect. 11.5.5).
2. For large ratios $\bar{L}/\xi$ (>10) ξ_e is about equal to the mesh size. However, one has to consider two geometrical situations depending on the ratio of the mean amplitudes $\langle u_\perp \rangle$ of the bending undulations to the mesh size. The equipartition theorem predicts that the mean square amplitude of the undulation is related to the length of the filaments by (cf. Lipowsky and Leibler 1986)

   $$u_\perp^2 \approx k_B\, T\, \bar{L}^3/B\,.$$

For small amplitudes $\langle u_\perp^2 \rangle \ll \xi$ one can assume $\xi_e \approx L_e$ (rod-like limit) and one finds with Eq. (43) $G_0' \propto 1/\xi_e^4$. Since ξ_e is related to the monomer actin concentration by $\xi_e \propto C_A^{-0.5}$ (cf. Schmidt et al. 1989)

$$G_0' \approx B \cdot C_A^2\,. \tag{44}$$

For $\langle u_\perp \rangle > \xi$ the bending undulations are partially truncated by the restrictions imposed by the mesh size. A law of the form

$$G_0' \propto B^{1/3}\, (k_B T)^{2/3} \cdot C_A^{5/3} \tag{45}$$

has been derived by Kroy and Frey (1996) while MacKintosh et al. (1995) proposed a scaling law: $G'_0 \propto C_A^{2.2}$. As demonstrated in Fig. 11.13 the power law $G'_0 \propto C_A^{1.7}$ is well fulfilled above the entanglement transition provided the networks do not undergo shear induced nematic alignment (cf. Fig. 11.19b below for such a case).

The above considerations show that there are many subtle effects which influence the elastic constants of networks of semiflexible macromolecules. They are most probably not very important for the biological role of actin. However, they are important for understanding the distinct behaviour of semiflexible polymer systems and they are very important in order to study the structural and dynamic properties of actin networks and their manipulation by actin regulating proteins by the highly sensitive viscoelastic techniques.

11.5.5
Dynamic Measure of Degree of Entanglement – The Entanglement Transition

As noted above for semidilute solutions the effective lattice constant or tube diameter, ξ_e, is not equal to the geometric mesh size, ξ. The reason is that free chain ends do not restrict the motion of neighbouring chains. Therefore, at low concentration many chains do not contribute to the formation of the tube determining the confinement of individual chains.

In order to account for this effect ξ_e is considered as an arbitrary length scale and the contour length L_e corresponding to ξ_e is called the "entanglement length". A more suitable parameter, the coordination number K has been introduced by Kavassalis and Noolandi (1988) which is defined as follows: a sphere of radius $\xi_e/2$ drawn about a point of entanglement is considered and K is the number of chains threading through this volume. In order to confine the singled-out test chain effectively, K must exceed a critical limit. Therefore the radius $\xi_e/2$ of the sphere is the larger fraction of free ends. Clearly, one expects $\xi_e = \xi$ if the chain length becomes infinite.

Based on the above considerations the effective lattice constant ξ_e has been calculated as a function of monomer concentration C_A for flexible chains by Kavassalis and Noolandi (1988) and for semiflexible chains by Kroy and Frey (1996). ξ_e diverges hyperbolically at small concentrations which points to a new type of transition from semi-entangled to a completely entangled state. The monomer concentration at this entanglement transition is

$$C_A^* \cong 27\,(1 + 0.6\,K)\,\tilde{C}_A/4\,,$$

where $\tilde{C}_A$ is the normal overlap concentration [cf. Eq. (7)] and the numerical analysis yields $K \cong 13$. The critical entanglement monomer concentration is thus expected to be a factor of 50 larger than the overlap concentration which is $C_A = 0.1\ \mu M$ for pure actin (Müller et al. 1991). The entanglement transition can also be expressed in terms of a critical chain length, $\bar{L}^*$, which is related to the geometrical mesh size ξ by

$$\bar{L}^* = 2.6\,\sqrt{3\,(1 + 0.6\,K)}\,\xi \approx 13\,\xi\,.$$

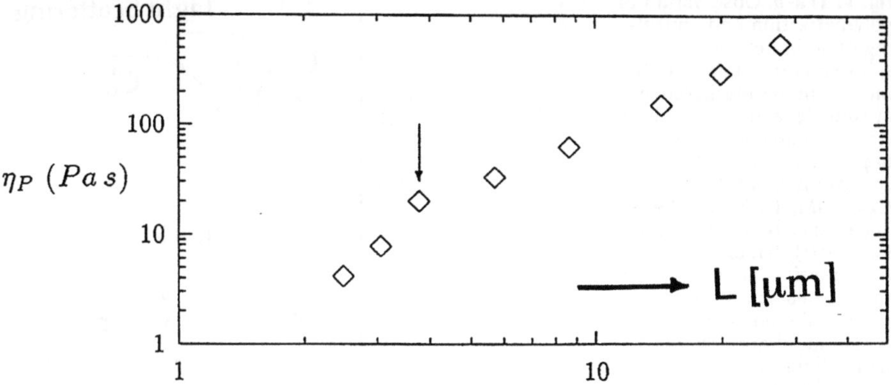

Fig. 11.16. Demonstration of entanglement transition of entangled actin network by measurement of zero shear viscosity as function of average chain length $\bar{L}$ for fixed mesh size. Actin concentration C_A=25 μM corresponding to mesh size $\xi \approx 0.3$ μm. The break (*arrow*) indicates an entanglement transition at $\bar{L}^* \approx 13\,\xi$

The entanglement transition is indeed observed for actin solutions by measuring the zero shear viscosity, η_z, as a function of chain length for a given mesh size. The experiment is shown in Fig. 11.16. The zero shear viscosity scales with the chain length as $\eta_z \propto \bar{L}^{1.6}$ for $\bar{L} \geq 4\,\mu$ and decays much faster with decreasing L below this limit pointing to an entanglement transition at the critical length $L^* \approx 12\,\xi_e$. This is in rather good agreement with the theoretical prediction.

The entanglement transition is further clearly demonstrated by the break of the concentration dependency of the storage modulus which was shown in Fig. 11.13b.

11.6 Practical Applications of Visoelasticity

In the following a few examples are presented showing that high sensitivity visoelasticity measurements can be useful to study even subtle effects of actin-binding proteins on the structural and dynamic properties of actin networks or on the actin polymerization kinetics.

11.6.1 Rheology as a Tool to Follow Actin Polymerization Kinetics

It is well established that the polymerization of actin is delayed due to the fact that a nucleus (of three monomers) has to be formed before the chain can grow. Since threefold collisions are very seldom, a delay of several minutes results unless nuclei are formed by actin-binding proteins. This behaviour is highly important for the control of the sol–gel transitions in cells, in particular since many of the actin-binding proteins can simultaneously act as nucleation promoters and can thus accelerate the polymerization. Examples

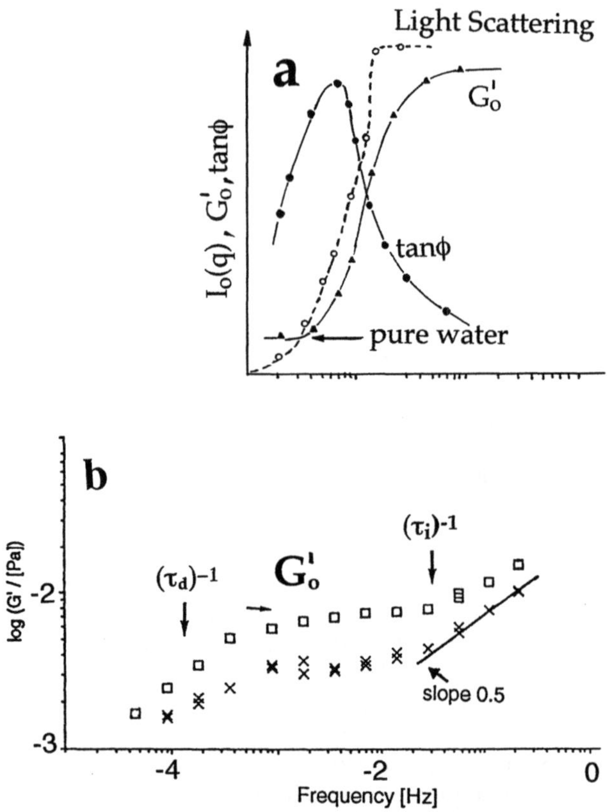

Fig. 11.17 a–b. Observation of polymerization and annealing of actin network.
a Observation of actin polymerization by measurement of time dependence of viscoelastic parameters G′ and $\mathrm{tg}\,\varphi = G''/G'$ at 0.7 Hz (*plateau regime*). Case of $C_A = 7\ \mu M$, $T = 25\,°C$. The appearance of a maximum for tg $\varphi(t)$ demonstrates that the viscosity increases faster than the elasticity. For comparison the polymerization is also monitored by dynamic light scattering by measuring the maximum initial intensity at $t = 0$ at scattering angle of 90° (after Ruddies et al. 1993).
b $G'(\omega)$ versus ω plots of entangled actin network ($C_A = 25\ \mu M$, 25 °C) measured 2 h and 20 h after starting the polymerization

are the severing protein gelsolin, profilin (which is also well established as a sequestering protein) and talin (an actin-membrane coupling protein).

Various methods of observing the polymerization process have been developed. Most prominent are fluorescence assays (Wegner 1982), static (Wegner 1982) and dynamic light scattering (Piekenbrock and Sackmann 1992). The first method is based on the change of the fluorescence quantum yield of fluorescence-labelled actin monomers after polymerization. The major advantage of this technique is that it can also monitor F-actin formation in cells. A disadvantage is that it does not yield information on the chain length. Light scattering techniques yield quantitative information on the chain length $\langle \vec{L} \rangle$ up to lengths comparable to the maximum value of the reciprocal scattering vector: (typically 1 μm) by measuring at small scattering angles of 20°.

These techniques are often not informative for the following reason: the delayed polymerization initially leads to the formation of a few very long filaments and many short chains since the very few initially formed filaments grow very rapidly and sequester most of the monomers before other chains start to grow. This leads initially to an exponential length distribution of the filaments (cf. Sackmann 1994).

Rheology is best suited to observing the later stages of the polymerization and is thus complementary to the other techniques. As shown in Fig. 11.17a, the plateau modulus G′ has reached only half its maximal value when the light scattering intensity has already reached saturation level. The appearance of a maximum of the tan φ-versus-t-curve shows the strong increase of the viscosity during the initial stages of the polymerization.

Even after completion of the polymerization (use-up of monomers) entangled actin networks exhibit structural changes for hours. This annealing process leads to more homogeneous networks as revealed by electron microscopy. This strange behaviour can be understood in terms of the vanishing of short chains in favour of longer filaments resulting in a more homogeneous length distribution. Clearly, if short filaments decay by fluctuation during the stationary state of treadmilling they cannot reform again at the very low stationary monomer concentration of about 0.2 μM (Piekenbrock 1992).

The annealing process is directly demonstrated by recording $G'(\omega)$-vs-ω plots during the later stages of polymerization as shown in the example of Fig. 11.17b. Both the plateau modulus and the terminal relaxation time increase by a factor of two after reaching the stationary state. This example shows that neglect of such remarkable differences between freshly prepared and annealed networks can lead to erroneous results.

11.6.2
Stiffening of Actin Filaments by Tropomyosin/Troponin

A distinct difference between semiflexible and flexible chains is that the plateau modulus of the former is not an universal function but depends on the filament bending stiffness: $G_0' \propto B^{1/3}$ in the limit $L_p > \xi$. Rheological measurements can thus be applied to measure modifications of the persistence length by actin-manipulating proteins provided one works at concentrations above the entanglement transition. Fig. 11.18 shows the effect of Tropomyosin/Troponin binding on the $G'(\omega)$-versus-ω curve. Two remarkable effects are found. The storage modulus increases and the slope of the ascending branch at the high frequency regime decreases remarkably yielding $G'(\omega) \propto \omega^{0.35}$. B increases by a factor of about 1.5 to 2 at an actin-to-T_m/T_n-ratio of 7:1, which is in good agreement with measurements by QELS (Götter et al. 1996).

The weaker frequency dependence of the ascending branch found for the stiffened chains is remarkable since the power law $G'(\omega) \propto \omega^{-0.35}$ (instead of $\omega^{-0.5}$) agrees well with that expected for semiflexible filaments provided the full spectrum of bending undulations contributes to the relaxation spectrum (Ruddies et al. 1993).

11.7
Non-Linear Viscoelasticity

Since the main emphasis was put on the application of rheological measurements for studies of structural and dynamic properties of actin and the ma-

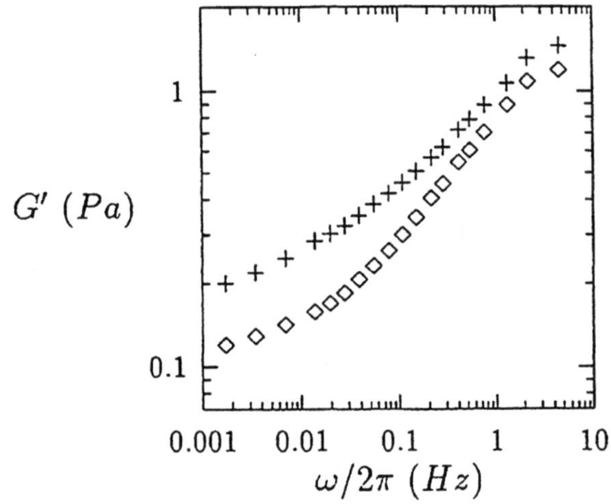

Fig. 11.18. Effect of Tropomyosin/troponin (Tr/Tn) binding on frequency dependence of storage modulus of entangled actin solution ($C_A = 8$ μM; molar ratio actin-to-Tr/Tn 7:1). Note increase in *plateau modulus* and decrease in *slope* of ascending branch at $\omega/2\pi > \tau_i$ from $\omega^{0.5}$ to $\omega^{0.35}$. *Diamonds* Absence of tropomyosin; *crosses* presence of tropomyosin/troponin

nipulation by actin-binding proteins we concentrated on weak shear rates where one deals with linear viscoelasticity. At high shear rates the viscoelastic response of actin networks becomes highly non-linear. Both viscoelastic parameters become strain dependent. There are two major reasons: rupture of chains and liquid crystal formation due to shear-induced alignment of the filaments.

Figure 11.19a shows an example of shear thinning by chain disrupture. The stress is measured as a function of shear strain with a torsional pendulum within the regime determined by internal chain dynamics. At small strains (≤ 10%) no stress is generated. It increases, however, steeply between 10 and 20% strain deformation and decreases again above 30%. Thus both strain hardening and strain thinning is observed. The latter effect is a consequence of chain rupture as demonstrated by the finding that no stress is built-up any more if the experiment is repeated immediately after application of the strain leading to shear thinning.

Shear thinning by nematic ordering is demonstrated in Fig. 11.19b. The storage modulus of a solution of actin filaments adjusted to a length of ≅5 μm by gelsolin is measured as a function of actin concentration C_A with the magnetically driven torsional rheometer. At concentrations $C_A \leq 50$ μM G_0' increases first nearly linearly with increasing C_A similarly to Fig. 11.13. In contrast, above 50 μM, G_0' remains nearly constant. Since the shear rate is below the rupture limit the only explanation for the above finding is nematic ordering. The crossover concentration $C_A \sim 50$ μM agrees well with the observation by microfluorescence (Käs 1996; Käs, this Vol.). The solution of filaments of 1.5 μm length behaves normally (cf. Fig. 11.13) since the tendency for nematic ordering increases with decreasing filament lengths (Semenov and Khoklov 1988).

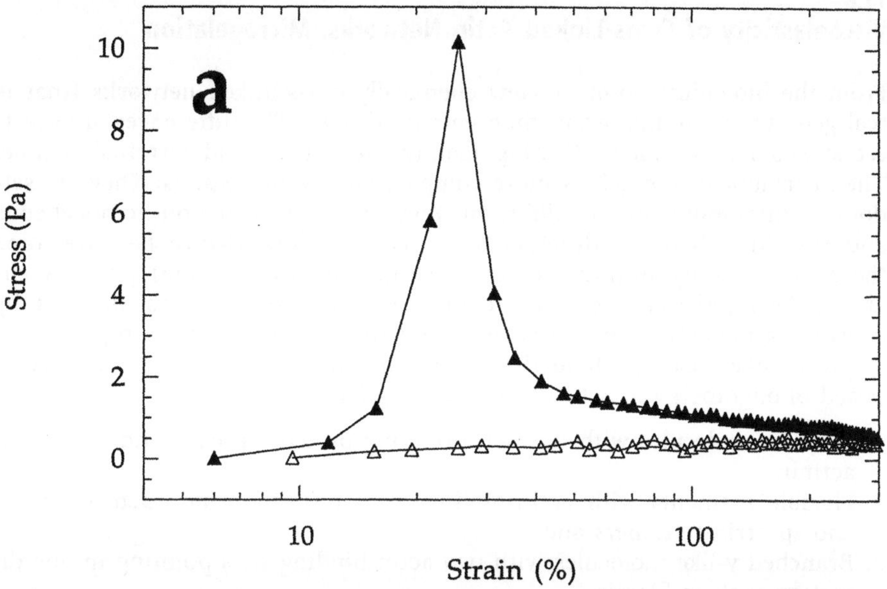

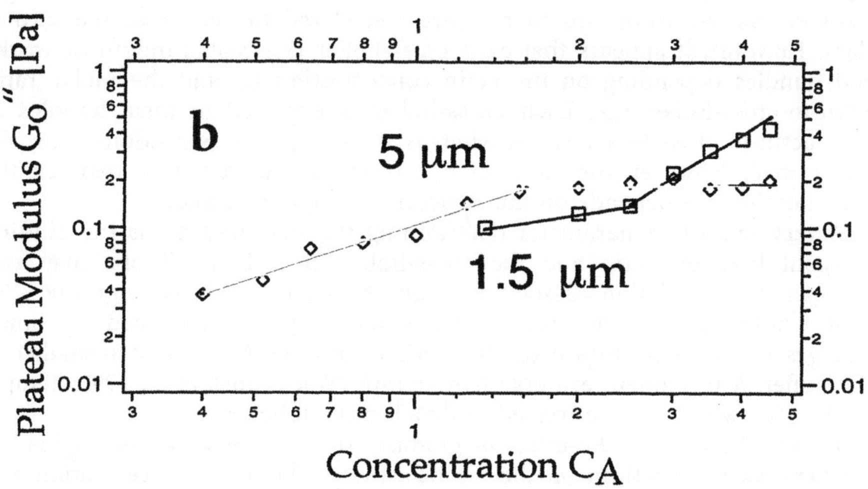

Fig. 11.19. **a** Demonstration of shear thinning with oscillatory rheometer by Janmey group (Janmey et al. 1994). The shear rate was 0.02 Hz. The *curve* marked by *full triangles* shows the first run while that marked by *open triangles* is observed during the second run demonstrating rupture of chains. **b** Apparent shear thinning by liquid crystal formation. Plot of storage modulus G_0' as function of actin monomer concentration for filament lengths of $\bar{L} = 5\,\mu M$ and $\bar{L} = 1.5\,\mu M$ (adjusted by gelsolin). Note the decrease of the *slope* of the G_0'-versus-C_A *curve* for $\bar{L} = 5\,\mu M$ at $C_A \geq 2$ mg/ml) ($\approx 50\,\mu M$) yielding $G_0' = \text{const}$. Note further the normal behaviour for $\bar{L} = 1.5\,\mu m$ (cf. Fig. 11.13)

11.8
Viscoelasticity of Cross-Linked Actin Networks: Microgelation

From the biological point of view chemically cross-linked networks (that is real gels) are more important than entangled ones. The differences are greatest at small shear rates affecting mainly the plateau and terminal regime. The behaviour of the gels is more complex for several reasons: They are seldom in thermodynamic equilibrium, they are in general non-homogeneous and transition between different states is very slow. Nature has overcome these problems by introducing the severing (or sequestering) proteins enabling the rapid cleavage of the connections between the cross-links or by controlling the activity of the cross-linkers through second messengers.

Cross-linkers can be divided into several subclasses which are often composed of oligomers (cf. Schleicher and Noegel 1992):

1. Rod-like molecules with two actin-binding proteins at each end such as α-actinin
2. Flexible filaments with actin-binding sites on either end such as fodrin and spectrin-tetramers and
3. Branched y-like molecules with two actin-binding sites pointing in one direction such as filamin

It is customary to divide cross-linkers into bundle formers and random gel formers and α-actinin and filamin are considered to belong to the former class. However, it appears that each cross-linker generates random networks and bundles depending on the actin concentration C_A and the molar ratio actin-to-cross-linker r_{AC}. Each cross-linker is expected to form bundles at high actin-to-cross-linker ratios whereas at low r_{AC}-values random networks are formed. However, the value of r_{AC} at which the transition between the two states occurs depends on the structure of the cross-linker.

A very important parameter determining the state of a gel is the binding constant between actin and the cross-linker. If it is small one may run through the association–dissociation equilibrium under physiologically relevant conditions and the sol–gel transition may be controlled by small changes of the temperature or the ionic conditions (pH, ionic strenght) of the buffer. A prominent example is α-actinin (Wachsstock et al. 1993; Tempel et al. 1996) which will be treated in detail in the following.

Figure 11.20 shows the effect of gradual cross-linking of an entangled actin network by α-actinin on the viscoelasticity. The actin concentration and the actin-to-cross-linker ratio are adjusted in such a way that the centre of the dissociation equilibrium is situated at about 20 °C. This is a consequence of the rather weak binding energy ($\Delta H \sim 3$ kJ/M) of the actin/α-actinin bond. It is a fortunate situation since it allows us to study the effect of the sol–gel transition on the structure and viscoelasticity in a reversible way.

We find three regimes of behaviour. In zone I the elastic modulus increases only slightly (although remarkably) with decreasing temperature (or increasing degree of cross-linking). Simultaneously the phase angle increases strongly reaching a maximum at the transition to a regime II where the storage modulus starts to increase steeply with decreasing temperature reaching

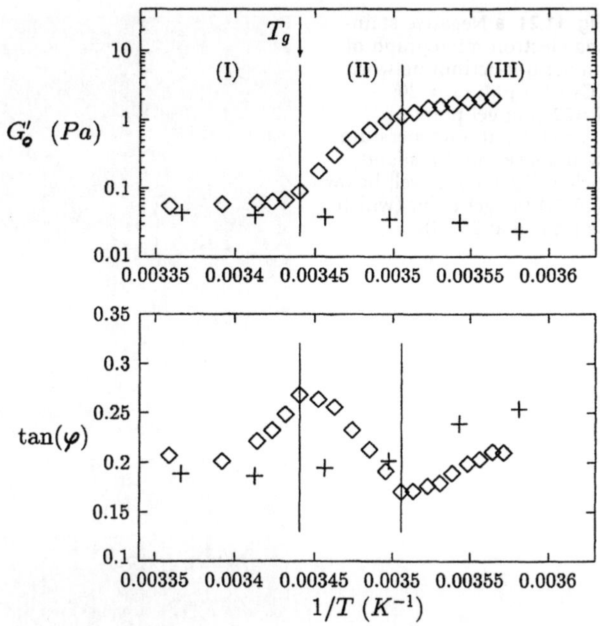

Fig. 11.20. Temperature dependence of storage modulus G'_0 and of phase angle $\tan\varphi = G''/G'$ plateau regime of actin network (C_A=9.5 μM, average chain length $\bar{L} = 22\,\mu m$) in absence (*crosses*) and presence (*diamonds*) of *a*-actinin at molar ratio r_{AC}=300. The temperature at which $\tan\varphi$ is a maximum is denoted as the gel point T_g

saturation within zone III. Tan φ decreases in regime II and increases again in III. Since $\tan\varphi = G''/G'$ is a measure for the loss modulus ($G'' = \eta \cdot \omega$) it follows that the viscosity (and therefore the energy dissipation) is a maximum at the transition from I to II where G'_0 starts to increase sharply.

As is well known from polymer physics (Colby et al. 1993) this behaviour is typical for a sol–gel transitions and the temperature T_g is therefore called the gel point.

The transition at T_g is associated with a remarkable change of the microstructure of the network as demonstrated by negative staining electron microscopy (Fig. 11.21 a). Whereas the network remains homogeneous on large scales it becomes heterogeneous on small scales exhibiting domains of more densely packed actin separated by regions of low filament density. The coarseness of this local segregation increases with decreasing temperature. We call this state a microgel in the following. At low actin-to-cross-linker ratios (r_{AC}< 10) and low temperature the microgel state becomes unstable and undergoes complete phase separation into densely packed clusters and highly dilute solution. The clusters consist of bundles. The bundles can become very long and can coexist also with randomly cross-linked gel as shown in Fig. 11.21 b for the case of an actin network chemically cross-linked by filamin.

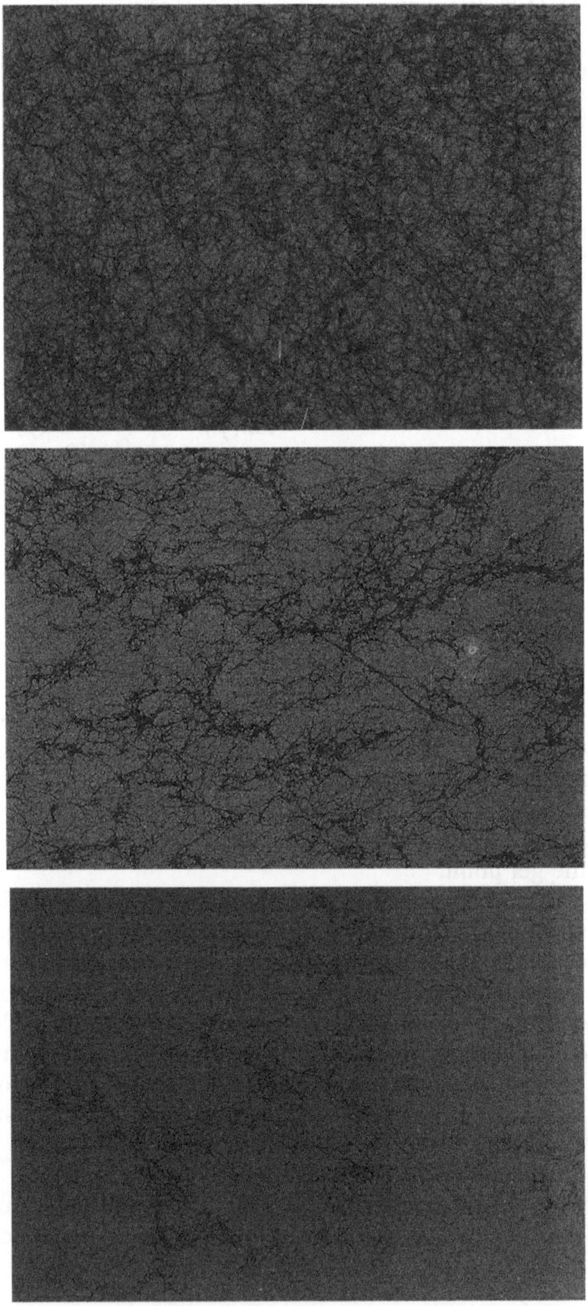

Fig. 11.21. a Negative staining electron micrograph of an actin-α-actinin network ($C_A=9.5$ μM; $r_{AC}=20$; $\bar{L}=22$ μm; gel point $T_g=21\,^\circ C$) at a temperature well above (30 °C), slightly below (19 °C) and well below (10 °C) the gel point (which is located at 20 °C)

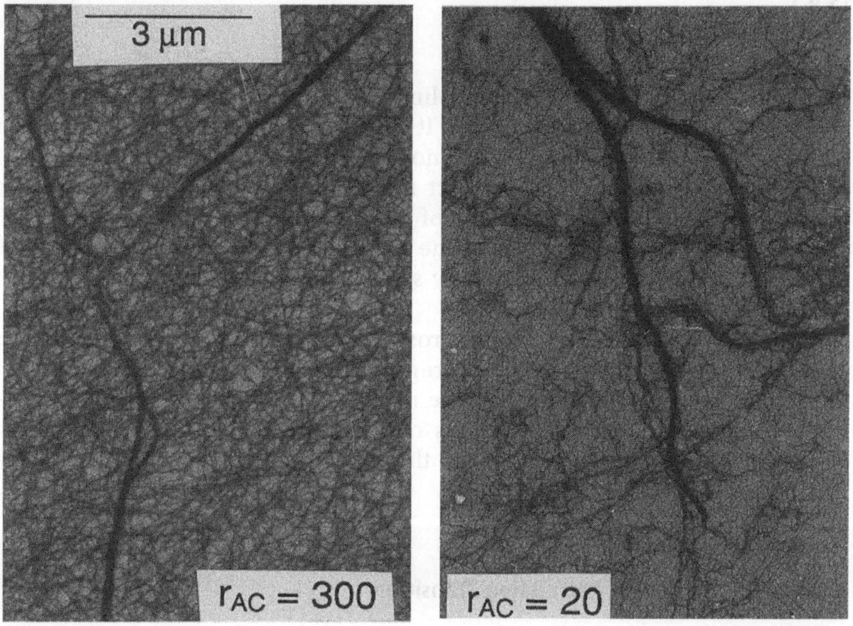

Fig. 11.21. b Negative staining electron micrograph of actin–filamin network at actin to cross-linker ratio $r_{AC}=300$ and $r_{AC}=20$ ($C_A=10$ μM, $r_{AG}=2000$, corresponding to $\bar{L}\approx 5$ μm)

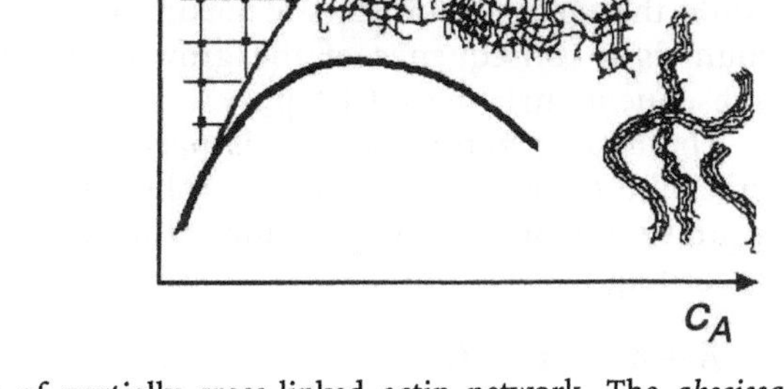

Fig. 11.22. Equivalent phase diagram of partially cross-linked actin network. The *abscissa* gives the actin concentration C_A (or the mesh size because of the law $\xi \approx C_A^{1/2}$). The *ordinate* gives the fraction of active cross-linker. Note that decreasing temperature is equivalent to increasing fraction of cross-linker activated. The *thin phase line* defines the phase boundary of the transition between homogeneous and the locally heterogeneous (microgel) state which can be described as a percolation transition. The *thick curve* defines the miscibility gap of the mixture below which the system decays into a highly dilute solution and a densely packed gel which is in general expected to be bundled

11.8.1
The Equivalent Phase Diagram

The behaviour of the actin-cross-linker system can be described by the phase diagram shown in Fig. 11.22. It is expected to be universal and to hold for all cross-linkers. If growing amounts of cross-linker (at fixed C_A) are added these are expected to connect first the natural points of entanglement leaving the equilibrium structure of the entangled solution essentially unchanged. The storage modulus in the plateau regime and the high frequency regime should therefore change only slightly as is indeed found. Only the terminal transition regime vanishes.

However, if the density of the cross-linkers exceeds that of the points of entanglement (that is when the average distance between two adjacent cross-linkers d_{cc} becomes smaller than the mesh size) further addition of cross-linker will result in local contractions of the network which remains homogeneous on a macroscopic scale. The thin phase line (where $d_{cc}/\xi < 1$) marks this transition.

11.8.2
Microgel Formation as Percolation Transition

The transition to the locally heterogeneous state has typical features of a percolation transition. This holds in particular for the sharp increase of the storage modulus at $T < T_g$. In this image the state of the network is determined by the fraction p, of cross-links formed. At increasing p, clusters of interconnected points of entanglement are formed within the network. These clusters grow in size with increasing p and become infinite at a critical value $p = p_c$, that is the whole system consists of chemical cross-linked network. The transition from $p < p_c$ to $p > p_c$ can be understood in terms of a percolation transition. The percolation theory predicts indeed that the storage modulus diverges if p exceeds $p_c (p > p_c)$ as according to the power law

$$G'_0 \cong (p - p_c)^t , \tag{46}$$

while the viscosity exhibits a maximum at the gel point ($p = p_c$). This maximum is a consequence of the growth of the clusters up to $p = p_c$ and their subsequent shrinkage at $p > p_c$.

The actin-a-actinin system is particularly interesting since the fraction of cross-links formed can be controlled through the association–dissociation equilibrium of the complex formation between actin (A) and a-actinin (a) according to

$$A + a^K (Aa) = C \tag{47}$$

$$K(T) = \frac{[C]}{[A][C]} = K_0 \cdot \exp\left(\frac{\Delta H}{k_B T}\right) , \tag{48}$$

where the brackets denote concentrations (in units of mol^{-1}) and ΔH is the heat of association. An important consequence is that the fraction ρ_a of a-ac-

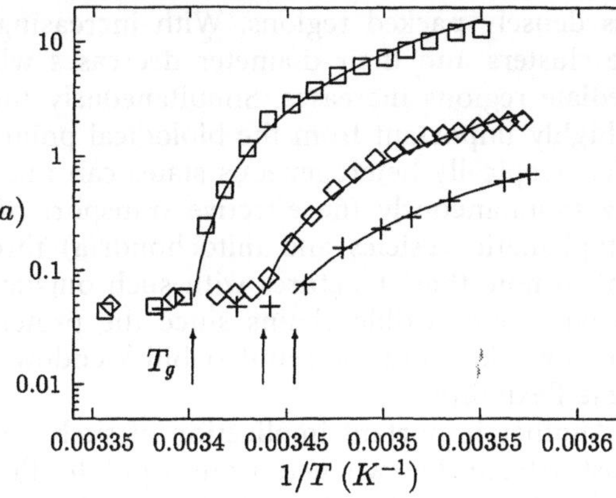

Fig. 11.23. Simulation of Arrhenius-like plots of plateau elastic constant G_0' (for r_{AC}=27, 20, 10, *left to right*, respectively) by percolation model assuming that degree of cross-linking is controlled by chemical equilibrium. All three fits are performed with nearly the same values of the free parameters ΔH, p_c and the exponent γ

tinin reacted (or the fraction of cross-links formed) becomes temperature dependent since (Tempel et al. 1996)

$$\rho_a\,(T) = \frac{[C]}{[C] + [a]} = \frac{[A]\,K\,(T)}{[A]\,K\,(T) + 1}\,. \tag{49}$$

Since α-actinin is bifunctional the fraction of cross-links formed is $p\,(T) = \frac{1}{2}\,\rho_a\,(T)$.

The percolation theory therefore predicts

$$G_0'(T) \cong [p(T) - p_c]^{\gamma}\,.$$

As demonstrated in the Arrhenius-like plots $G_0'\,(T)$–versus-T^{-1} in Fig. 11.23 the divergence of the elastic modulus at $T \geq T_g$ is well represented by the law predicted by the percolation theory (Tempel et al. 1996). A remarkable and from the point of view of physics surprising finding is that the exponent of the power law, γ=1.5–1.8, agrees rather with the value predicted by classical mean field theory of percolation than with the result of the rigorous percolation theory for which $\gamma \cong 3.8$ is expected (Stauffer and Aharony 1992).

11.8.3
Biological Implications and Consequences of Heterogeneous Gelation

An important conclusion following from the above study is that the viscoelastic behaviour and microstructure of actin gels depend critically on the ratio of the average distance d_{cc} between active cross-linkers to the mesh size ξ:

$$\zeta = \xi / d_{cc}\,.$$

ζ is an order parameter characterizing the state of the gel. ζ=1 characterizes a geometrical limit where all initially formed points of entanglements are cross-linked.

The microgel state is characterized by three length scales: the diameter of the clusters, the mesh size of the dense domains and the mesh size of the less densely packed regions. With increasing parameter ζ the mesh size in the clusters and their diameter decreases while the mesh size in the intermediate regions increases. Simultaneously the network becomes stiffer. This is highly important from the biological point of view since networks in such microscopically heterogeneous states can maintain high elastic forces and allow simultaneously the effective transport of rather large particles (such as cytoplamatic vesicles and mitochondria) through the network. It is important to note that structures with such outstanding features can only be built up with semiflexible chains since the protein molar fractions in such networks are low and the semiflexibility endows the cross-linked networks with some flexibility.

Another important implication of such microgel-states is that they can be most effectively switched from rigid to fluid states by cleavage of a few chains between the clusters. The clusters become separated and can flow freely with respect to each other under stress.

An important parameter characterizing the effectivity of cross-linkers is the binding constant and confusion may arise if this is ignored. Thus for filamin and the 120 kDa cross-linker of Dictyostelia cells the percolation transition occurs at $r_{AC} \approx 500$ (for $C_A = 7\ \mu M$) which corresponds to a parameter $\zeta \approx 1$ while for α-actinin $r_{AC} \cong 10$ is required in order to reach $\zeta \approx 1$.

The weak binding constant of α-actinin has distinct advantages for the control of the sol–gel transition in cells. The state may be sensitively tuned by variation of the concentration of (non-sequestered) actin monomers when the environmental conditions are biased in such a way that the actin-cross-linker system is at the midpoint of the association–dissociation equilibrium. According to Eq. (49) the midpoint of the reaction is determined by the condition

$$[A]\, K_0 \exp\{\Delta H/RT\} = 1\,. \tag{50}$$

For α-actinin $\Delta H \sim 1.8$ kJ $\approx k_B T$. Thus, an increase of [A] by a factor of ten would shift the absolute temperature of the midpoint from 300 K to 600 K and therefore to the side of complete association. In cells the equilibrium could thus be easily shifted between dissociation and association by slight variation of the fraction of actin sequestered by profilin under conditions where the actin-based cytoskeleton is maintained near the percolation instability (cf. also Forgacs 1995).

11.8.4
Rheological Redundancy and the Compensation of Mutation-Induced Deficiencies of the Cytoskeleton

An enigma of cell biology is why the behaviour of cells (e.g. adhesion and locomotion) is often astonishingly insensitive towards drastic mutations of the actin-based cytoskeleton.

Thus it has been found that removal of three actin-binding proteins thought to be essential (such as severin, the 120 kDa gelation factor and α-actinin) from cells of the slime mould *Dictyostelium discoideum* does not affect remarkably the viability and actin dependent chemomechanical pro-

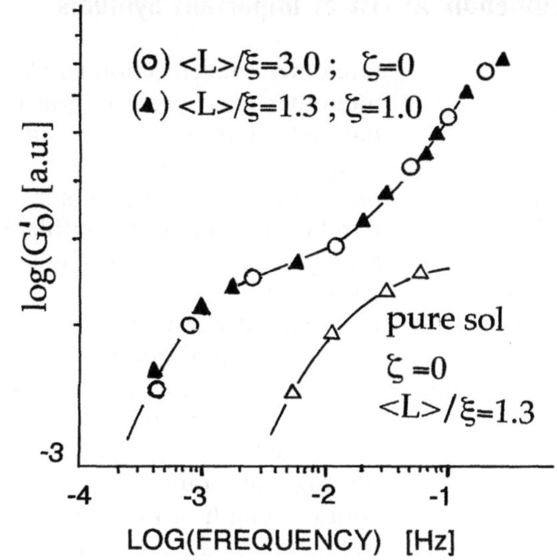

Fig. 11.24. Rheological redundancy of actin networks. Frequency dependencies of storage modulus of two purely entangled actin solutions exhibiting ratio of length-to-mesh-size of $L/\xi=1.3$ (*data points* Δ) and $L/\xi \approx 3$ (*data points* 0) and of one partially cross-linked network with $L/\xi \approx 1.3$ (*data points* Δ). By adjusting the ratio of the mesh size to the average distance between cross-linkers (120 kD gelation factor of Dictyostelia cells) to $\xi/d_{cc} \approx 1$ the $G'(\omega)$-versus-ω curve fits exactly with that of the purely entangled solution with $L/\xi=3$ over the whole frequency range

cesses (such as locomotion on substrates) of the cells (Schindl et al. 1995). The same has been found for myosin II deficient cells (Pastemak et al. 1989).

One natural and likely explanation is the redundancy of the actin-binding proteins, such as the presence of several types of equivalent cross-linkers. This is indeed suggested by the universal effects of the cross-linkers postulated above. Another explanation is a rheological redundancy based on the fact that the same frequency dependency of the viscoelastic parameters can be realized by a manifold of sets of the three variables: actin concentration, filament length and actin-to-cross-linker ratio. As noted above, the mechanical impedance $G^*(\omega)$ of cross-linked networks is characterized by the ratio of the mesh size to the average distance between cross-linkers (ξ/d_{cc}) while the impedance of entangled solutions is determined by the ratio of the filament length to the mesh size. An example of rheological redundancy is shown in Fig. 11.24. It shows that the same frequency dependence of the storage modulus over several frequency decades can be realized for a purely entangled and a partially cross-linked network by choosing the right order parameters (L/ξ and $\zeta=\xi/d_{cc}$), respectively.

Acknowledgements. Helpful discussions with Erwin Frey, Klaus Kroy and Bernhard Hinner are acknowledged. I would also like to thank B. Hinner and K. Kroy for providing results from their Diploma theses prior to publication.

Appendix A: List of Important Symbols

C_A	monomer concentration (moles per litre)
ρ_M	monomer mass density (g/ml)
n_M	monomer number density (monomers per unit volume)
N_L	Avogadros number
M	molecular weight of polymer
N	number of monomers per filament
$\bar{L}$	average contour length of ensemble of filaments
<L>	end-to-end distance
$G'(\omega)$, $G''(\omega)$	frequency-dependent storage and loss modulus, respectively
$G(t)$	shear relaxation modulus
$J'(\omega)$, $J''(\omega)$	frequency-dependent compliances
$\eta(\omega)$	frequency-dependent viscosity ($\eta=G''/\omega$)
η_z	zero shear viscosity ($\eta_z=\eta(\omega\to 0)$)
γ	tilt angle of strain
s	contour length of filament
$\vec{R}(s)$	radius vector defining position of contour site s on filament
$t(s)$	tangent to filament at contour position s
$H(s)=\partial^2 R(s)/\partial s^2$	total curvature of filament

References

Bastide J, Leibler L, Prost J (1990) Enhancement of inhomogeneities in gels upon swelling and stretching. Makromolecules 23:1821–1828

Colby RH, Gillmor JR, Rubinstein M (1993) Dynamics of near critical polymer gels. Phys Rev E 48:3712–3715

deGennes P (1979) Scaling concepts in polymer physics. Cornell University Press, Ithaca

Doi M, Edwards SF (1986) The theory of polymer dynamics. Clarendon Press, Oxford

Evans EA (1993) New physical concepts for cell amoeboid motion. Biophys J 64:1306–1322

Ferry JD (1980) Viscoelastic properties of polymers. Wiley, London

Forgacs G (1995) On the possible role of cytoskeletal filamentous networks in intracellular signaling: an approach based on percolation. J Cell Sci 108:2131–2143

Fuller GG, Mikkelsen K (1988) Optical rheometry using a rotatory polarization modulation. J Rheol 33:701–769

Fung YC (1981) Biomechanics: Mechanical properties of living tissue. Springer, Berlin Heidelberg New York

Götter R, Kroy K, Frey E, Bärmann M, Sackmann E (1996) Dynamic light scattering from semidilute actin solutions: a study of hydrodynamic screening, filament bending stiffness and the effect of tropomyosin/troponin binding. Macromolecules 29:30–36

Hinner B (1997) Messung viskoelastische Eigenschaften halbvedünnter und konzentrierter Aktinlösungen. Diploma Thesis, Technical University Munich

Isambert H, Maggs AC (1996) Dynamics and rheology of actin solutions. Macromolecules 29:1036–1040

Isenberg G (1991) Actin binding proteins - lipid interactions. J Muscle Res Cell Motility 12:136–144

Janmey P (1995) Cell membranes and cytoskeleton. In: Lipowsky P, Sackmann E (eds) Structure and dynamics of membranes. Handbook of biological physics, vol 1B. Elsevier - North Holland, Amsterdam

Janmey PA, Hvit S, Lamb J, Stossel TP (1990) Resemblance of actin-binding protein/actin gels to covalently cross-linked networks. Nature 345:89–92

Janmey P, Käs J, Sackmann E et al. (1994) The mechanical properties of actin gels elastic modulus and filament motions. J Biol Chem 269:32503–32513

Kaufmann S, Käs J, Goldmann WH, Sackmann E, Isenberg G (1992) Talin anchors and nucleates actin filaments at lipid membranes: a direct demonstration. FEBS Lett 314:203–205

Kavassalis TA, Noolandi J (1988) A new theory of entanglements and dynamics in dense polymer systems. Macromolecules 21:2869–2879
Käs J, Strey H, Sackmann E (1993) Direct measurement of the wave-vector dependent bending stiffness of freely flickering actin filaments. Europhys. Lett 21:865–870
Käs J, Strey H, Sackmann E (1994) Direct imaging of reptation for semiflexible actin filaments. Nature 368:226–230
Käs J, Strey H, Tang JX et al. (1996) F-actin, a model polymer for semiflexible chains in dilute/semidilute and liquid crystalline solutions. Biophys J 70:609–625
Korn ED, Carlier MF, Pantaloni D (1987) Actin polymerization and ATP-hydrolysis. Science 238:638–644
Lipowsky R, Leibler S (1986) Unbinding transition of interacting membranes. Phys Rev Lett 56:2541–2544
Kroy K, Frey E (1996) Force extension relation and plateau modulus for wormlike chains. Phys Rev Lett 77:306–309
MacKintosh F, Käs J, Janmey PA (1995) Elasticity of semiflexible biopolymer networks. Phys Rev Lett 75:4425–4428
Müller O, Gaub H, Bärmann M, Sackmann E (1991) Viscoelastic moduli of sterically and chemically cross-linked actin networks in the dilute to semidilute regime: measurements by an oscillating disc rheometer. Macromolecules 24:3111–3120
Onsager L (1949) The effect of shape and interaction of colloidal particles. Ann NY Acad Sci 51:627–650
Pasternak C, Flicker PF, Ravid S, Spudich JA (1989) Intermolecular versus intramolecular interactions in dictyostelium myosin: possible regulation by heavy chain phosphorylation. J Cell Biol 109:203–210
Piekenbrock T, Sackmann E (1992) Quasielastic light scattering study of thermal excitation of F-actin solutions and of growth kinetics of actin filaments. Biopolymers 32:1471–1489
Ruddies R, Goldmann WH, Isenberg G, Sackmann E (1993) The viscoelasticity of entangled actin networks: the influence of defects and modulation by talin and vinculin. Eur Biophys J 22:309–321
Sackmann E (1994) Intra- and extracellular macromolecular networks: physics and biological function. Macromol Chem Phys 195:7–28
Saito N, Takahashi K, Yunoki Y (1997) The statistical mechanical theory of stiff chains. J Phys Soc Jpn 22:219–225
Schiessel H, Blumen A (1995) Mesoscopic pictures of the sol–gel transition: ladder models and fractal networks. Macromolecules 28:4013–4019
Schindl M, Wallraff E, Deubzer B, White W, Gerisch G, Sackmann E (1995) Cell substrate interaction and locomotion of *Dictyostelium* wild-type and mutants defective in three cytoskeletal proteins: a study using quantitative reflection interference contrast microscopy. Biophys J 68:1177–1190
Schleicher M, Noegel AA (1992) Dynamics of the dictyostelium cytoskeleton during chemotaxis. New Biol 4:461–472
Schmidt C, Bärmann M, Isenberg G, Sackmann E (1989) Chain dynamics, mesh size and diffusive transport in networks of polymerized actin: a quasielastic light scattering and microfluorescence study. Macromolecules 22:3638–3648
Schmidt FG, Ziemann F, Sackmann E (1996) Shear field mapping in actin networks by using magnetic tweezers. J Eur Biophys 24:248–253
Semenov AN, Kokhlov AR (1988) Statistical physics of liquid-crystalline polymers. Sov Phys Usp 31:988–1014
Stauffer D, Aharony A (1992) Introduction to percolation theory. Taylor and Francis, London
Tempel M, Isenberg G, Sackmann E (1996) Temperature-induced sol–gel transition and microgel formation in α-actinin cross-linked actin networks: a rheological study. Phys Rev E 54:1802–1810
Wachsstock DH, Schwarz WH, Pollard TD (1993) Affinity of α-actinin for actin determines the structure and mechanical properties of actin filament gels. Biophys J 65:205–214
Wachsstock DH, Schwarz WH, Pollard TD (1994) Cross-linker dynamics determine the mechanical properties of actin gels. Biophys J 66:801–809
Wegner A, Savko P (1982) Fragmentation of actin filaments. Biochemistry 21:1909–1913
Zaner SK, Stossel TP (1983) Physical basis of the rheological properties of F-actin. J Biol Chem 258:11004–11009
Ziemann F, Rädler J, Sackmann E (1994) Local measurement of viscoelastic moduli of entangled actin networks using an oscillating magnetic bead microrheometer. Biophys J 66:2210–2216

Subject Index

MIX
Papier aus verantwortungsvollen Quellen
Paper from responsible sources
FSC® C105338

If you have any concerns about our products,
you can contact us on
ProductSafety@springernature.com

In case Publisher is established outside the EU,
the EU authorized representative is:
Springer Nature Customer Service Center GmbH
Europaplatz 3, 69115 Heidelberg, Germany

Printed by Libri Plureos GmbH
in Hamburg, Germany